The Sand Wasps

The Sand Wasps

Natural History and Behavior

Howard E. Evans
Kevin M. O'Neill

Harvard University Press
Cambridge, Massachusetts
London, England
2007

Printed in the United States of America

ISBN-13: 978-0-674-02462-5
ISBN-10: 0-674-02462-1

Cataloging-in-Publication Data available from Library of Congress

For Mary Alice Evans
and Ruth Pettinga O'Neill

Contents

Foreword

Although my husband, Howard, had studied spider wasps as a graduate student at Cornell University, his first job took him to Kansas, where sand wasps were a natural for him and his students. When he came back to Cornell as a member of the faculty in 1952, he brought this interest with him. I met him not long after his return, and some of our first trips into the field were to teach me what a good sand wasp site looked like, how to take meaningful biological notes on the wasps, and how to dig out their nests without destroying them. After we had our three children my job changed, as I sought to keep them from "helping" their father dig in the sand. And as far as I know, no other family traveled with a bug net for each member. As our children reached school age and as Howard's interests broadened beyond the continental United States, we could not always travel with him. We did, however, spend a year in Australia. How well I remember how thrilled Howard was at finding wasps that used other wasps as prey, and how his three "assistants" dashed around trying to help him.

Over the years, Howard did a lot of writing, both scientific and popular, but as he aged he came to believe that a person, and especially a scientist, should write no more books after reaching the age of eighty. But when he reached that age he continued to write, so I asked him why. His reply was, "What else would I do?" What he was doing was putting his thoughts in order. Some years earlier, he had started an autobiography, intended only for his family, and from time to time he added to it. Then he summarized what he thought had been his contribution to science. Finally he began writing about his favorite sand wasps and what he thought still needed to be learned. Attached to the manuscript on the ethology of sand wasps was a note that said that this was not a publication, but should be sent to Kevin O'Neill, who might find the information useful. This book is a testament to Howard's belief in Kevin's ability to use the material well.

Mary Alice Evans

Preface

By 1950 the field of ethology, led by Niko Tinbergen and other Europeans, was maturing rapidly (Kruuk 2004). In that year Howard Ensign Evans, a recently hired assistant professor at Kansas State University, published his 340-page *Taxonomic Study of the Nearctic Spider Wasps Belonging to the Tribe Pompilini (Hymenoptera: Pompilidae),* Part I. A casual browser, if not first put off by the ominous title, might expect a dry, specialized tome on the classification of some obscure group of wasps. Indeed that is what it is, of course, but tucked in with the standard anatomical descriptions are synopses of the biology of these solitary wasps: records of their spider prey, brief notes on nesting habits, and lists of the flowers they visit. And important for the future of wasp studies, some of the notes were original observations of the author, making clear that his interests extended beyond taxonomy and insects stuck on pins to behavior and insects observed in the field.

In the 1950s the main stars of ethology were animals such as gulls, geese, and stickleback fish. But solitary wasps had also been important in ethology, through studies of homing and hunting behavior of the beewolf *Philanthus triangulum* by Tinbergen (1932, 1935) and of nesting behavior of *Ammophila campestris* by Baerends (1941). And van Iersal's study of homing by *Bembix rostrata* would appear in 1952. These three Dutch scientists demonstrated how solitary wasps could serve ethology as subjects of experimental field studies of behavior. But ethology is also about comparative analyses in which related species are observed in different environments to gain insight into how environment shapes the evolution of behavior. For example, Esther Cullen's classic study of gulls and kittiwakes showed how nesting habitat was associated with specific behavioral and anatomical adaptations (Cullen 1957). Insects, of course, are attractive for comparative studies. One can get a fuller picture of the behavior of each species more quickly, because insects are simpler than gulls. Insects are often easier to study without inserting oneself too intrusively into their lives. Entomologists also have available a much greater diversity of species, and that means a greater sample size in a comparative study as well as tremendous behavioral and structural diversity if the researchers have chosen the appropriate insects. Here it certainly helps to be a taxonomist who can identify species encountered in the field. With his taxonomic skills and ethological inclinations, Howard Evans was perfectly positioned to contribute to the field of comparative ethology. He saw his work in systematics and in ethology as inseparable aspects of comparative biology, as reflected in his statement that "one cannot intelligently discuss behavior and structure separately. Behavior is what an animal does with its structure; structure is what an animal uses to behave" (Evans 1966a). And earlier he had noted that the "nests of insects, birds, and other animals are morphological expressions of behavior patterns and as such are particularly interesting to the comparative ethologist" (Evans 1957b).

At Colorado State University in the late 1970s, Howard's graduate students, who at that time included Darryl Gwynne, Al Hook, Rob Longair, Bill Rubink, and myself, joked that he described a new species or revised a genus before breakfast each morning. Although it may have been the truth, the real fun for Howard was not sitting at a microscope but lying on his belly in the dirt in some remote location, digging out and diagramming a wasp's nest, and extracting eggs, cocoons, prey, and parasites. His enthusiasm for such work was mirrored by another of my teachers, Frank Kurczewski, who once asserted that opening a wasp's nest is like opening a Christmas present, because you never know exactly what is inside. In my years of doing fieldwork with Howard, I never saw him more surprised and pleased than when he discovered lacewing prey in a *Bembix stenebdoma* nest in New Mexico (the uniqueness of this finding is discussed in Chapter 7). On the other hand, I never saw him more displeased in the field than when I ran out of gas on a field trip in eastern Colorado, having stupidly failed to note that the busted gas gauge had not moved for several hundred miles. He at least smiled when I pointed out that, if we had stalled a few miles farther up the road, we would have been stranded in front of a gigantic feedlot, the odor of which was often stupendously repellent on hot summer days (possibly even for a man who once studied the behavior of wasps around piles of dog excrement [Evans 1989]). All was not lost, however. When I returned from hitchhiking to get gasoline, he was contentedly collecting wasps from flowers in a weedy ditch—and, as I recall, even found one of his treasured bethylid wasps.

Through the 1950s and 1960s Howard published ever-longer comparative analyses of wasp behavior, as he moved from Kansas State back to Cornell University (in 1952), on to Harvard University (in 1960), and finally to Colorado State University. Notable among these were works on the behavior of spider wasps (1953; 17 pages), digger wasps of the genus *Astata* (1957a; 27 pages), sand wasps of the genus *Bembix* (1957b; 248 pages), and sand wasps of the entire subfamily Bembicinae, then called the Nyssoninae (1966a; 526 pages). Later, it would be books on Australian *Bembix* (1973, with Robert Matthews; 387 pages) and North American beewolves of the genus *Philanthus* (1988, with myself; 278 pages). And these were interspersed with hundreds of shorter scientific papers. In 1976, "for his work over a 25-year span on the biology and evolution of behavior in wasps," he was awarded the Daniel Giraud Elliot Medal by the U.S. National Academy of Sciences, a honor he shares with such biological luminaries as William Morton Wheeler, Theodosius Dobzhansky, Sewell Wright, Ernst Mayr, George Gaylord Simpson, and George C. Williams.

Although many entomologists can boast of a body of distinguished scientific works, few are also as adept as Howard was at writing for a general audience. Over a nearly 40-year span, he produced a number of general books on entomology, natural history, and the history of science, some of which have reached wide audiences (H. E. Evans 1963, 1969, 1984, 1985, 1993, 1997, 2001; H. E. Evans and M. A. Evans 1983, 1991; M. A. Evans and H. E. Evans 1970; Evans and West-Eberhard 1970). Recently his often-reprinted *Life on a Little-Known Planet* was even released as a book-on-tape (New Millennium Audio, 2002), a rarity for a work on insects. But then again, the book has the hallmarks of a best-selling thriller: bloodshed (by bedbugs), violence (among crickets), deception (by insect mimics), code-breaking (by entomologists studying insect communication), and, of course, sex in its many insectan manifestations. It is no coincidence that Howard was an avid reader of detective mysteries (a habit he shared, along with a talent for popularization and a love of beewolves, with Niko Tinbergen [Kruuk 2004]).

As one might expect, someone like Howard does not find it easy to put a pen down, even in retirement. Along with the five books he published following his official retirement in 1983, we can now add a recent collection of his essays, edited by his wife, Mary Alice Evans, and published under the title *The Man Who Loved Wasps: A Howard Ensign Evans Reader* (Evans 2005). But that is not the last word. After Howard died in 2002, Mary Alice, herself an accomplished naturalist, passed to me his unpublished 20,000-word manuscript entitled "Recent Research on the Ethology of Sand Wasps (Sphecidae: Bembicini)," which he apparently intended as a partial update to his 1966 *Comparative Ethology and Evolution of the Sand Wasps.* To complete this manuscript, and to make it parallel with that earlier book, I have expanded the material on the tribe Bembicini prepared by Howard (Chapters 6–7 here). I also added material on the remaining four tribes (Chapters 2–5) covered in the 1966 book, but not in his manuscript. Taken along with his 1957 *Bembix* and the 1973 Australian *Bembix* book he coauthored with Bob Matthews, this work should provide a comprehensive review of the biology of sand wasps. Mary Alice Evans approved the general format of the book, but because Howard had no say over its final form, it is appropriate to be explicit about which authors were responsible for each chapter: 1–5 (O'Neill), 6–7 (Evans, with additions by O'Neill), and 8 (both authors). Needless to say, I take responsibility for any errors, whether of inclusion or exclusion.

In preparing this work, the intent was not to repeat all of the information in the 1966 review. The bulk of this work will be a tribe-by-tribe, species-by-species review of biological information appearing since 1966, though we make frequent reference to earlier studies and provide a comprehensive summary in Chapter 8 that integrates old and new information. We include a few previously unpublished observations at appropriate places in the text (designated as "HEE and/or KMO, previously unpublished"), as well as an appendix containing some general questions that might help guide future research. Chapter 1 provides an overview of the classification and behavior of sand wasps. Chapters 2–5 are each devoted to a single tribe of the subfamily Bembicinae, whereas coverage of the large tribe Bembicini is split between Chapters 6 and 7. For each genus, we first summarize reports appearing since 1966 on individual species, which are listed alphabetically. Our summaries of information on genera and/or tribes include a mix of new information and that available to Evans (1966a). Readers should consult that book, along with Evans (1957b), for references to older accounts.

I would like to thank Ruth Pettinga O'Neill, William Wcislo, and an anonymous reviewer for comments on the manuscript; Thomas O'Neill for assistance with proofing; Mary Jane West-Eberhard for encouraging me to pursue the project; and Ann Downer-Hazell and Elizabeth Gilbert of Harvard University Press. Several colleagues and friends helped by translating papers into English: Byron Alexander, Richard Miller, and Caroline Von Meyenfeldt. Al Hook and Greg Palmer allowed us to cite some of their unpublished information on *Bembecinus neglectus*. Richard Hurley, Gabriel Melo, Richard Miller, and Martin Villet provided advice on various taxonomic issues.

Kevin M. O'Neill

The Sand Wasps

— 1 —

Introduction

A Sand Wasp Sojourn

In the Chihuahuan desert, at the LaJoya State Game Refuge in central New Mexico, there is a habitat that is incongruous in its surroundings: 30 acres of natural wetlands within the flood plain of the Rio Grande. Adjacent to the wetlands is a habitat more commonly associated with deserts, a series of dry sand dunes that include a mix of sparse vegetation and open sand. Naturalists visiting the refuge's marshes and dunes typically come for the sandhill cranes, double-crested cormorants, turkeys, Gambel's quail, pyrrhuloxia, black-throated sparrows, and vermillion flycatchers, among other attractions. Our main interest in the area, however, and that of our colleague Bill Rubink, was the ground-nesting solitary wasp fauna.

Solitary wasps do not live in highly integrated, all-female colonies like the more familiar hornets and yellow jackets that most people fear for their stings. Rather, each female constructs her own nest, most often in the ground, and provisions it with food for her larvae. The stock of food necessary for each offspring to reach its own adult stage may be provided all at once, before the young wasp larva hatches from its egg. Or, as in the case of many of the wasps we studied, the mother continues feeding the larva after it hatches, sometimes monitoring its progress until it is fully grown and ready to spin a cocoon. Although solitary wasps do not, by definition (O'Neill 2001), cooperate with one another, nests of one or more species may be bunched in large, often closely spaced aggregations. The aggregations may simply be a by-product of the females' shared preferences for particular habitats, although there is also some evidence that clustering has its own virtues (Wcislo 1984; Larsson 1986). The tendency for many solitary wasps to aggregate is one of their features that appeals to ethologists.

Also, unlike many of their social relatives, solitary wasps do not attack and sting human intruders, even when the human has the gall to unearth their nests in search of data. The more important lures to study, however, are the fascinating behaviors of these often large and attractively colored insects.

Within a relatively restricted area of the LaJoya dunes and surrounding alluvial flats, we found a taxonomically and behaviorally diverse group of solitary wasps. One was *Cerceris bicornuta* Guérin, which nested in small aggregations near the dunes and fed paralyzed adult weevils to its young (Evans and Rubink 1978). Its nests occurred in the firm sandy roadside at the base of the dunes, along with those of *Tachytes aurulentus* (F.), a predator of katydids (Evans and Rubink 1978), and *Plenoculus cockerelli* Fox, a tiny predator of even tinier caterpillars (Rubink and O'Neill 1980). In the sparse vegetation surrounding blowouts on the dunes, we found nests of one of the smallest North American species of beewolf, *Philanthus psyche* Dunning, a predator of small bees and wasps, including *Plenoculus* (Evans and O'Neill 1988). Not only are beewolves interesting in themselves, but examination of their prey can aid in surveys of wasp and bee faunas because they are at least as good as entomologists in finding these insects. To be honest, they are probably better at it. Another small caterpillar predator, *Plenoculus boregensis* Fox, is a true dune inhabitant at the refuge, digging its nests in expanses of open sand (Rubink and O'Neill 1980).

But the 6 mm long *Plenoculus* can be easily overlooked amid aggregations of the most impressive wasp dweller of the dunes, *Bembix pallidipicta* F. Smith (Figure 1.1), whose activities were studied at LaJoya by Bill Rubink (1978, 1982). The black and pale white *B. pallidipicta* are not only much larger than other solitary wasps at LaJoya, being several centimeters long, but they nest in conspicuous aggregations of thousands of individuals. Early in the nesting season, the activity of males provided an almost chaotic spectacle. While searching for virgin females, thousands of males alternated long streaming flights just above the dune surface with "hopping dances" that, in this species, make groups of males look like "aggregations of very small toads" (Evans 1957b). Mass swarms such as these are famous among wasp biologists as "sun dances," a term first used by Phil and Nellie Rau (1918). But as the supply of virgin females dwindled during the season, males disappeared and females got down to the business, now without male harassment, of digging and provisioning nests. To nest in the unstable sand of the dunes, females construct an amazing but short-lived structure (described earlier for this species by Evans 1957b). The ability of the

Figure 1.1. Female *Bembix pallidipicta* at nest entrance. Photo by H. E. Evans.

female to dig a shallow, half-meter-long tunnel just below and essentially parallel to the surface of the dry sand, and not have it collapse, is enough to make a devoted builder of sand castles green with envy. And the wasps do it without the use of moisture to bind sand particles together. The single brood cell sitting 16–56 cm beneath the surface in each nest is stocked with a variety of flies. Hunting females probably benefit from the great quantities of flies available in the nearby marsh. We saw this as vicarious revenge for the benefit that many of the marsh insects received from feeding on our blood while we camped at LaJoya.

Bembix pallidipicta and its relatives are members of the subfamily Bembicinae, traditionally placed in the hymenopteran family Sphecidae, to be discussed further below. Apoid wasps are also often referred to as *digger wasps,* even though not all of them dig nests. Wasps of the Bembicinae subfamily are themselves loosely referred to as *sand wasps,* despite the fact that many species do not nest in sand. Such is the imprecise nature of common names used for higher taxonomic categories that include a behaviorally and ecologically diverse set of organisms. But, like the name "grasshoppers" (many of which do not live on or eat grasses) and "scorpionflies" (only some of which have scorpion-like tails), the use of common names is dictated more by tradition than by biological appropriateness. We continue to use the term *sand wasp* here to refer to all members of the subfamily Bembicinae, and will give a more precise taxonomic definition below.

Two other sand wasps inhabiting the LaJoya dunes were *Trichogorytes cockerelli* (Ashmead) and *Microbembex hirsuta* Parker. *Trichogorytes cockerelli* drew our attention because its biology was unstudied, and it belongs to a genus that occurs only in the southwestern United States and Sonora, Mexico. It turned out to be a run-of-the-mill leafhopper predator like its closest relatives, but certain aspects of its nest-digging behavior were somewhat unique (Evans 1976a), as will be described later. Not to be surpassed by our collecting skills, of course, the beewolf *Philanthus psyche* also included *T. cockerelli* among its prey (Evans and O'Neill 1988).

Microbembex hirsuta, a member of a better-known and widespread New World genus, was of interest because it had been studied only briefly in the past (Evans 1976b). The "hunting" technique of *Microbembex* females is unique among all solitary wasps. Rather than prey on living insects, they provision nests with the dead bodies or body fragments of arthropods scavenged from the soil surface. The dry, often partially decayed fragments they feed to their offspring must have a much lower water content than that provided by solitary wasps that hunt living arthropods. In most digger wasps, not only is the food fresh and moist when placed in the nest, but the prey have often been injected with a venom that keeps them alive, but subdued, until consumed. Nevertheless, *Microbembex* obviously do quite well on their dry, crunchy diets, because they are commonly encountered in arid and semi-arid habitats of the western United States. Though "only" scavengers, *Microbembex* should be admired for their ability to compete with those most ubiquitous and efficient of scavengings insects, the ants. In fact, they often pick up dead ants to feed their offspring.

Places like LaJoya provide valuable, relatively undisturbed locations for sand wasp studies, but the variety of wasps at the site, while interesting, is fortunately not unique. Sand wasps occur worldwide and are particularly diverse in dry habitats. But some Bembicinae occur in relatively moist streamside habitats, and others can even be found on suburban lawns or urban playgrounds. The areas of the southwestern United States and northern Mexico surrounding LaJoya contain a rich fauna of sand wasps that includes little-studied (or unstudied) genera like *Arigorytes, Hapalomellinus, Hyponysson, Trichogorytes, Xerogorytes,* and *Xerostictia* that occur nowhere else in the world (Bohart and Menke 1976). Endemic representatives of geographically widespread genera also occur there, but the southwestern United States itself is not unparalleled. The genera *Editha, Neoplisus, Selman, Trichostictia,* and *Zyzzyx* are found only in South America, and *Carlobembix, Hemidula,* and *Liogorytes* are restricted to Argentina and *Neonysson* to Chile. Only in Australia can we find *Acanthostethus, Austrogorytes,* and *Exeirus,* and South Africa boasts a unique set of species that has received increased scrutiny in recent years. But not all sand wasps are local curiosities. The group also includes genera that have wide geographic distributions, like *Argogorytes, Bembecinus, Bembix, Gorytes, Hoplisoides, Nysson, Sphecius,* and *Stizus,* though any given area may have its unique set of species with unique behaviors. This will be evident when we review, for example, the adaptive radiation of *Bembix* that has occurred in Australia (Evans and Matthews 1973), and the remarkable unique nests and prey of several *Bembecinus* in South Africa (F. W. Gess and S. K. Gess 1975).

Sand Wasp Natural History / Sand Wasp Science

The final chapter of the *Comparative Ethology and Evolution of the Sand Wasps* (Evans 1966a) noted that much had been learned since 1945 when E. T. Nielsen published "Moeurs des *Bembex*" ("The Habits of *Bembex*"); "*Bembex*" in Nielsen's usage included a large portion of the group we now call sand wasps. The last 40 years has seen a similar increase in our knowledge. We know much more about certain species and genera that had received little or no attention by 1966. The sand wasps of particular geographic regions, especially Australia, South Africa, and South America, are much better studied. Some aspects of sand wasp biology that previously garnered little consideration have been examined in detail. The data gathered are often more quantitative and amenable to statistical analyses. And sand wasp behavior is interpreted in the light of theoretical developments and refinements stimulated by such scientists as William D. Hamilton (1964), George C. Williams (1966), Willi Hennig (1966), and Robert Trivers (1972).

The trend toward greater empirical and theoretical rigor in wasp biology is to be welcomed, but there is still much good to be said about publications that present basic records of wasp natural history with little, if any, statistical treatment of data and with no explicit testing of major theories in ethology and evolution. E. O. Wilson (1992) defines natural history as "ecology expressed in the details of the biology of individual species," and Greene (2005) refers to it as "descriptive ecology and ethology." Natural history studies help fulfill what Stephen Jay Gould (2001) called one of the two major goals of science, "to determine, as best we can, the empirical character of the natural world." Unfortunately, natural history "has earned the pejorative of 'alpha' ecology'" to equate it with, most likely in a denigrating way, alpha taxonomy or the description and naming of species; and natural history "has often been considered to have little or no potential for generating ideas" (F. C. Evans 1985, quoted in Noss 1996). Mary Jane West-Eberhard (2001) took the opposite stand in her criticism of the "belief that meaningful questions can be, or even must be, formulated without knowing something about the organism first." We agree, and argue that carefully conducted natural history studies not only help provide that "something about the organism," but contribute much to our knowledge of the biological diversity of the planet.

Piecemeal accounts of solitary wasp behavior, including those that provide accounts of a few individuals of one species at one place and time, continue a strong and important tradition of scientific natural history studies of Hymenoptera. Natural history accounts of the type that form the core of this book have two important functions in ethology, or three if you count their value as entertainment. One is that, even when spare and nonquantitative, they have heuristic value if they stimulate curious skeptics to look deeper into the biology of an organism. As West-Eberhard (2001) notes, in her praise of organism-centered research, "the more you know about a particular species or a particular group of organisms the more you can find out about it, and the more valuable it is as a resource for research of general interest." Admittedly, many natural history accounts have so little rigor that they are no more than suspect anecdotes, and may not lead to further research. One of us (KMO) was once asked by a bartender to identify an insect that, according to the hastily drawn figure on a napkin, had four legs and a distinct smile on its face. That amateur naturalist was apparently serious and his curiosity was admirable, despite his suspect observational skills. But the question stimulated no further research, though it certainly had the above-mentioned entertainment value. An example of a more productive trigger to research, one that we relate in Chapter 7, was an anecdotal report by Wheeler and Dow (1933) of a species of *Bembix* preying upon damselflies; their finding was surprising at the time, because all North American and European *Bembix* studied prior to the 1970s preyed only on true flies. Wheeler and Dow's report, based on a brief description of nest cells excavated in a dry lake bed in Australia, was subsequently confirmed by Evans and Matthews (1973), who trekked to the remote site specifically to follow up on the earlier report.

Another value of natural history accounts is that they provide raw data that can be integrated into reviews and syntheses of the biology of a group, and a chance to test ecological and evolutionary hypotheses. In a sense, we can consider natural-history data gathering to be an endeavor by a loosely organized community of naturalists who communicate and coordinate mainly through publications (and more recently the Web). No meetings are held to organize the project, and the resulting data may be uneven in quality because individual participants vary in outlook, training, resources, and luck (e.g., having the good fortune to find an interesting population to study). Perhaps an organized, concerted effort would be more productive, but it seems highly unlikely that funding agencies and politicians will support a Wasp Ethology Project as they did the Human Genome Project. As Greene (2005) points out, the annual budget for the Systematic Biology Program at the U.S. National Science Foundation is only one-third of what NASA spent for one toilet on a space shuttle. Even so, members of the wasp natural history community eventually accumulate enough data so that certain members can pause to review and synthesize. But wasp researchers do not record just any old observation, they are guided by past research and by evolutionary theory. A researcher might focus on recording the manner in which a wasp carries its prey, since this provides information that could be used to test hypotheses about the phylogeny of behavior. Or she could study interactions that wasps have with their neighbors in a nesting aggregation, as a means of testing hypotheses concerning the evolution of social behavior. On the other hand, it would probably be a waste of time to record the color and texture of every rock on which a wasp perched during the day (though if you are clever enough, you might even be able to deduce something relevant from that—perhaps some rocks provide better camouflage or remain cooler on hot days). Reviews of solitary wasp behavior appearing since 1966 include Evans and West-Eberhard (1970), Iwata (1976), Evans (1977), Alcock et al. (1978), Wcislo (1981), Evans and O'Neill (1988), Cowan (1991), Field (1992), and O'Neill (2001). Some analyses have even crossed major taxonomic boundaries, as is the case of

Itô's (1978) analysis of the evolution of reproductive rates of plants, insects (including solitary wasps), and vertebrates, and Brockmann's (1997) comparison of wasp and bird social systems.

Why study sand wasps in particular? Is the choice of sand wasps (rather than, say, thrips or velvetworms) just a smaller-scale version of the decision to fund research on zero-gravity toilets rather than on the taxonomy of liverworts? Of course, all natural history is fun and all knowledge is useful to some extent, but ethologists looking for a focal group of study organisms (and companions on the road to fame, fortune, and tenure) are not likely to be swayed by such answers. One important part of our answer, to which we have already alluded, is practical: the subfamily Bembicinae includes some of the largest solitary wasps in the world, and many species nest in dense aggregations that often include a mix of species. Large size and gregarious habits increase the ease of observation. This should get the attention of anyone who thinks that insects (such as thrips) are too small to observe easily in the field. But if you are up to the challenge of making field observations on insects that are just several millimeters long, you can also find sand wasps in that size range. Equally important is the diversity of sand wasps, which far exceeds that of velvetworms, so scientists making themselves familiar with the local sand wasp fauna have a wide variety of subjects available. Along with the diversity of species, sand wasps exhibit diversity in behaviors that facilitate comparative analyses. Although most studies have been strictly observational, bembicines are also amenable to experimental studies, as demonstrated, for example, by studies of homing behavior (van Iersal 1952), deceptive pollination (Ågren and Borg-Karlson 1984; Borg-Karlson 1990), thermoregulation (Ghazoul and Willmer 1994), sexual interactions (O'Neill et al. 1989; Eason et al. 1999), and digging behavior (Coelho and Wiedman 1999).

Sand Wasp Classification: A Short Course

To appreciate the diversity of sand wasps and to revel in the intricacies of each species behavior and ecology, one needs to know very little about how hymenopterists classify sand wasps and their closest relatives. Here we will summarize the current state of sand wasp classification, mentioning along the way a few issues that are still unresolved.

Sand wasps are members of the insect order Hymenoptera, suborder Apocrita, which includes a lineage of insects informally known as the

Aculeata: the ants, bees, and various groups of wasps. Among the superfamilies making up the Apocrita is the Apoidea, which includes the bees and certain wasps (called apoid wasps), but not the social wasps which are placed in the superfamily Vespoidea. Most of those "certain wasps" have been traditionally placed in the family Sphecidae, the family name still found in many textbooks. Recent phylogenetic analyses, however, now support splitting the Sphecidae into multiple families. Whatever the final verdict of entomologists on this matter, it seems likely that the sand wasps covered in this book will be placed in the subfamily Bembicinae of the appropriate family. Evans (1966a) and Bohart and Menke (1976) actually used the term Nyssoninae rather than Bembicinae, but here we follow Menke (1997). So, from here on, we usually use the term Bembicinae even when referring to earlier works that used Nyssoninae. The Bembicinae, as constituted in Bohart and Menke (1976), is the second-largest subfamily of the Sphecidae, containing (at the time of their treatment) 71 genera and nearly 1,500 described species. In Pulawski's (2006a) updated count, the same tribes listed by Bohart and Menke contain >80 genera and >1,700 species, or nearly one-third of the genera and ~18% of the species of apoid wasps.

Depending upon which authors one consults, the Bembicinae is listed under one of three family names: Sphecidae *(sensu lato),* Nyssonidae, or Crabronidae. The Sphecidae, as traditionally and broadly defined, is now thought to be a paraphyletic group, meaning a taxon (i.e., group of organisms at some particular taxonomic rank) that encompasses organisms that share a common ancestor, but which excludes related organisms that also share that ancestor. A vertebrate equivalent would be a taxon that included wolves, coyotes, and tigers, but excluded lions and other cats and dogs. The paraphyletic group we are interested in is sometimes called by the informal name "spheciforme wasps." Spheciforme wasps would form a monophyletic taxon, one that represents a complete phylogenetic lineage, only in combination with their closest relatives: all of the bees and, perhaps, the small wasp family Heterogynaeidae (Alexander 1992; Brothers and Carpenter 1993; Brothers 1999; Prentice 1998; Melo 1999; Michener 2000; Ohl and Bleidorn 2006) (Figure 1.2). The exact timing of the evolutionary divergence of bees from their "wasp" relatives is subject to debate, but bees, themselves monophyletic, were certainly diverse by the late Eocene (Michener 2000). The evolutionary divergence of bees from their wasp relatives was associated with a switch to feeding offspring with pollen

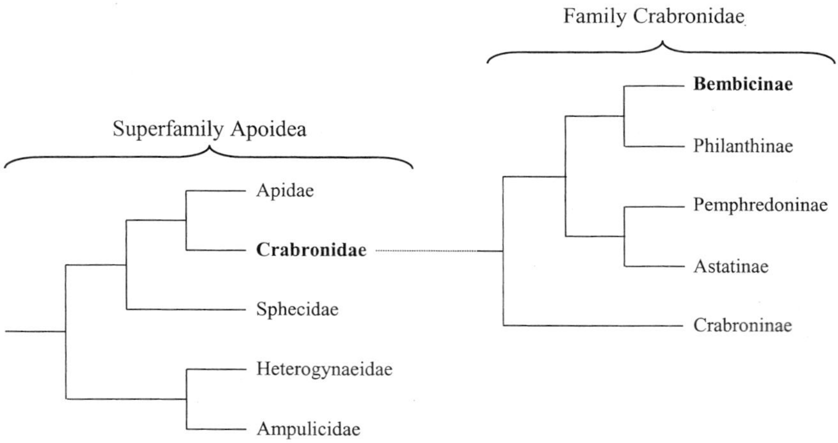

Figure 1.2. Phylogenetic relationships among families of the superfamily Apoidea (left) and subfamilies of the family Crabronidae (adapted from Melo 1999). Melo's Apidae refers to a monophyletic "Apidae" *(sensu lato);* bees are divided into numerous families by some authors. The "Sphecidae" here is the Sphecidae *(sensu stricto)*, referring to the wasps that Bohart and Menke (1976) included in the subfamily Sphecinae of their family Sphecidae.

and nectar, and the acquisition of correlated morphological traits (e.g., branched body hairs and an enlarged hind basitarsus for collecting and carrying pollen in some groups of bees).

In response to the recognition that the family Sphecidae *(sensu lato)* is paraphyletic, some authors have raised its component subfamilies to the level of monophyletic families (Table 1.1). In the most recent published analysis, for example, Melo (1999) places the Bembicinae in the family Crabronidae, along with the Astatinae, Crabroninae, Pemphredoninae, and Philanthinae. Other traditional sphecid subfamilies (i.e., as in Bohart and Menke 1976) are placed elsewhere, the Sphecinae in the Sphecidae *(sensu stricto)* and the Ampulicinae in the Ampulicidae. Analyses by Prentice (1998) and Melo (1999) support the conclusion that the Bembicinae considered in Evans (1966a) and here is a monophyletic group (along with the rare and little-studied South American genus *Heliocausus*).

Moving down the Linnaean hierarchy, we get to the question of which of the traditional tribes of the Bembicinae should be included in the subfamily and which genera should be included in each tribe, assuming that the goal is to construct monophyletic subfamilies and tribes. The standard reference on the Sphecidae, Bohart and Menke (1976), further subdivided the Bembicinae into seven tribes: Mellinini, Heliocausini, Alyssontini, Gorytini, Nyssonini, Stizini, and Bembicini (Table 1.1). Two of those seven are not further considered in this book. The Mellinini are excluded because recent analyses suggest that they are not closely related to the five tribes covered here. The Heliocausini are left out because of a lack of biological information on this small group of wasps (Alexander 1992; Prentice 1998; Melo 1999). The taxonomic and phylogenetic status of the five tribes considered in this book will be discussed individually in Chapters 2–6.

The foregoing should provide the reader with a general idea of the protean state of the classification of sand wasps and the correspondence of familial, subfamilial, and tribal names used in different taxonomic and biological accounts (Table 1.1). Future reshuffling and renaming of the subfamilies, tribes, and genera of apoid wasps may modify the details. For now, we will use the tribal affiliations given in Bohart and Menke (1976) and Menke (1997). The species names of Bembicinae we use here have been checked against those in the Catalog of Sphecidae *sensu lato* (Pulawski 2006a), which also contains reference to specific taxonomic works (including keys) on individual taxa. In most cases, species names used for prey, brood parasites (except when they are other apoid wasps), parasitoids, and predators of the wasps are those given in the cited publications.

Biology of the Bembicinae: A Primer

In this section we provide a general overview of the behavior of sand wasps, in order to place the individual species accounts that follow into a general context. In the final chapter, we will discuss differences among species in more detail. A general review of the behavior of both female and male solitary wasps, including sand wasps, can be found in Evans and West-Eberhard (1970) and O'Neill (2001), which contain numerous references to the material presented here.

Female Bembicinae are either solitary nest provisioners or brood parasites of other wasps. In the nest-provisioning species, each female constructs one or more ground nests using the mandibles and front legs as the primary digging tools. Digging with the legs is made more efficient in most species by the presence of a row of stout "rake spines" (also called a pecten) on each foreleg. A nest consists of a *main burrow* and one or more *brood*

Table 1.1 Alternative classification schemes for sand wasps. In zoological systematics, family names end with -idae, subfamilies end with -inae, tribes end with -ini, and subtribes end with -ina.

Bohart and Menke (1976)	Krombein et al. (1979); Finnamore and Michener (1993)	Menke (1997)	Melo (1999)	Prentice (1998)
Sphecidae: Nyssoninae; Mellinini	Mellinidae:	Sphecidae: Bembicinae; Mellinini	Crabronidae: Crabroninae	Crabronidae: Mellininae
Sphecidae: Nyssoninae; Heliocausini	Nyssonidae: Heliocausinae[a]	Sphecidae: Bembicinae; Heliocausini	Crabronidae: Bembicinae	Crabronidae: Bembicinae; Bembicini (Heliocausina)
Sphecidae: Nyssoninae; Alyssonini	Nyssonidae: Alyssoninae	Sphecidae: Bembicinae; Alyssontini	Crabronidae: Bembicinae	Crabronidae: Bembicinae; Alyssontini
Sphecidae: Nyssoninae; Gorytini	Nyssonidae: Gorytinae	Sphecidae: Bembicinae; Gorytini	Crabronidae: Bembicinae	Crabronidae: Bembicinae; Bembicini (Clitemnestrina,[b] Gorytina, Handlirschiina)

Sphecidae: Nyssoninae; Nyssonini	Nyssonidae: Nyssoninae	Sphecidae: Bembicinae; Nyssonini	Crabronidae: Bembicinae	Crabronidae: Bembicinae; Nyssonini
Sphecidae: Nyssoninae; Stizini	Nyssonidae: Stizinae	Sphecidae: Bembicinae; Stizini	Crabronidae: Bembicinae	Crabronidae: Bembicinae; Bembicini (Bembecinina, Stizina)
Sphecidae: Nyssoninae; Bembicini	Nyssonidae: Bembicinae	Sphecidae: Bembicinae; Bembicini	Crabronidae: Bembicinae	Crabronidae: Bembicinae; Bembicini (Bembicina)

[a]Because Heliocausinae occur only in southern South America, the subfamily is not included in Krombein et al. (1979), which covers only North American taxa; [b]subtribe Exeirina of Pulawski (2006a).

cells, which may be just off the main burrow or at the end of side burrows (cell burrows) (Figure 1.3). The number of cells varies among species, and only one nest is maintained at a time by a female. The type of soil substrate inhabited by females is generally species-specific, and across the subfamily includes dry or damp soil, which may be hard clay, or sand and loam of various grain sizes.

A nest burrow with an open entrance leaves the brood accessible to predators and parasites (see below). Thus in many species, females construct a *temporary closure* at the nest entrance while they are away hunting. Although a temporary closure may prevent entry of natural enemies, it also delays reentry of the female when she returns with prey. Following completion and provisioning of the final brood cell, the female soon leaves the nest for good. At this time, females of most species construct a *final closure,* packing soil into the burrow using the tip of the abdomen. Along with the *outer closure* at the level of the nest entrance, females of some species construct a deeper *inner closure* at the level of the brood cells. Most Bembicinae leave the mound of soil that accumulates at the nest entrance intact both during and after nest construction and provisioning. Some species, however, level the mound at various times so that the nest entrance

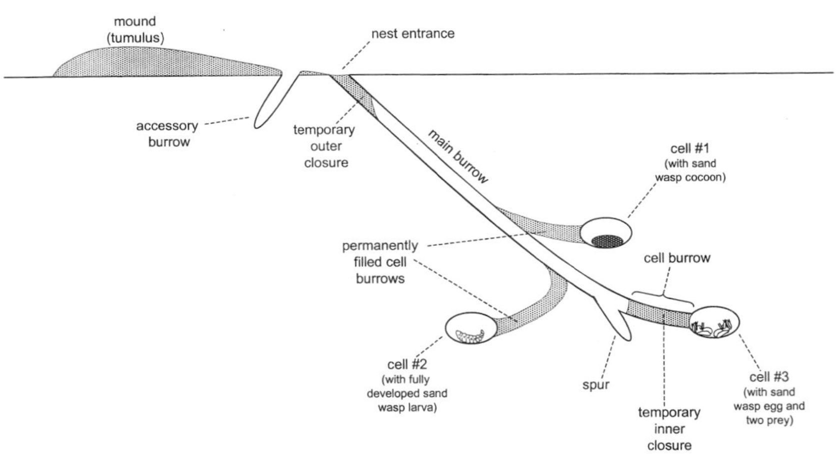

Figure 1.3. Diagrammatic representation of the nest of a sand wasp (see text for explanation). Note that all of the features illustrated (e.g., multiple cells, accessory burrows, and spurs) may not be present in all species' nests, or in all nests of the same species; nor will such a wide range of developmental stages necessarily be present simultaneously in one nest.

is concealed, often going through a set of stereotypical leveling movements that may leave tell-tale patterns in the sand. Females sometimes also build *accessory burrows* (or *false burrows*) within several centimeters of their real nest entrance. These short burrows do not lead to brood cells and are not used for prey storage or sleeping. Although soil from the accessory burrows is sometimes used to provide fill for closing the main burrow, their primary function may be to provide visual cues that distract the attention of natural enemies away from the nest.

Non-soil nesting is common in other subfamilies of solitary aculeate wasps, but no Bembicinae are definitively known to nest in preexisting cavities in plant stems or wood, and none build free-standing nests. The intriguing possibility, based on scant evidence, that one species of the genus *Argogorytes* excavates nests in rotten wood and reports that some species nest in preexisting cavities in rock or soil are discussed later. Although communal nest sharing and even eusociality have evolved in other subfamilies of apoid wasps (Matthews 1991), nest sharing in the Bembicinae has been reported in just one species of *Bembix* (S. K. Gess and F. W. Gess 1989).

Each brood cell in a sand wasp nest houses a single offspring, along with prey provided by the mother for its sustenance. Females of nest-provisioning Bembicinae range from specialist predators that take prey of a single family or genus to extreme generalists (opportunists) that take prey from multiple, often only distantly related, orders of insects. The range of prey used in any population and in any year is determined by innate preferences and hunting methods, as well as by local prey availability. The subfamily Bembicinae as a whole preys on families of the insect orders Odonata (damselflies), Orthoptera (grasshoppers and katydids), Mantodea (mantids), Homoptera (leafhoppers, treehoppers, spittlebugs, and so on), Hemiptera (true bugs), Neuroptera (antlions and lacewings), Lepidoptera (butterflies and moths), Diptera (true flies), and Hymenoptera (bees and wasps). Several groups of arthropods notable for their absence from this list are preyed upon by other solitary wasps. Coleoptera (beetles), the largest order of insects, is not exploited by female bembicines, though they are used by other solitary wasps, most notably *Cerceris, Eucerceris* (Philanthinae), and many eumenine wasps (Vespidae). Nor do sand wasps prey on spiders, a group utilized by solitary wasps such as *Chalybion, Sceliphron* (Sphecinae), *Miscophus, Pison, Trypoxylon* (Larrinae), and the entire family Pompilidae. *Microbembex* occasionally include dead beetles and spiders among the materials they scavenge for food, but they do not hunt living members of these groups.

Several other aspects of female behavior associated with hunting, nest provisioning, and parental care are often noted in publications on sand wasp behavior. One is the manner in which they carry prey to the nest while returning from their hunting grounds. The method of prey carriage varies among species of solitary wasps, but tends to be relatively constant within genera (being affected by the shape and relative size of the prey). Alyssontini carry prey in their mandibles, sometimes while walking part of the way. But species of Gorytini, Stizini, and Bembicini carry prey in flight all the way back to nests, holding the prey primarily with their middle legs (Figure 1.4), though the hind legs may help with maintaining a stable grip on larger prey. Females of all species carry just one prey at a time.

A second commonly reported aspect of parental behavior is the timing and location of egg laying. The egg may be laid in the empty cell prior to provisioning (sometimes on a prepared pedestal), in a particular location on the first or last prey in the cell, or on top of the mass of prey after all prey have been provided (Figure 1.5). The rate of egg laying is likely to be variable among species. Brood parasitic sand wasps, which forgo nest building and provisioning, have a higher potential rate of egg laying, which

Figure 1.4. Female *Rubrica nasuta* carrying prey. Female has just landed at the nest entrance and is removing temporary nest closure. Photo by H. E. Evans.

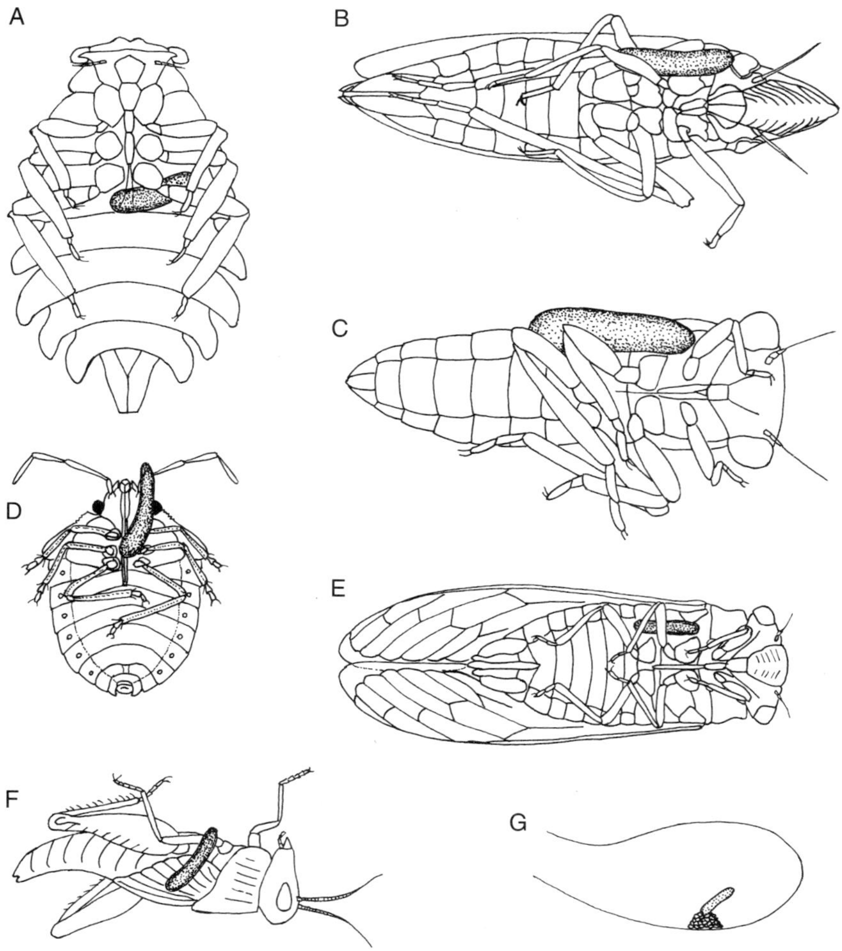

Figure 1.5. Position of sand wasp eggs (stippled sausage-shaped objects) on prey or in an empty cell (not drawn to same scale): (A) immature treehopper prey of *Hoplisoides nebulosus*, bearing egg of *Nysson daeckei*, a brood parasite; (B) adult leafhopper prey of *Alysson melleus*; (C) immature leafhopper prey of *Gorytes canaliculatus;* (D) immature stinkbug prey of *Bicyrtes ventralis;* (E) adult cicada prey of *Sphecius speciosus;* (F) immature short-horned grasshopper prey of *Stizus fasciatus;* and (G) empty nest cell of *Bembecinus neglectus* in cross-section. All redrawn from Evans (1966a).

is reflected in the fact that they carry more mature eggs (up to six) in their ovaries than do nest provisioners. At the other extreme in the subfamily are the progressive provisioners (see below), which may lay only one egg per week and carry no more than a single mature egg with them at any time.

The third feature is the number of prey per cell (i.e., per offspring), which varies, as might be expected, as a function of the size of the prey relative to the size of the offspring. A few sand wasps that take large prey may provide some offspring with just a single prey as the offspring's entire sustenance, whereas other species provide many dozens of prey to each offspring.

The fourth is the duration and timing of provisioning relative to the developmental schedule of egg and larva. At one extreme, *mass provisioners* bring in all of the prey and close the cell before the egg hatches. *Delayed provisioning* (also called *slow mass provisioning*) is basically mass provisioning that occurs over a long enough interval that the egg hatches prior to cell closure. Evolutionarily, it was likely the precursor of *progressive provisioning* (Evans 1966a), where the provisioning period overlaps substantially with the larval period of the offspring. In *truncated progressive provisioning*, the female closes the cell before the larva completes feeding. In its *fully progressive* form, progressive provisioning continues until the larva is ready, or nearly ready, to spin its cocoon. Females of fully progressive provisioners have the opportunity to monitor the development and needs of their offspring, and the mothers in a few species even clean refuse from the cell during the process. Among solitary aculeate wasps, the Bembicinae probably show the greatest range of types of provisioning, though some *Ammophila* of the Sphecinae are also progressive provisioners.

Other less commonly reported details of bembicine nesting and foraging behavior that will occasionally be treated in this book are the homing and orientation of females going to and from nests, the location of their hunting grounds, host-searching behavior, stinging behavior, and the form and duration of prey paralysis. The latter three are often difficult to study because it may be difficult to find wasps hunting, which may take place at some distance from their nest sites. Species of sand wasps that prey on flies can often be seen hunting in the vicinity of large animals, animal droppings, and garbage. Sand wasps that take herbivorous insects undoubtedly hunt on or around the host plants of their prey.

After consuming its provisions, the larva spins a cocoon within its cell. The result is a quite remarkable structure, considering that it is constructed by an animal lacking appendages, using only a pair of mandibles to manipulate objects. The cocoon is an elongated egg-shaped shroud of silk, whose walls are lined with sand grains chosen from within a particular size range. Along the equatorial line around the major axis of the cocoon, the larva also constructs a series of minute pores, sometimes with cone-shaped rims. The function of the pores is unstudied, though they perhaps aid gas exchange, while preventing entry of liquid water. At some time after capping the cocoon, the larva enters a resting and quiescent *prepupal* phase. Pupation does not occur until much later, not until the following spring in temperate zone species.

Adults of the following generation emerge from their subterranean natal cells over a period of several weeks, the initiation and peak of male emergence usually preceding that of females (a phenomenon referred to as *protandry*). In species having compact or relatively discrete nest aggregations, the area in which adults appear from the ground the following generation is termed the *emergence area.* Most, if not all, temperate zone bembicines go through a single generation per year *(univoltinism),* but some species in warmer areas have two generations per year *(bivoltinism).* A few bembicines are known to be *parsivoltine,* meaning that adults of one generation have a bimodal emergence period, with the second peak occurring up to one year after the first.

Adult sand wasps of most species probably live from several weeks to several months. They spend much of their time in an inactive state referred to as *sleeping,* which in honey bees has many physiological and behavioral similarities to vertebrate sleep (Kaiser 1995). Sleep may be forced upon wasps by lower temperatures and light levels during the night or intervals of inclement weather. Many wasps spend these periods within nests, though we cannot always be sure that females are completely inactive at those times—because they could be building cells or arranging prey, for example. Other wasps build short, temporary sleeping burrows, whereas many sleep on plants or under rocks, sometimes singly, sometimes in groups that may include more than one species.

The nests of sand wasps are attacked by a variety of natural enemies, many of whom specialize on bembicine wasps and their relatives. The legless, soft-bodied sand wasp larvae are fairly helpless creatures, their only line of defense being the nest constructed by the mother and the measures she undertakes that help prevent access of parasitoids and predators to the nest. Parasitoids of the bembicine larvae, that is, those that gain all or most of their sustenance from feeding on the wasp larva itself, include flies of the family Bombyliidae, beetles of the family Rhipiphoridae, and aculeate wasps of the families Chrysididae and Mutillidae. The other major enemies that attack the nest exploit the prey stores in cells. These *brood parasites* (also called *cleptoparasites*) often kill the wasp's offspring, mainly to remove it as a competitor for the provisions. Brood parasites of bembicines include flies of the family Sarcophagidae (subfamily Miltogramminae) and some cuckoo wasps of the family Chrysididae. Obligate brood parasitism, where a species never prepares and provisions its own nests, has also evolved independently at least twice among apoid wasps, both times in the Bembicinae and both times originating from nest-provisioning ancestors. The tribe Nyssonini (chapter 4), with 18 genera, is the larger of the lineages; the second lineage comprises the single genus *Stizoides* of the tribe Stizini (chapter 5). In addition, adult females of some Bembicinae may steal prey from one another, and place them in their own nests.

Even adult sand wasps, with their heavier armor and strong flight capabilities, are susceptible to predators, parasitoids, and parasites. Predators of adults, most of which are generalists that occasionally feed on sand wasps, include birds, lizards, robber flies (Asilidae), antlions (Myrmeliontidae), and velvet ants of the genus *Pseudophotopsis* (Mutillidae). Flies of the genus *Physocephala* (Conopidae) are parasitoids that lay their egg between the abdominal segments of the adult sand wasp, which is usually a male. Finally, a few species of Stylopidae (Strepsiptera) have been recently identified as parasites of adult bembicines.

As is true of the overwhelming majority of apoid wasps, males do not participate in nest building or provisioning. Males, which are usually smaller than conspecific females on average, focus their activities on obtaining mates. The behaviors by which they achieve this are often as intriguing as female behaviors associated with nest building and provisioning, though they are considerably less complex. Mating strategies of male sand waps include territorial defense, "hilltopping," and patrolling in mass aggregations where they often engage in intense competitions for available females. The latter includes the "sun dances" that we described earlier for *Bembix pallidipicta.*

Although there are some strong commonalities in the biology of bembicines, the devil is in the details of the exact forms of nesting behavior, nest structure, prey selection, hunting behavior, prey carriage, and defensive and mate-seeking strategies. Of course, the details are also a delight, or this would be a very short book. The next six chapters recount, species by species, information on the biology of sand wasps that has appeared since the mid-1960s. We have included even the briefest of behavioral notes reported in the literature, including a few records of bembicines as prey of other insects. In individual species accounts, we have primarily restricted ourselves to newer literature, though we sometimes present pre-1966 data to provide a context for discussing recent work. In each species heading, we have provided a brief notation on the geographic range of the species, based primarily on information in Bohart and Menke (1976) and Pulawski (2006a). Following our treatments of the species in each genus, we include an overview of the genus, but only when relatively substantial information is available and there are more than just a handful of species in a genus. In the summary sections of chapters, we provide only citations to information not covered earlier in the chapter.

— 2 —

Cool Wasps of the Alyssontini

Members of the tribe Alyssontini, all of which are in the genera *Alysson*, *Didineis*, and *Analysson*, are "small, slender wasps, most of which are taken only infrequently by collectors" (Bohart and Menke 1976). Thus it is not particularly surprising that the group is poorly known, or that we found just a single brief report on the biology of *Didineis* since 1966. New data on *Alysson*, however, permit a somewhat broader comparison of the ethology of that genus.

Phylogenetic status. Alysson, with 42 described species, is the largest genus of Alyssontini, whereas *Didineis* has 28 species and *Analysson* one (Pulawski 2006a); the tribe constitutes just 4% of the Bembicinae. The tribe Alyssontini was included in the subfamily Bembicinae by Evans (1966a) and Bohart and Menke (1976). No consensus has been reached, however, on the relationship of the Alyssontini to other apoid groups of wasps. Three recent phylogenetic analyses alternatively suggest that Alyssontini are most closely related to wasps of the (1) subfamily Larrinae (Crabronidae) (Alexander 1992), (2) tribe Gorytini (Melo 1999), or (3) tribe Nyssonini (Prentice 1998; see also Bohart and Menke 1976). Finally, note that in previous works, including Evans (1966a), the name of this subfamily was spelled Alyssonini, but Menke (1997) corrected the name on grounds of grammar.

Alysson

Alysson includes both Palearctic and Nearctic species, including eight in North America, but the genus ranges as far south as Madagascar and Australia. Species of *Alysson* vary in color, *Alysson melleus*, for example, being brightly colored with a strongly contrasting stripe on the abdomen, whereas *Alysson triangulifer* Provancher is drab brown. The best-studied North American species are *Alysson conicus* and *A. melleus.* Ten or so species in the genus have been studied in various parts of the world. *Alysson* are apparently found at sources of honeydew, but are not collected on flowers (Bohart and Menke 1976).

Alysson cameroni Yasumatsu and Masuda—Palearctic (Japan)

At an altitude of ~1300 m in the mountains of central Honshu, Japan, a "crowded colony" of about 25 females nested on a path within an area of ~0.5 × 3.0 m (Tsuneki 1969), confirming an earlier report of high nest density in *A. cameroni* (i.e., 50/m^2 by Yasumatsu and Masuda 1932). Nest entrances were surrounded by mounds of soil, and burrows entered the ground vertically, later bending and branching (Figure 2.1A, B). The 4–9 cells in each nest sat at depths of 6–24 cm, and each housed 2–9 prey; this contrasts with Yasumatsu and Masuda (1932), who reported unicellular nests with 10–20 prey per cell. All 13 prey in one nest were identified as leafhoppers of the species *Cicadella viridis* (L.) (Cicadellidae); identities of prey in four other nests were not given. Females carried prey in flight

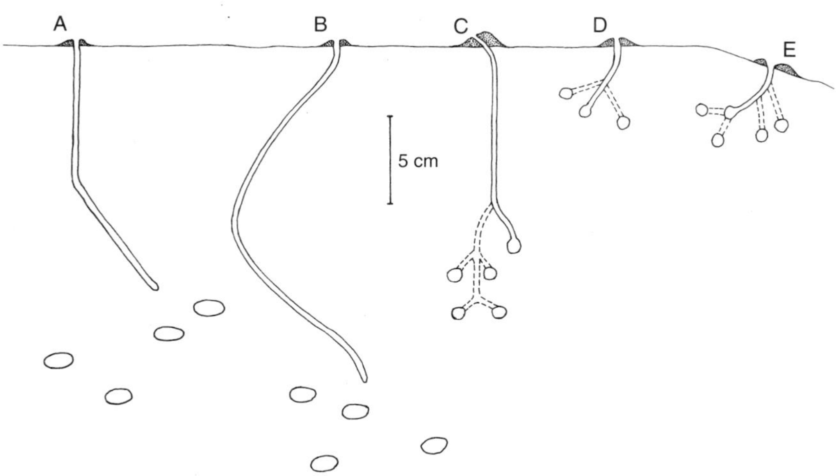

Figure 2.1. Cross-sections of *Alysson* nests: (A, B): *A. cameroni* (Tsuneki 1969); (C) *A. melleus* (Evans 1966a); (D, E) *A. conicus* (O'Brien and Kurczewski 1982). All redrawn from sources indicated.

venter-to-venter, using only their mandibles to grasp prey (by their beaks). Because the nest's entrance was left open during foraging, females could enter directly without dropping the prey. The wasp laid an egg on the ventral side of the thorax of the last prey brought into a cell.

Alysson conicus Provancher—Nearctic (eastern Canada, northeastern United States)

The account by O'Brien and Kurczewski (1979, 1982) of the nesting biology of this species from the vicinity of Cranberry Lake, New York, is one of most detailed to have appeared for *Alysson.* Females were active in moist shady habitats where air temperatures ranged from 15 to 22°C (though the exact location of the measurements was not reported). Although nests were within sunlit areas, females remained continuously active on overcast days and were inactive during the middle of sunny days. Thus, like other *Alysson,* they seem not just to tolerate cooler conditions than other sand wasps, but to prefer such conditions.

Nests occurred at a density of about 15/m^2, in a "firm, moist, gravelly road" along a stream. Females "formed pellets of sand with their forelegs, transferred them back to the midlegs, and pushed them into the entrance with hindlegs. The pygidium (at the tip of the abdomen) often assisted in pushing pellets into the opening." A small conical tumulus (Figure 2.2) formed at the nest entrance as "pellets were then pushed to the side of the entrance, the female rotating in a clockwise direction." However, the mound of soil was easily weathered away, so that nest entrances were often inconspicuous. The vertical or slightly oblique burrows led to 2–5 cells at depths of 1.6–6.0 cm deep in moist soil (Figure 2.1D, E).

Females left nests open while provisioning 5–12 prey per cell, carrying prey venter up and grasping the leafhopper's beak in their mandibles (Figure 2.2). After landing near the nest, a female walked the remaining distance, dropped the prey, turned around, and pulled it inside. The 246 prey (all Cicadellidae) consisted of 230 *Empoa albicans* Walsh, 2 *Empoa venusta* (McAtee), 2 *Empoa querci* Fitch, 2 *Empoa latifasciata* Christian, 8 *Empoasca atrolabes* Gillette, 1 *Typhocyba persephone* McAtee, and 1 *Ribautiana* sp. The egg was laid on the topmost leafhopper in a cell, and many prey seemed dead or deeply paralyzed. No nest parasites were found, although nearly two-thirds of the leafhopper prey one year were parasitized by Dryinidae, which perhaps made them easier for the wasp to capture.

Alysson melleus Say—Nearctic (eastern North America)

Evans (1966a) presented a detailed account of the nesting biology of *A. melleus*, a species that constructs multicellular nests in moist soil (Figure 2.1C). That report included records of nine cicadellid species as prey from Pottawatomie County, Kansas, and nine species of cicadellids and

Figure 2.2. Female *Alysson conicus* carrying prey (top); tumulus of soil formed around nest entrance of *A. conicus* (bottom). From O'Brien and Kurczewski (1982); used with permission.

one Delphacidae from along Fisheating Creek in Florida. Evans (1968) later presented further prey records from both sites. For Kansas, he reported three further cicadellids: *Agallia constricta* Van Duzee, *Neokolla hieroglyphica* (Say), and *Scaphytopius* sp., as well as numerous unidentified nymphs of Cicadellinae. For Florida, he added the cicadellids *Graminella nigrifrons* (Forbes) and *Macrosteles fascifrons* Stål, along with the delphacid *Delphacodes basivitta* (Van Duzee), which had been listed in Evans (1966a) as an unidentified delphacid. Kurczewski and Kurczewski (1971) reported one adult *Keonolla dolobrata* (Ball) (Cicadellidae) as prey in Kansas.

Alysson oppositus Say—Nearctic (eastern North America)

Near Fort Collins, Colorado, a single female of this species was taken as prey by *Philanthus barbatus* Smith (Evans 1982b). Bohart and Menke (1976) listed the distribution of this species as being "N. America, e. of the Rocky Mountains," so this record comes from the western edge of its range. As in 1966, apparently nothing is known about its nesting biology, although it is "common around sandbanks and along watercourses in the northeastern United States," so it may also be a "hygrophile" like other *Alysson* (Evans 1966a).

Alysson pertheesi Gorski—Palearctic (Eurasia)

In Japan, females nested in sandy soil among grasses along a river (Itami 1967). Like other *Alysson,* females dug nests with mandibles and legs, using their abdomens to clear soil from the burrow. Prey, which are stung ventrally, and carried venter-to-venter with the mid-legs, were Homoptera of three subfamilies of Cicadellidae. A female returning with prey deposited it at the nest entrance, entered, rotated, and pulled the prey into the nest. Dollfuss (1994, in Pulawski 2006a) lists this species and *Alysson tricolor* Lepeletier and Serville as endangered in Austria.

Alysson spinosus (Panzer)—Palearctic (Europe)

Near Ceva in northwest Italy, Pagliano and Alma (1997) captured a single female carrying an adult *Cicadella viridis* in an uncultivated grassy area, but gave no other details.

Alysson triangularis Krombein—Oriental (Sri Lanka)

Krombein (1985) made several brief observations of this species in the Udawattakele Sanctuary of the mountains of southcentral Sri Lanka. He collected two prey, both *Idioscopus clypealis* (Lethierry) (Cicadellidae), from females thought to be nesting in "sloping to nearly vertical" soft sandstone banks in a shaded, moist site.

Didineis

Didineis includes seven North American species, and one from Cuba, the other 20 species being in the Old World as far south as Bangladesh and Kenya. Most *Didineis* are poorly known ethologically, and Evans (1966a) and Bohart and Menke (1976) cite no reports on *Didineis* later than 1945. On the basis of older reports, Evans tentatively concluded that the biology of *Didineis* was similar overall to that of *Alysson.*

Didineis lunicornis (F.)—Palearctic (Europe)

In a paper that recommends techniques for collecting this species and provides a list of captures in Great Britain, Packer (1987) also recounts a few behavioral observations (for a species he refers to as *Alysson lunicornis*). Males emerged before females and were found flying several centimeters above the ground, often alighting on grasses, presumably while searching for females. Females apparently nested within cracks in bare patches of soil which ranged from "clay substrate" to "a mixture of sand and clay" to "brickearth of almost gravelly quality." Dricker (in Packer 1987) observed females carrying the prey into cracks; the single prey identified was *Aphrodes* sp. (Cicadellidae), a genus not recorded in a much earlier study by Ferton (1911).

Overview of the Tribe Alyssontini

Habitat. Unlike many Bembicinae whose activity is strongly, if not strictly, dependent on warm air temperatures and direct sunlight, *Alysson* are adapted to relatively low temperatures in shaded, moist habitats. *Alysson cameroni* nested in cool, moist habitats on Mt. Santoge in Japan at 1700 m elevation (>5500 feet) (Yasumatsu and Masuda 1932), whereas *A. triangulifer* was recently collected in the Little Belt Mountains of Montana at over 2100 m (7000 feet). Here, it was the only bembicine found among more than 30 species of apoid wasps collected during a four-year period (KMO and J. E. Fultz, unpublished). Even in the tropics, Alyssontini tend to inhabit wet habitats. In Sri Lanka, Krombein (1985) found *A. triangularis* in "shaded, damp situations in the Wet Zone" in the mountains. In southeast Sri Lanka, *Analysson rufescens* Krombein was an inhabitant of the Dry Zone, but was collected "only after rainy periods." Among apoid wasps, tolerance of low temperatures is not unique to *Alysson,* as it has also been seen in some *Crabro, Mellinus,* and *Bembix* (O'Neill 2001). However, tolerance of low temperatures appears to be a consistent feature of the tribe Alyssontini as a whole. Even those species for which we have no nesting observations are commonly collected near water.

Nesting. Females often nest at moderate to high densities. Reported values for nest density (which seem sometimes to be based on rough estimates) include 15/m^2 for *A. conicus,* 17/m^2 and 50/m^2 for *A. cameroni,* and >250/m^2 for *A. melleus* in one Kansas aggregation (Evans 1966a). Observations on *A. melleus* indicate that density may vary considerably among aggregations of the same species. *Alysson* nest in relatively firm, often moist soil, loosening it first with their mandibles and clearing it with their legs and abdomens. There are several obvious morphological correlates of this type of nesting behavior. Female *Alysson* have stout mandibles and a well-developed pygidium (for pushing soil from the burrow), but weakly developed rake spines on the foretarsi (Evans 1966a; Bohart and Menke 1976). Nests are multicellular, with vertical or near vertical burrows, and sometimes occur in fairly dense aggregations. Nests contain as many as nine cells in *A. cameroni,* five in *A. conicus,* and five in *A. melleus.* Mounds of soil accumulate at nest entrances that are left open during foraging, but may be easily dispersed by rain. Alyssontini apparently do not maintain either inner or outer closures during provisioning, although *Alysson melleus* completely fill the burrow upon completion of a nest.

Foraging. Prey are carried to the nest in the mandibles, females holding prey by their beaks and flying most of the way to the nest. In the final approach, females carry prey on the ground for the last several centimeters or even meters. Cells are mass provisioned, with 2–20 prey by *A. cameroni,* 5–12 by *A. conicus,* and 3–23 by *A. melleus,* and the egg is laid on the last (topmost) prey in the cell. Most prey records of *Alysson* are Cicadellidae, but Cercopidae, Delphacidae, and Issidae are also reported. In both *A. conicus* and *A. melleus,* it seems that prey are at first deeply paralyzed, but many die before consumption; Evans (1966a) reported that prey condition was variable in *A. melleus.* Like *Alysson, Didineis* are predators of Homoptera, the few records being of Cicadellidae (*Agallia, Aphrodes, Chiasmus, Eupelix, Thamnotettix*), Cixiidae (*Cixius*), and Delphacidae (see Evans 1966a).

Natural enemies. Spofford and Kurczewski (1992) reported no parasitism among 22 cells of *A. melleus,* contrasting with rates of 9–46% parasitism in six other species of bembicine they studied. Evans (1966a) had reported just a single miltogrammine in 30 cells of *A. melleus,* and we have found no other reports of brood parasites.

— 3 —

Cicada and Hopper Hunters of the Gorytini

Biological information has accumulated on more than half of the genera of Gorytini, and certain species, such as *Gorytes canaliculatus, Hoplisoides nebulosus,* and *Sphecius speciosus,* are well studied. Within the Gorytini, we have biological information on over one-quarter of the species for just one genus, *Sphecius*). For several medium- to large-sized genera of Gorytini, a small minority of species have been studied (according to our survey): 1 of 27 *Ammatomus,* 5 of 24 *Argogorytes,* 1 of 31 *Austrogorytes,* 5 of 67 *Clitemnestra,*10 of 46 *Gorytes,* 8 of 73 *Harpactus,* 14 of 79 *Hoplisoides,* 1 of 18 *Lestiphorus,* and 2 of 33 *Pseudoplisus.* And we have no biological information on 11 genera with a total of >40 species. Nevertheless, since 1966, new biological information lesser extent, has appeared on at least 40 species of Gorytini.

Phylogenetic status. The tribe Gorytini, as constructed by Bohart and Menke (1976), includes 41 genera and nearly 550 species, or approximately one-third of all Bembicinae (Pulawski 2006a). Bohart (2000) states that "many of the characteristics which describe the Gorytini are negative ones, that is, they represent the absence of characters which determine related tribes." Thus although the Gorytini apparently belongs in the subfamily Bembicinae, its component genera may not all belong together in a discrete tribe, an issue that we defer to later writers. Despite likely paraphyly of the traditional tribe Gorytini (Alexander 1992; Prentice 1998; Melo 1999), all species considered here do have some broad behavioral similarities: all are ground-nesters that mass provision with Homoptera.

Clitemnestra

Bohart and Menke (1976) listed 12 species of *Clitemnestra*, all from Chile and Australia. The addition of newly described species (e.g., Pulawski 1997) and the lumping of *Clitemnestra* and *Ochleroptera* (Bohart 2000) have expanded the number to 67 (Pulawski 2006a). Three-quarters of *Clitemnestra* are neotropical. Although *Clitemnestra* is now the sixth-largest genus of Bembicinae, we know very little of its biology. The single well-studied species, *Clitemnestra bipunctata* (= *Ochleroptera bipunctata*), is the only species in America north of Mexico. Janvier (1928) studied two of the three Chilean species, *C. chilensis* (Saussure) and *C. gayi* (Spinola). Both nested in dense aggregations in clay soils, constructed multicellular nests, and preyed on small Homoptera. Evans (1966a), however, questioned the accuracy of several of the details in Janvier's report and stressed the need for further studies.

Clitemnestra bipunctata (Say)—Nearctic and Neotropical (North America, Cuba)

In the United States, this species often nests in vertical banks (Figure 3.1A) and preys on Homoptera (from five families). To the prey records in Evans (1966a), Evans (1968) added two species, *Colladonus clitellarius* (Say) and *Prescottia lobata* (Van Duzee) (Cicadellidae), for a population in Ithaca, New York. Kurczewski and Kurczewski (1971) reported one adult *Japananus hyalinus* (Osb.) (Cicadellidae) as prey from Fayetteville, New York.

Genaro (1994) studied *Clitemnestra bipunctata* in Cuba (as *Clitemnestra jamaica* Pate). Nests were in vertical banks, and had burrows that were 7.5–10 cm long and led to as many as seven cells, each mass provisioned with 5–20 prey. The 424 prey recorded were from eight families of Homoptera: Cicadellidae (4 species), Cixiidae (4), Delphacidae (5), Dictyopharidae (1), Flatidae (3), Membracidae (1), Psyllidae (1), and Tropiduchidae (1, immatures). The most common prey species were *Oliarus complectus* Ball (Cixiidae, 32% of prey), *Euidella* sp. (Delphacidae, 16%), *Agallia* sp. (Cicadellidae, 10%), *Neurotmeta sponsa* Guérin (Tropiduchidae, 7%), and *Scaphytopius* sp. (Cicadellidae, 6%). Overall, the 389 adult prey consisted of 63% females. Just 8.3% of the prey were immatures, four Membracidae and 31 Tropiduchidae. Pupae of the miltogrammine fly *Metopia argyrocephala*

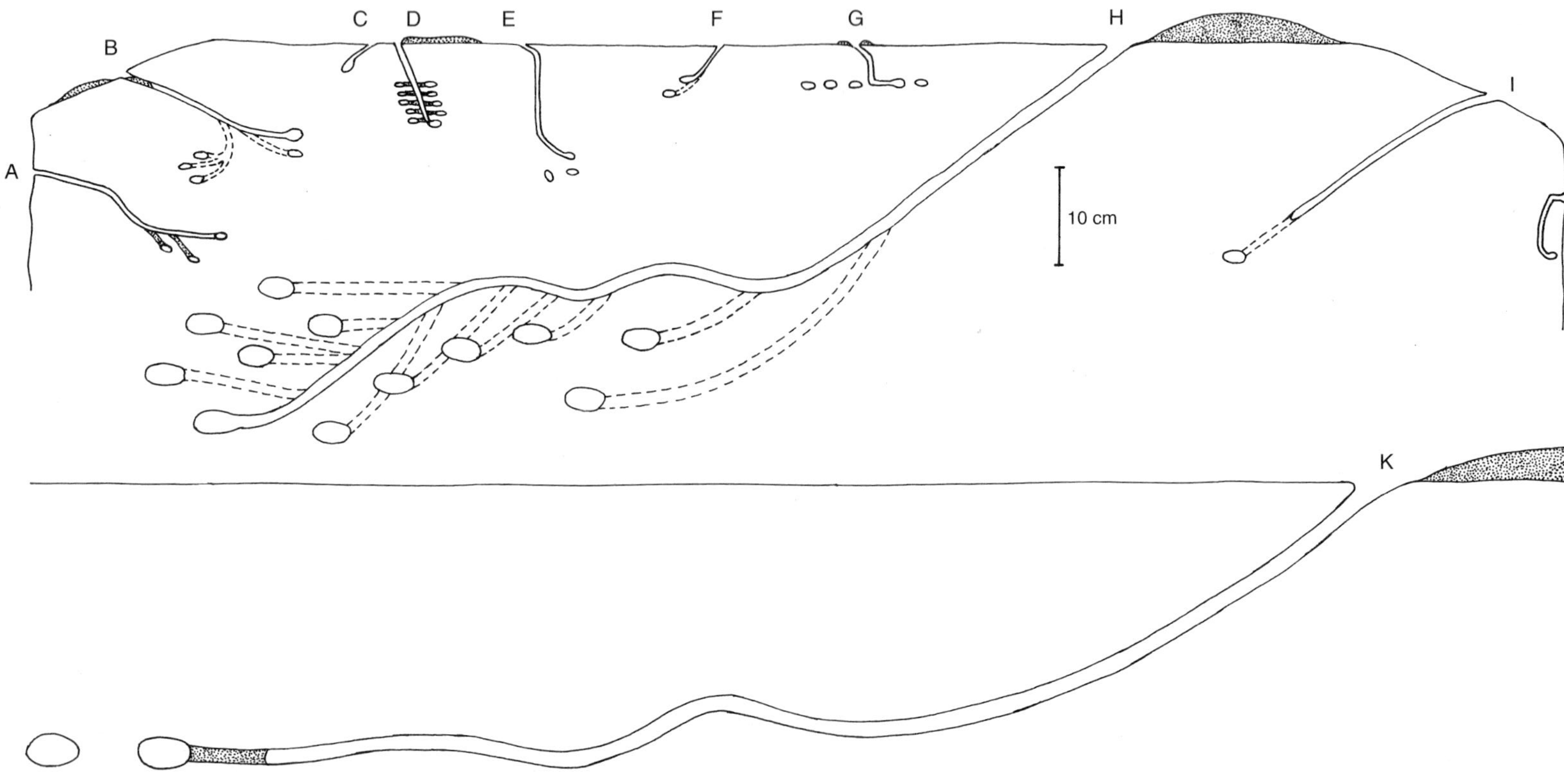

Figure 3.1. Nests of Gorytini: (A) *Clitemnestra bipunctata* (Evans 1966a); (B) *Austrogorytes bellicosus* (Evans and Matthews 1971); (C) *Gorytes tricinctus* (Tsuneki 1982); (D) *Argogorytes carbonarius* (Harris 1994); (E) *Tanyoprymnus moneduloides* (Hook 1981a); (F) *Hoplisoides hamatus* (Evans 1970); (G) *Ammatomus icaroides* (Hook 1981a); (H) *Sphecius pectoralis* (Evans and Matthews 1971); (I) *Trichogorytes cockerelli* (Evans 1976a); (J) *A. carbonarius* (Harris 1994); (K) *Sphecius speciosus* (Evans 1966a). All redrawn to same scale from cited sources).

were found in cells. Spofford and Kurczewski (1990) reported that *C. bipunctata* nests were parasitized by the miltogrammine fly *Phrosinella aurifacies* Downes (1 of 3 cells examined). In Wyoming, Lavigne and Holland (1969) reported *C. bipunctata* among the prey of the robber fly *Diogmites angustipennis* (Loew).

Clitemnestra plomleyi (Turner)—Australasian (Australia)

In southeast Australia, Evans and Matthews (1971) observed females carrying prey and entering nests in sloping banks of soil with high gravel content that unfortunately made it impossible for the researchers to excavate and diagram nests. The nests appeared to be preexisting holes, perhaps made by beetles or other species of wasps. The single confirmed prey record (from a museum specimen) was a cicadellid leafhopper. Observations were not sufficient to compare the behavior of this species with those made on congeners by Janvier, other than that they are all predators of Homoptera and that they nest in banks.

Exeirus

The single species of this genus, *Exeirus lateritius* from Australia, is the ecological counterpart of the cicada-killer wasp, *Sphecius speciosus*, in North America—to which it is not particularly closely related (Prentice 1998). Froggatt referred to this species as a member of the genus *Priocnemis*, which is now a name used for species of the family Pompilidae.

Exeirus lateritius (Shuckard)—Australasian (Australia)

Exeirus lateritius is a large species known for its habit of drinking sap from wounds made in trees by cicadas and for its mousehole-sized nest burrows (Froggatt 1903). Females dig multicellular nests up to 60 cm deep in gardens and sandy areas. Cells are provisioned with cicadas, usually one per cell (reviewed in Evans 1966a). Alcock (1980) observed male-female interactions in Warrumbungle National Park. One to six males patrolled at the center of an opening in a dry eucalyptus forest, flying about 50 cm high. Such flights were alternated with flights up into trees, where females hunted cicadas. Although males chased each other and other insects, they were not aggressive toward one another. A single 26 min–long copulation ensued after a male intercepted a female on a low plant.

Sphecius

Cicada-killers of the genus *Sphecius* rank among the largest solitary wasps, rivaled only by other giants such as *Editha magnifica* (Bembicini) and tarantula hawks of the genera *Hemipepsis* and *Pepsis* (Pompilidae). *Sphecius* is a widely distributed genus and contains 21 described species, four of which occur in the North America (Pulawski 2006a). In 1966 the biology of just one species, *Sphecius speciosus,* was well known. Our knowledge of *S. speciosus* has since expanded considerably since then, and data on a few other species have begun to appear, along with a new key to New World species (Holliday and Coelho 2006).

Sphecius antennatus Klug—Palearctic (Eurasia)

According to Pulawski (2006a), Kazenas (2001) gives *Cicadatra querula* Pallas, as prey of this species.

Sphecius grandis (Say)—Nearctic (Central America, western United States)

For this, the western cicada-killer wasp, Evans (1966a) reported just a single prey record, *Tibicen dealbata* (Davis), garnered from a female carrying prey (its venter up) in Kansas. Later, Hastings (1986) reported *Tibicen duryi* Davis and *Tibicen parallela* Davis as prey. In a sample of 355 prey (both species combined), 55% were males, a significant sex bias. Measurements of the dry weight of *T. parallela* obtained from prey-carrying wasps showed that prey averaged about two-thirds heavier than the provisioning females themselves (0.42 g vs. 0.25 g). Such large prey were difficult to carry, with the result that the largest females in the population had the highest provisioning success because they were better able to carry large prey. Fourteen of the largest wasps in the population (9% of all females) accounted for 38% of the prey brought into nests during Hasting's study, whereas the smallest females (13% of all females) accounted for <1% of prey. A second correlate of provisioning success was the date on which the female wasps emerged as adults, the overall emergence period being from 25 July to 15 August. Because the abundance of prey changed during 1983, females that emerged in the middle of the emergence period were more successful at provisioning. However, such stabilizing selection on emergence date may not occur every year.

After they emerged as adults, males established territories within the emergence area during times of day in which virgin females emerged (Alcock 1975a, Hastings 1989a,b; Figure 3.2). During battles among males that included butting and grappling, wings and other appendages were sometimes damaged, perhaps partially explaining why the mean adult male lifespan was just 11–15 days (Hastings 1989b). A male sometimes grasped another male in midair and carried it upward, a feat that Hastings (1989a) suggests could be a show of strength. The most successful males were those that occupied areas of high female density at times when the rate of female emergence was highest. Although these large "primary residents" were in the best position to intercept virgins, smaller males also had opportunities to mate. In the afternoon, when large males left the emergence area, they were sometimes replaced by intermediate-sized, "secondary residents" that mated with the late-emerging females, which were fewer in number. (Is it perhaps the case that the largest males need to retire early each day to replenish their reserves at flowers?) The smallest males in the population often eschewed territoriality altogether and perched in trees adjacent to the

Figure 3.2. Mating pair of *Sphecius grandis*. Photo by Chuck Holliday; used with permission.

emergence area, where they attempted to intercept females missed by territory holders.

The date on which a male emerged had a significant effect on his potential mating opportunities, determining both the number of receptive females he was likely to encounter and the number of male competitors he was likely to face. Males emerged 1–7 days before females each year, so many males were present when females first appeared. However, expected male mating success varied with emergence date in two of three years examined (Hastings 1989b). One year, males that emerged 1–2 weeks prior to the peak of female emergence had the highest mating opportunities. Another year, females emerged over a longer period and males had shorter life spans, so that late-emerging males had the highest mating opportunities. Thus both female and male reproductive success varied with emergence date.

Sphecius hogardii (Latrielle)—Nearctic (Florida), Neotropical (West Indies)

Sphecius hogardii is distinguished by its overall bright orange coloration, excluding its black terminal abdominal segments (Stange 2000). Its behavior has been studied only in Cuba, where 63 nests occupied a 9 × 21 m area of hard stony soil, along with nests of *Sphex jamaicensis* (Drury); nest entrances were from 70 to 200 cm apart (Genaro and de Varona 1998). The oblique nest burrow entered the substrate at 30–45° and was apparently difficult to map because of its great length and because of the compacted rocky soil. However, cells were found at depths of 15–30 cm; three completely provisioned cells contained 4–6 prey. Prey, which were carried venter upward and head forward, consisted of 42 female and 22 male cicadas, all *Uhleroides walkeri* (Guérin), so sex bias among prey was the opposite of that observed for *Sphecius grandis.* The egg was laid longitudinally on the ventral side of the thorax near the mid and hind coxae on the first prey deposited in the cell.

Sphecius pectoralis Smith—Australasian (Australia)

In eastern Queensland, a single nest was excavated in a disturbed area with a layer of fine grain sand 20 cm thick above "black and loamy" soil (Evans and Matthews 1971). The nest descended at about 45° for ~40 cm to about 30 cm depth, where it then angled to the left and extended for another 35 cm (Figure 3.1H). When excavated, the single 50 cm–deep cell at the end of the main burrow was empty, but 11 other provisioned cells were found near the empty burrow and along the burrow; the side tunnels leading to the cells had been filled by the female wasp. The cells were up to 4 cm long and contained 7–12 small cicadas of the genus *Tamasa,* along with larvae of various ages (9 cells) or an egg (one cell); the egg was in the same position observed for other *Sphecius.* The ages of the larvae suggested that the female had first provisioned cells closest to the entrance, perhaps extending the main burrow as later cells were added.

Sphecius speciosus (Drury)—Nearctic (United States east of Rocky Mountains, Mexico)

The eastern cicada-killer is one the few solitary wasps that North American entomologists commonly deal with in their interactions with the public. The reasons include the wasps' large size (up to 4 cm in length) and their penchant for nesting in urban areas, including playgrounds and lawns (Evans 1966a; Tashiro 1987). Another problem is that territorial males act as if they are aggressive toward humans when they are seeking females, though they are really only aggressive toward each other (see below). As long ago as 1935, H. C. Will reported an "epidemic" of cicada-killers nesting on the grounds of an electric plant in Virginia, where the wasps annoyed plant personnel and defaced the "immaculate grounds." Amid other reports of "pests" in lawns, Romel and Dykstra (1991) reported *S. speciosus* nesting in a playground in Ontario, Canada. Coelho (2002) noted that "one Indiana women chose to move rather than tolerate the annual invasion of her yard, and an Illinois town closed its park in 1996 due to a large wasp [cicada-killer] infestation." Our informal survey of a few extension entomology sites on the Internet revealed that potential control measures include insecticides, mulching and fertilizing to cover the soil in nest sites, and tennis rackets (!), none of which are really necessary for the nonentomophobic among us who can tolerate a few brown spots in our lawns and occasional libidinous, dive-bombing male wasps. Despite their negative image, cicada-killers can also be considered beneficial, because they prey upon cicadas, which can be pests of ornamental trees. Those confronted with a cicada-killer crisis on their property may need to balance their entomophobia against their landscaping goals, which ought to include sharing space with wasps anyway. For entomologists, cicada-killers have provided subjects for some of the most intriguing and imaginative studies of solitary wasps.

Females dig multicellular nests (Figure 3.1K) that in one location contained an average of nearly 16 cells (Dambach and Good 1943). The size of the soil mound outside the nest (up to 15 × 25 cm in extent and 10 cm in depth), the width of the nest entrance (up to 4 cm), the length of the burrows (up to 100 cm long), and the size of the cells (up to 3 cm in diameter) attest to the amount of soil potentially moved by a nesting female. This amount was more directly assessed by Coelho and Wiedman (1999), who estimated that the weight of the soil was up to one thousand times a female's body mass. Evans (1966a) had noted that a female *S. speciosus* loosened soil with her mandibles, raked it beneath her body with her forelegs, and then backed out of the burrow. As she did so, "she pushed the soil behind her with her hind legs, which opened and closed in a somewhat scissor-like fashion." He also noted the presence of very large spurs on the apex of the hind tibiae of *S. speciosus,* but suggested (on the basis of others' observations) that they function to hold prey in flight. Coelho and Wiedman (1999; see also Coelho 2002), however, demonstrated that the spurs are actually tools for excavating nests. As a digging female backs out of the nest, she flexes her hind tarsi, causing the spurs to extend. In resisting extension, the spurs produce sufficient force to aid in moving soil out of the burrow, perhaps reducing the "energetic and temporal costs of burrowing."

Female *S. speciosus* mass provision cells with 1–4 cicadas (Figure 3.3) from at least 12 species. Three records come from publications appearing since 1966: *Tibicen figurata* Walker and *T. resh* Haldeman (N. Lin 1979b) and *T. superba* (Coelho 1997). Evans (1966a) listed nine other species of *Tibicen,* but one on that list, *Tibicen vitripennis,* is now named *Diceroprocta vitripennis* (Say). Coelho (1997) found no evidence that females choose larger individuals or one or the other sex among potential prey, which is also true for *S. grandis* and *S. hogardii.* Reported biases in prey sex ratio could be due to skews in the sex ratios of available prey.

When fully provisioned, cells with female offspring received 2.3–3.9 g of prey, whereas males receive 0.9–2.1 g. N. Lin (1979a) noted that all daughters received two prey, all sons one. Females can afford to provision cells with so few prey only because of the large size of prey. Female bembicine wasps generally carry prey smaller than themselves (see *Hoplisoides vespoides* below for another exception). However, the weight of prey carried by female *S. speciosus* ranged as high as 265% of the provisioning wasp's

Figure 3.3. Female of *Sphecius speciosus* dragging prey cicada to burrow. Photo by Chuck Holliday; used with permission.

body mass in Lin's study and up to 300% in Coelho's (1997; see also Balduf 1941). In Lin's study, the largest prey weighed >900% of the mass of the smallest adult male in the population, whereas in Coelho's study the value was >1400%. Carrying huge prey places a tremendous burden on females, who are unable to take off in flight with average-sized prey. They can carry prey in flight only by first walking with them to a higher location on a tree and then flying downward to the nest, often reaching it only after several repetitions. Selection on body size, because of its relationship to load-carrying ability, may explain why females average nearly 2.5 times greater body mass than males (Coelho 1997). Because of the their size, larger females not only provide more provision than small females but are less likely to be evicted from nests during aggressive encounters among females (Pfennig and Reeve 1989).

Evans (1966a) placed doubt on Riley's (1892) claim that cicadas stung by *Sphecius* remain paralyzed for a year or more under some conditions. Because feeding by the larva is completed within several weeks following oviposition (Dambach and Good 1943), such a degree of prey preservation would seem unnecessary. Bringer (1996), in an unpublished thesis supervised by Coelho, discusses her work on venom-induced paralysis of *Tibicen*. Bringer compared the life spans of cicadas collected from provisioning females with those collected from trees; both groups were placed in simulated nest cells and kept under controlled conditions. Paralyzed *Tibicen* lived 147.5 h on average, 66% longer than live-collected cicadas, which also lost mass at a greater rate during storage.

In large nesting aggregations studied by Pfennig and Reeve (1989), female *S. speciosus* sometimes entered the burrow of another female, causing an aggressive encounter. Although two females may temporarily cohabit a burrow, only rarely do they both provision it. Why should a female tolerate another female in her burrow, even for a short period of time, when there exists the potential danger of having the intruder lay an egg in a cell? Two patterns emerged. First, females were less likely to engage an intruder in a violent fight if it was a neighbor whose nest was within 1 m of her own. Because neighbors may commonly enter each other's burrow accidentally, it may not pay to fight when the intruder has come in by mistake (such mistakes would perhaps be common). Non-neighbors, however, are more likely to enter a nest distant from their own in order to usurp it. Second, it also turns out that neighbors are more closely related genetically than non-neighbors, so tolerance may be a form of kin-selected nepotism.

Spofford and Kurczewski (1990, 1992) reported that 38% of 24 *S. speciosus* nest cells were parasitized by the miltogrammine flies (Sarcophagidae), five by *S. trilineata* (Van der Wulp) and four by undetermined miltogrammines. One cell contained maggots of both *S. trilineata* and *Muscina stabulans* (Muscidae). *Senotainia trilineata* females are called *satellite* flies because, after locating a prey-carrying female, they follow closely behind, awaiting an opportunity to attack and attach a maggot to the prey. Female cicada-killers attempted to avoid a trailing *S. trilineata* with "freeze-stops," wherein they halt during a prey carriage flight and remain motionless (during which time the fly may be drawn off to chase another female). If the fly contacted the prey (i.e., attempted to larviposit), some wasps abandoned the prey rather than risk provisioning with fly-infested cicada. Females also abandoned burrows or individual prey items to avoid parasitism.

The basic picture of male territorial behavior in *S. speciosus* was first outlined by N. Lin (1963a, 1966, 1967). Working along paths bordering two baseball fields in Brooklyn, New York, he described how males defended territories within an emergence nesting area. Briefly, males emerged before most females (as in *S. grandis* and as confirmed for *S. speciosus* by N. Lin 1978) and established territories that they returned to for up to 12 days. Territories centered around groups of emergence holes, the density of which was probably a good general predictor of future emergence locations. Males defended their plots by chasing, head butting, grappling, and biting opponents. Mating occurred after males pursued newly emerged virgins, coupled with them, and left the emergence area in tandem. Copulation, a relatively prolonged affair, for wasps, lasted from 29 to 51 min.

Territories are, by definition, plots of ground from which intruders are excluded via aggression or advertisement (Brown 1975). In practice, however, it may be difficult for researchers to delineate territory boundaries. N. Lin (1963a) roughly estimated the area of male cicada-killer territories to be 1.5–9.0 m^2, but did not give the methods by which this was determined. Furthermore, males often left territories in pursuit of females (and potential females). Eason et al. (1999), in a study of a population in Louisiana, defined boundaries precisely by noting the location of clashes between males on neighboring territories. Thus, in their study, unmanipulated territories were estimated to average 0.9 m^2 in area, and none (judging by the maps they provide) exceeded 2 m^2. Following initial determinations of territory location, the researchers then laid dowels 90 cm in length on the ground in areas not corresponding to territory boundaries. The next day, territory boundaries shifted so that they were aligned with the dowels in every case; originally, there were no obvious visual landmarks to demarcate territories. After the dowels became the visual reference points for the boundaries of territories, the cost of territorial defense declined by over five times; cost was measured as the duration of time spent in interactions with neighbors.

Joos and Casey (1992) and especially Coelho (1998, 2001) have also refined our knowledge of male cicada-killers. Coelho notes that, because of their large size and high levels of activity, and because they are active on and near hot soil surfaces, males potentially suffer high, detrimental heat loads. Part of their ability of avoid lethal high temperatures, he observed, was due to behavioral responses. At hotter times of sunny days, males perched on plants and tended to face the sun (to reduce the amount on radiation intercepted), whereas those on the soil surface often perched in shade or perched for shorter durations. The latter has also been observed in the sand wasps *Bembecinus strenuus* (Evans and O'Neill 1986) and *B. quinquespinosus* (O'Neill et al. 1989), and in the beewolf *Philanthus psyche*

Dunning (O'Neill and O'Neill 1988). Unlike bumble bees (Heinrich 1993), cicada-killers "have almost no ability to cool themselves by shunting hot hemolymph from thorax to abdomen" (Coelho 1997). They sometimes regurgitated fluids to cool their heads, but this ability may be limited by the availability of free liquid in their guts. Early and late in the day, when they were trying to raise body temperature, males tended to bask with their bodies perpendicular to the direction of the sun (to increase radiative heat gain); they also may shiver without wing movement to generate heat endothermically in their flight muscles (see Chapter 7 for a description of this in *Bembix rostrata*).

Overview of *Sphecius*

Even among entomologists, big animals are easier to study. So cicada-killers have been the subjects of a greater diversity of biological studies that any other bembicine, studies not only of female nesting biology but of pest control, functional morphology, thermoregulation, kin selection, landmark recognition, territoriality, and the reproductive consequences of body size and protandry.

Mass provisioning seems to be the rule, and some species (*Sphecius hogardii, S. pectoralis,* and *S. speciosus*) are known to construct multicellular nests. Cicada-killers have among the narrowest prey preferences of any sand wasp. All ten species for which records are available provision cells with cicadas, including species in a dozen genera (Table 3.1). The large size of the wasps and their narrow diet breadth are linked; homopteran predators as large as *Sphecius* have no prey options other than cicadas, unless they were to stock cells with hundreds of smaller prey. Because there is a very large size gap between cicadas and most other Homoptera, these wasps' choices are quite limited. Thus the evolution of large size in this genus is likely linked to its expansion into a niche not open to many invertebrate predators. Both of the well-studied North American species (*S. grandis* and *S. speciosus*) are recorded (with one exception) as taking only dog-day cicadas (*Tibicen* spp.). Dog-day cicadas have multiyear life cycles, but broods are not synchronized, so some adults appear each year (Borror et al. 1989) and provide relatively predictable sources of prey in late summer. Not surprisingly, periodical cicadas *(Magicicada)* are not reported as prey, because they appear as adults in early summer and have locally synchronized, multiyear broods that do not produce adults every year.

Table 3.1 Genera of Cicadidae reported as prey of *Sphecius*. See text and Evans (1966a) for references and details.

Specius species	*Afzeliada*	*Cicada*	*Cicadatra*	*Diceroprocta*	*Munza*	*Oxypleura*	*Soudaniella*	*Tamasa*	*Tettigades*	*Tibicen*	*Uhleroides*	*Yanga*
S. antennatus			X									
S. convallis Patten				X						X		
S. grandidieri Saussure[a]												X
S. grandis										X		
S. hogardii											X	
S. milleri R. Turner[b]	X				X	X	X					
S. nigricornis		X										
S. pectoralis								X				
S. speciosus				X						X		
S. spectabilis Taschenberg									X			

[a] *Yanga brancsiki* (Distant) (originally reported as *Poecilopsaltria brancsiki* Distant) and *Yanga pulverea* (Distant) (originally reported as *Platypleura pulverea* Distant); [b] *Afzeliada lindiana* (Distant) (originally reported as *Platypleura lindiana* Distant), *Oxypleura quadraticollis* (Butler) (originally reported as *Platypleura quadraticollis* Butler), *Soudaniella marshalli* (Distant) (originally reported as *Platypleura marshalli* Distant).

Tanyoprymnus

The sole species in this genus, called *Ammatomus moneduloides* in Evans (1966a), is restricted in distribution to North America, whereas all remaining *Ammatomus* are Old World species. Prior to the account related below, all that was known of *T. moneduloides* was that it preyed on Dictyopharidae and nested in sandbanks (Krombein 1959).

Tanyoprymnus moneduloides (Packard)—Nearctic (United States, Mexico)

On St. Catherine's Island, Georgia, females nested in a sandy road and sandbank adjacent to salt marsh and live oak forest (Hook 1981a). The single nest examined was apparently in an abandoned *Bembix* burrow. A burrow 21 cm long led to a single cell, 15 cm deep (Figure 3.1E). It was provisioned with 13 *Thionia simplex* (Germar) (Issidae) (Figure 3.4), which were oriented head inward; the egg was laid on the coxa of one prey.

Figure 3.4. Female *Tanyoprymnus moneduloides* carrying prey. From Hook (1981a); used with permission.

One cell also contained a second egg, apparently of *Nysson* sp. (see Chapter 4). In Big Bend National Park, Texas, an excavated nest in a sandy silt bank had burrow 14 cm long that also appeared to have begun in an existing hole. Each of the three cells, which were 14–17 cm deep, contained four *Cyrpoptus vanduzeii* Ball (Fulgoridae); eggs were laid along the thorax of one prey, over the coxae. In both nests, the entrance was left open during foraging.

Ammatomus

Pulawski (2006a) lists 30 species, all Old World. Evans (1966a) discussed only *Ammatomus moneduloides,* which has since been moved to *Tanyoprymnus* (see above). Since then, Hook (1981a) added observations on one Australian species and concluded that "nest architecture, prey carriage, prey arrangement, oviposition site, and final closure were similar in the two genera."

Ammatomus icarioides (Turner)—Australasian (Australia: Queensland)

Females nested (often under leaf litter) in a disturbed habitat near Blunder Creek, Brisbane, Australia (Hook 1981a). When initiating a nest, females loosened the ground with their mandibles before raking with their legs. A small tumulus, 3 cm wide, formed around the nest entrance. In the eight nests excavated, a 1–2 cm oblique burrow led to side burrows that were 1–8 cm long with 2–7 cells per nest (1.8–6.5 cm deep) (Figure 3.1G). The nest entrance was left open during foraging trips, but there was a final closure, for which females obtained soil from the burrow wall (using their mandibles to loosen soil) and the tumulus.

Nineteen of the 25 prey identified were from five species of Flatidae: *Colgar rufostigmatum* Distant (12 specimens), *Parasalurnis roseicincta* (Walker) (2), *Euphanta* sp. (2), *Sephenia* sp. (2), and *Massila sicca* Walker (1). The remaining prey were Eurybrachidae of the species *Dardus abbreviatus* Guérin-Menneville. Bohart and Menke (1976) also report seeing a specimen of this wasp from Townsville, Queensland, that was "pinned with the presumed prey," an adult Flatidae. Hook found that the 3–6 prey per cell were all oriented head inward when stored. Prey on which eggs were laid (placed as noted for *Tanyoprymnus*) sat venter-up in the cell. The only evidence of natural enemies in the 32 cells examined was several maggots feeding on prey in one of the cells. Hook concluded that because "cells of the same nest often had fresh prey and cocoons . . . most nests were provisioned over several days." During "cocoon construction, a silken mesh was spun and lined with a sand layer, then an additional silk layer was deposited inside. A single pore with a raised rim was produced at the cocoon's anterior end."

Argogorytes

Just two species (*Argogorytes nigrifrons* (F. Smith) and *Argogorytes sapellonis* (Baker)) of the 31 species of the widespread genus *Argogorytes* occur in North America, and nothing is known of their biology. Evans (1966a) reported information on the biology of two European species (*Argogorytes fargei* [*A. campestris*] and *Argogorytes mystaceus*). We now have biological information on five *Argogorytes,* but comparatively little evidence on nest structure and prey; and data on *A. nipponis* nests is so at variance with what we know of other Bembicinae that it needs confirmation.

Argogorytes carbonarius (F. Smith)—Australasian (New Zealand)

Gourlay's (1964) report on this species, the only bembicine in New Zealand, was not discussed in Evans (1966a). Gourlay observed it in a garden constructing multicellular nests mass provisioned with up to three large spittlebug (Cercopidae) nymphs per cell. All prey were *Carystosterpa fingens* (Walker), a member of a genus endemic to New Zealand. Prey were captured on nearby trees, and attacked within masses of spittle produced by the nymphs. The egg was laid dorsally between the developing wings of the nymphs.

Harris (1994) provides a comprehensive account of the nesting behavior of *A. carbonarius.* Nests were generally found in "slightly moist, vertical clay and clay/sand/loam banks" though they were sometimes made in "sloping soil." The structure of the nests was quite variable in the sample of 65 nests excavated. At two sites, Haast and Wilton Bush, nests had one or two cells (Figure 3.1J), with main burrows 45–180 cm long. At Outram, near Dunedin, one- and two-celled nests were also found, but in an aggregation of over 60 females, Harris documented nests of 6–10 cells. In these nests, cells were arranged along main burrows at the ends of side burrows that were 0.5–2.5 cm long (Figure 3.1D). A 6–9 cell nest took 9–15 days to complete, and individual females constructed up to five nests per season.

Females hunted prey on various native and exotic plant species, carrying prey to nests "venter up, facing forwards, with the sternum appressed to that of the wasp" (Figure 3.5). Nest were left open during foraging, though inner cell closures were present prior to the final closure of the nest. Both the inner and the final closures were made of clay. Prey at Harris's sites included not only *C. fingens* but *Carystoterpa vagans* Hamilton and Morales and *Philaenus spumarius* (L.). The latter is a European cercopid first found in New Zealand in 1960, and has since spread to many areas of both North and South Island; it is a known prey of *Argogorytes mystaceus* in Europe. Since its arrival in New Zealand, *P. spumarius* has provided an abundant source of prey for *A. carbonarius,* and, according to Harris, may account for the recent tendency of the wasps to construct nests with more than two

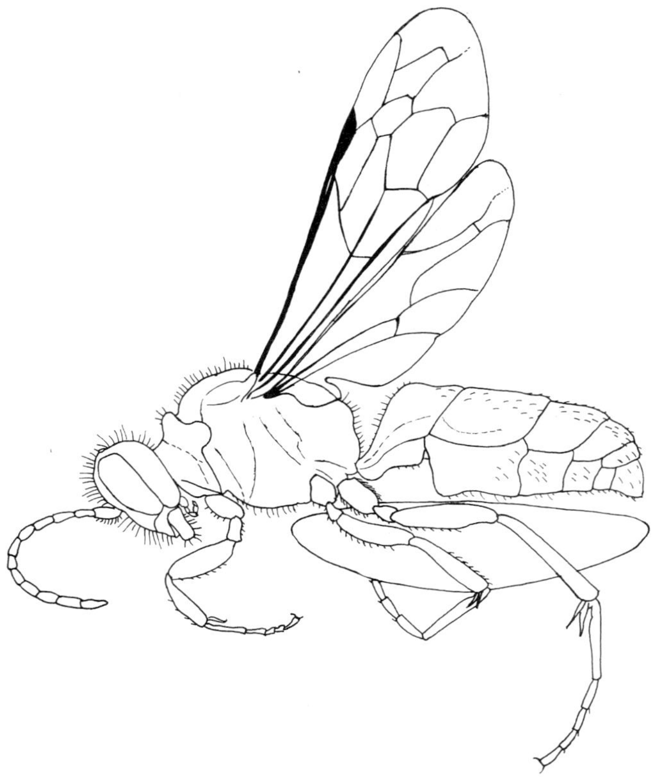

Figure 3.5. Relative position of legs and prey for prey-carrying females of *Argogorytes carbonarius* carrying prey. Adapted from Harris (1994).

cells (two cells being typical prior to the mid-1980s). Each nest cell was stocked with 14–26 prey, always permanently paralyzed nymphs of instar II or beyond. Provisioning often occurred rapidly, with as little as 7 min passing between visits with successive prey; one female completely filled a cell with 17 prey in ~4 h.

Schöne et al. (1993) examined homing behavior of female *A. carbonarius* at a site near Dunedin, New Zealand, where 20–25 females nested within a 4 × 6 m area on a sloping, northeast-facing, bank about 30 m from the Tairei River. Nest-provisioning females that had already oriented to the positions of their nests were transported from the nesting area to sites 500 m away, where they were released after a maximum of about 50 min following capture. Some were transported in "closed" containers (i.e., in a plexiglass tube in the researcher's pocket), others in "open" containers (i.e., in a plexiglass tube held so that the wasp could scan its surroundings during transport). Females returned to nests after intervals of from several minutes to two hours. Curiously, they did not fly directly toward the nest site when released. Their initial orientation was generally displaced to the right of the true direction, and a few females flew directly away from the nest site. The researchers found no differences between wasps in open or closed containers, either in return time or initial orientation upon release. However, those wasps transported in closed containers had shorter return times and more accurate initial orientations when tested a second time; this suggests an improvement in their homing capabilities, perhaps because of learning that occurred at the earlier release. A later study, in which females were released within larger arenas and videotaped, produced similar results (Schöne et al. 1994).

Argogorytes fargeii (Shuckard)—Palearctic (Eurasia)

This is one of two species of *Argogorytes* studied in Sweden in their role as pollinators of orchids of the genus *Ophrys* (see under *A. mystaceus* below). In a study of the mandibular contents of *Argogorytes fargei* and *A. mystaceus,* Borg-Karlson and Tengö (1980) found 2,5 dimethyl-3-isopentylpyrazine in both males and females, a chemical also present in the secretions of many ants, bees, and wasps (El-Sayed 2006).

Argogorytes hispanicus (Mercet)—Palearctic (Europe)

In southern Spain, Janvier (1974, summarized in Callan 1981) observed females constructing and mass provisioning nests having burrows 10–12 cm long and 5–7 cells. All prey in the two cells examined were *Hysteropterum reticulatum* (Herrich-Schaeffer) (Issidae). In Callan's interpretation of Janvier's report, the nest entrance is left partly open during foraging trips. The egg was apparently laid on the first prey in the cell and then further prey were brought in over several days. Janvier estimated that 40–50 prey were required to provision the 5–7 cells in a nest. One prey had an egg attached to the upper mesopleuron.

Argogorytes mystaceus (L.)—Palearctic (Eurasia)

Females have long been known to dig multicellular nests stocked with up to 27 prey per cell (Cercopidae, *Philaenus*) (Hamm and Richards 1930).

Pagliano and Alma (1997) recently added a few observations for an Italian population of *A. mystaceus*, confirming the use of *P. spumarius* nymphs that were attacked inside spittle masses. Following stinging, nymphs were carried venter-to-venter in the middle legs of the female wasp to nests located in a bank of soil. A report originally said to document the nesting biology of *Argogorytes mystaceus grandis* Gussakovskij (Tsuneki 1965a) was later corrected, as the species turned out to be *Gorytes tricinctus* (Tsuneki 1982) (see under that species below).

Recently, *A. mystaceus* (along with *A. fargeii*) has been the subject of a very different kind of study. The flowers of many species of orchids mimic females of particular species of insects in order to attract males of those species, which normally search for females in nesting areas (Kullenberg 1956; Alcock 2005). Deceptive pollination is a common form of parasitism by orchids, species of which are known to recruit males from two families of beetles, one species of ichneumonid wasp, four families of bees, and four families of solitary wasps, including *A. mystaceus and A. fargeii* (see review in O'Neill 2001). A male *Argogorytes* deceived by an orchid gains nothing (because the flower produces no nectar), but the flower stands to receive pollination services. The process begins when male *A. mystaceus*, flying in an agitated manner, approaches the flower of *Ophrys insectifera* from downwind and lands on the labellum, a modified petal that provides a landing platform. He is first attracted by volatile chemicals released from the flower that resemble those in the Dufour's glands of female *A. mystaceus*. As a male nears a flower, he responds visually to the shape, size, and color of the flower, which resembles the female wasp (at least to the eyes of the male wasp). Upon landing, he is probably stimulated by the fine hairs on the labellum that match the length, density, and pattern of those on the abdomen of a female *A. mystaceus*. He then attempts to copulate and, while engaged, a pollinium (pollen package) becomes attached to his head. On a later visit to another *O. insectifera* flower, pollen grains may be transferred to its stigma (Ågren and Borg-Karlson 1984; Borg-Karlson 1990). Bradbury and Creswell (1979), however, question whether the activities of male *A. mystaceus* are sufficient to explain the level of pollination observed among *Ophrys* in Great Britain.

Argogorytes nipponis Tsuneki—Palearctic (Japan)

In Japan, Nambu (in Tsuneki 1982) documented a remarkable set of observations which appear to apply to nests of this species. We say "appear" because, although Nambu saw *A. nipponis* flying about nests, females were never actually observed at nests and attempts to rear wasps from cells were unsuccessful. One piece of evidence in favor of the argument that the nests belonged to *A. nipponis* was that cells were provisioned with Cercopidae (281 records). However, the four nests are very different from those of other *Argogorytes* and, in fact, different from any other known in the Bembicinae. First, the nests were within very soft material in rotten logs, whereas other Bembicinae are known to nest only in soil. Second, some of the cells were aligned in a linear sequence within the burrows. However, Janvier (1974) reports a similar configuration of cells in the ground-nesting *A. hispanicus*. In the illustration provided by Tsuneki, the entire bifurcated nest is about 16 cm long and similar in structure to that of many wood-nesting Pemphredoninae (O'Neill 2001).

Independent of these observations, Krombein (1985) notes that species of the apoid wasp genus *Lestica* (Crabroninae) having a broad pygidium (a flattened dorsal plate on the last segment of the abdomen) nest in soil, whereas those with a narrow pygidium nest in rotten wood. He also notes that *Argogorytes* with a broad pygidium are soil nesters, but that the biology of those with a narrow pygidium is unknown. Perhaps a similar pattern holds for both *Lestica* and *Argogorytes,* because Tsuneki (1965a) emphasizes that *A. nipponis* has a narrow pygidium. As for *A. mystaceus grandis,* he only notes that its pygidial plate is broader than that of the European form of the species. *A. nipponis* and other *Argogorytes* with narrow pygidia warrant further work, as they may provide unique examples of non-soil nesting in the Bembicinae.

Harpactus

Eight of the 73 species of *Harpactus* listed by Pulawski (2006a) occur in America, north of Mexico. Both the composition and the name of the genus have changed (Pulawski 1985) since it was covered as *Dienoplus* by Evans (1966a) and Bohart and Menke (1976). One species reviewed in Evans (1966a) has since been moved to *Oryttus* (*O. concinnus* (Rossi)), and another, *Harpactus houskai,* has been moved from *Oryttus.* Nineteen new species have been described, most from central Asia, but also four from the United States (Bohart 1980). Data are available on one North American and seven Palearctic species, but the level of taxonomic activity since 1966 has far outstripped the increase in biological information; *Harpactus* prey upon Cicadellidae, Cercopidae, and Issidae (Evans 1966a).

Harpactus affinis (Spinola)—Palearctic (Europe, Middle East)

Kazenas (2001, in Pulawski 2006a) gives *Graphocraerus ventralis* (Fallen) as prey of this species.

Harpactus formosus (Jurine)—Palearctic (Europe, Middle East, northern Africa)

Gayubo and Sanza (1986, in Pagliano and Alma 1997) report Cercopidae as prey of this species.

Harpactus laevis (Latreille)—Palearctic (Eurasia)

In Mongolia, Tsuneki (1969) excavated a single four-celled nest with a near vertical tunnel 18 cm long, dug into a steep bank. At the time of its examination, the tunnel leading to the fourth cell was still open, but the other side tunnels were packed with soil along their entire lengths. (Cercopidae as well as Cicadellidae were recorded as prey in previous reports.) The egg in one cell was "flatly attached to the underside of the prey at the outside of the legs along the coxae, from the neck region to the hind coxae." Tsuneki notes that an earlier report concerning a Japanese population of *H. laevis* in Japan (Iwata 1937), which was cited in Evans (1966a), actually concerned *Harpactus tumidus.*

Harpactus pictifrons (W. Fox)—Nearctic (western North America)

During Evans's (1970) surveys of the wasps of Jackson Hole, Wyoming, he found no nests of this species, but did find 14 males and one female *H. pictifrons* among the diverse prey of *Philanthus pulcher* Dalla Torre; one female was later found among 93 prey of *Philanthus pacificus* Cresson in Teton County, Wyoming (Evans and O'Neill 1988).

Harpactus tumidus (Panzer)—Palearctic (Europe)

Earlier reports, summarized in Evans (1966a), that females of this species prey on Cercopidae and Cicadellidae were confirmed by Pagliano and Alma (1997), who captured a female carrying a female *Aphrodes* sp. (Cercopidae).

Trichogorytes

The two described species of *Trichogorytes,* both small wasps, are mainly restricted to the southwestern United States and Sonora, Mexico. *Trichogorytes argenteopilosus* Rohwer, which remains unstudied, has been collected in both Arizona and Arkansas, whereas *T. cockerelli* is known only from Arizona (Pulawski 2006a). The biology of this genus was apparently unknown until the appearance of the report cited below.

Trichogorytes cockerelli (Ashmead)—Nearctic (southwestern United States)

At LaJoya, New Mexico, four nests (3–10 m apart) were located on a dune in gently sloping, fine-grained sand among sparse plants (Evans 1976a). Nests of *Ammophila, Bembix, Plenoculus,* and *Philanthus* were nearby. In fact, two female and six male *T. cockerelli* were among prey in nests of *Philanthus psyche* Cresson (Evans and O'Neill 1988). The nests and digging behavior differ from those of related genera such as *Gorytes* and *Hapalomellinus.* During burrow construction, the actions of the female in dispersing soil resulted in a circular 3 × 3 cm mound about 3 cm from the nest entrance. Most unusually, a groove made by the female as she dispersed soil led from the nest entrance through the mound. In two nests excavated (Figure 3.1I), 1–2 cells sat at a depth of 23–28 cm and each contained 8–10 prey of three species of Cicadellidae: *Circulifer tenellus* (Baker), *Exitianus exitiosus* (Uhler), and *Norvellina* sp.

Austrogorytes

In a revision of the genus first erected by Bohart (1967), Bohart (1984) listed 31 species, a substantial increase over the 15 in Bohart and Menke (1976). All of the species, of which only one has been studied, occur in Australia.

Austrogorytes bellicosus (F. Smith)—Australasian (southern Australia)

In southeast Australia, most females nested in a mound of sand and a gravel slope in a disturbed area several meters from the edge of the Cotter

River (Evans and Matthews 1971). While working at the nest entrance, a female walked backward, scraping soil with her front legs and clearing it with her hind legs. The soil that accumulated at the entrance was not leveled, so it formed a mound approximately 6 × 3 × 1 cm. The nest consisted of an oblique burrow, 15–30 cm long, and up to six cells in one or two clusters 7–15 cm deep; within a cluster, cells tended to be 2–4 cm apart, but at the same depth (Figure 3.1B). When leaving to hunt, the females erected a temporary outer closure and sometimes made an orientation flight in circles and figure eights.

Among the 218 prey found in 10 nests were seven species of the Eurymelidae (an endemic family of Australian homopterans that feed on *Eucalyptus*): 16 *Eurymelella tonnoiri* Evans, 18 *Eurymelessa moruyana* (Dist.), 20 *Eurymeloides bicincta* (Erichs.), 7 *Eurymeloides marmorata* (Burm.), 5 *Eurymeloides pulchra* (Sign.), 40 *Pauroeurymela amplicincta* (Burm.), and 112 *Platyeurymela semifascia* (Walk.). Because of variation in prey size, the number of prey per cell varied from five large *E. pulchra* to 27 immatures and adults of smaller species. The egg was laid longitudinally on the side of the venter and on the topmost prey in the cell when it was fully provisioned. A chloropid fly (*Lasiopleura* sp.) emerged from one cell (the only one parasitized of 31 cells examined).

Gorytes

Gorytes are widely distributed in the Palearctic and Afrotropical regions, but are apparently absent from Australia and South America. About one-third of 47 species of *Gorytes* (Pulawski 2006a) occur in North America (Evans 1966a). Evans (1966a) reviewed information on eight species, including four from the United States. With the exception of one study of *G. tricinctus,* additional data on *G. canaliculatus,* and a few other scattered reports, little has been done on *Gorytes* since 1966. F. X. Williams's (1928) earlier reference to *G. brasiliensis* no longer refers to *Gorytes* (as the species has been moved to the genus *Sagenista*).

Gorytes atricornis Packard—Nearctic (North America)

At Littleton, Massachusetts, six females of this wasp, which is known to take both Cercopidae and Membracidae as prey, were themselves taken as prey by *Philanthus sanbornii* Cresson, a generalist predator of both bees and wasps (Stubblefield et al. 1993).

Gorytes canaliculatus Fox—Nearctic (North America)

Evans (1966a) was able to report data from 10 sites in eight states in the United States. These reports indicate that *G. canaliculatus* builds nests (Figure 3.6) of 1–4 cells stocked with Cicadellidae and, in one location, Cixiidae (*Haplaxius pictifrons* Stål). A further note on nest structure was provided by Evans (1970), who excavated a unicellular nest with a burrow 8 cm long and a cell 6 cm deep provisioned with leafhoppers. Similarly, Powell (1974) excavated nests with burrows 9 cm long and 4 cm deep on a west-facing slope of "uniform, textured, fine-grained sandstone." Three recent reports confirm that cicadellids of the genus *Idiocerus* are the most common prey; Beirne (1956) lists *Populus* and *Salix* as the most common host of *Idiocerus,* so hunting likely occurs on those tree species. Evans (1968) added records for Jackson Hole, Wyoming, of *Idiocerus apache* Ball and Parker and *I. perplexus* Gillette and Baker. Kurczewski and Kurczewski (1971) recorded 92 Cicadellidae as prey from Presque Isle, Pennsylvania: 89 adult *Idiocerus perplexus* Gillette and Baker, 1 adult *Idiocerus snowi* Gillette and Baker, 1 nymphal *Idiocerus* sp., and 1 adult *Macropsis viridis* (Fitch). Powell (1974) found that all prey at a site in Arcata, California,

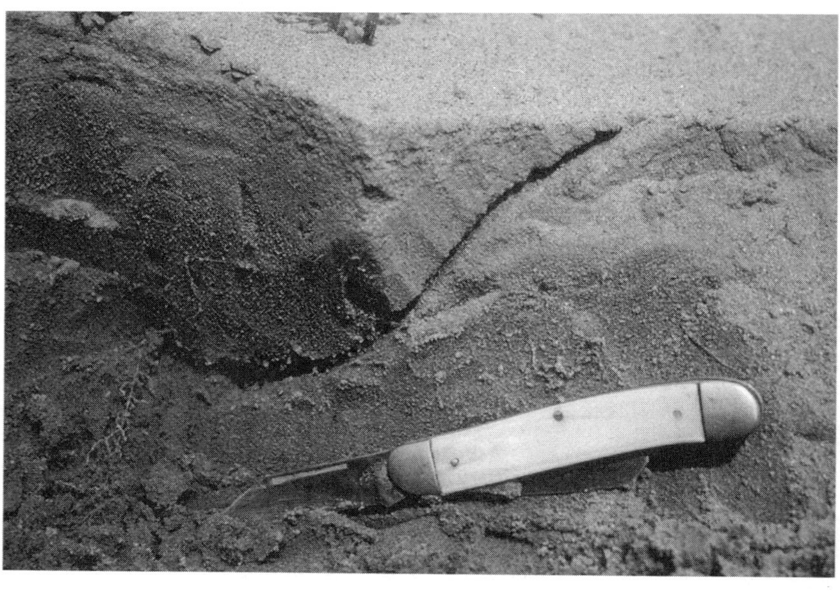

Figure 3.6. Unicellular nest of *Gorytes canaliculatus.* Photo by H. E. Evans.

were *Idiocerus* sp. Three completed cells contained 13–20 prey, all but one of which were nymphs. Fourteen percent of the prey were parasitized by dryinid wasps. Also confirming earlier observations, Powell noted that prey were carried venter-to-venter, and that the nest was closed during foraging trips. Egg placement (along the side of the prey) and the degree of prey paralysis (prey showed no signs of movement or recovery after the first day) were similar to earlier reports by Evans (1966a).

The miltogrammine fly *Metopia argyrocephala* (Mg.) was reared from a nest by Kurczewski and Kurczewski (1971). Spofford and Kurczewski (1990, 1992) reported that 45% of 38 *G. canaliculatus* nest cells were parasitized by the miltogrammines that included *Phrosinella aurifacies*, *Senotainia trilineata* (Van der Wulp), and *Senotainia vigilans* Allen, as well as undetermined species. When trailed by a *S. trilineata*, some females made diversionary flights in an attempt to elude their pursuers. Other females abandoned prey if contacted by a fly.

Gorytes laticinctus (Lepeletier)—Palearctic (Europe, Middle East)

Earlier prey records of females preying only on the spittlebug *Philaenus spumarius* (Cercopidae) have been supplemented by a single female caught carrying the cercopid *Aphrophora corticea* (Germar) in Holland (Felton 1987), and another carrying a male *Aphrophora alni* (Fallén) (Pagliano and Alma 1997). In Japan, Tsuneki (1969) provided a brief note on one female *Gorytes laticinctus koreanus* Handlirsch nesting on a "heap of earth at the road side under the dense foliage of the tall trees." The female carried the prey (Cercopidae) venter-to-venter and quickly removed a temporary closure to enter her nest. In the cell in the shallow nest, he found a single prey that had a maggot under its wing.

Gorytes quadrifasciatus (F.)—Palearctic (Eurasia)

Jacobs and Oehlke (1990, in Pagliano and Alma 1997) report *Philaenus spumarius* as prey.

Gorytes quinquecinctus (F.)—Palearctic (Europe, Middle East, North Africa)

Schmidt (1979, in Pagliano and Alma 1997) report *Philaenus* sp. as prey.

Gorytes simillimus F. Smith—Nearctic (North America)

Kurczewski and Kurczewski (1971) identified two Cicadellidae as prey from Presque Isle, Pennsylvania: one adult each of *Gypona melanota* Spångberg and *Gyponana flavolineata* (Fitch), a genus previously reported as prey. At Littleton, Massachusetts, one male of this wasp was found among over 3,000 prey of *Philanthus sanbornii* Cresson (Stubblefield et al. 1993). Waldbauer et al. (1977) list *G. simillimus,* as well as other *Gorytes,* among a complex of the models hypothesized to provide protection to a group of mimetic flies of the families Conopidae and Syrphidae.

Gorytes tricinctus Pérez—Palearctic (Japan, Korea)

In Japan, females dug and provisioned nests in the soil in plant pots within a garden (Tsuneki 1982). Five nests had slightly curved oblique burrows, 3–5 cm long, leading to single cells at depths of 3–6 cm; two pots housed four and five nests each. The five cells contained 32 adult *Aphrophora maritima* Matsumura (Cercopidae), 5–10 per cell. Tsuneki cites unpublished observations of S. Kubota who also observed nests within a plant pot, in this case stocked with 36 *Aphrophora flavomaculata* Matsumura and *A. maritima.* The egg was laid in the middle of the prey mass, but the timing of oviposition was not determined. Females constructed thick temporary closures that took from 30 to 60 s to clear when they returned with prey, which were carried in flight, beneath the body and venter-to-venter. Returning females eventually dropped the prey, entered the nest, turned around, and pulled the prey in with their mandibles. At final closure, the mound was completely leveled. Earlier, Tsuneki (1965a) observed two nests on the sandy shore of a lake, but the wasps were misidentified as *Argogorytes mystaceus,* when they were actually *G. tricinctus.* The nest burrows were slightly longer (8–12 cm). Cells in the nests, one of which was unicellular (Figure 3.1C), the other bicellular, were 2.5–3.7 cm deep. Prey were a large species of *Aphrophora,* four per cell. Tsuneki (1982) provides details on cocoon-spinning behavior.

Overview of *Gorytes*

Considering the size of this genus, our knowledge is deficient. Four species of *Gorytes* (*G. canaliculatus, G. planifrons, G. simillimus,* and *G. tricinctus*) are known to nest in sand. Both *Gorytes tricinctus* and *G. laticinctus* have been reported nesting in flowerpots. The report of *Gorytes koreanus* Handlirsch nesting in a pile of earth along a road does not mention the soil type, but this is now considered a subspecies of *G. laticinctus*. Nests are relatively shallow, and only *G. canaliculatus* is reported to make multicellular nests, providing as many as 10 prey per cell. Among the 10 species for which prey records are available, half are predators of spittlebugs (Cercopidae); no species is known to take more than two families as prey (Table 3.2).

Pseudoplisus

This former subgenus of *Gorytes* includes 32 species (Pulawski 2006a), mostly in North America. The first detailed reports on their biology did not appear until 1991 (and those were for South African species).

Pseudoplisus natalensis (F. Smith)—Afrotropical (southern Africa)

In Eastern Cape Province, South Africa, females nested in friable soil in flowerpots. Both the mandibles and the legs were used for digging (Callan 1991a). The single nest excavated had two cells at about 10 cm depth at the end of a vertical burrow. The one completed cell was stocked with four froghoppers, *Ptyelus grossus* (Cercopidae), which at 22 mm length were longer than the adult female wasps. Prey were carried to the nest venter-to-venter, with the middle legs holding the prey and the hind legs supporting it. Callan also notes that Fred Gess found this species nesting in flowerpots in Grahamstown, South Africa.

Pseudoplisus phaleratus (Say)—Nearctic (North America)

At Littleton, Massachusetts, a single female was taken as prey by *Philanthus sanbornii* (Stubblefield et al. 1993).

Pseudoplisus ranosahae (Arnold)—Malagasy (Madagascar)

In Madagascar, females nested in large aggregations in damp clay banks with a northeast exposure. Nest entrances were 5–10 cm apart (Callan 1991a). Males patrolled for females in the emergence/nesting area, where several copulations were observed. Like those of *P. natalensis,* females dug nests with the mandibles and forelegs, sometimes carrying lumps of damp soil in their mandibles. The nest burrows entered the bank horizontally or

Table 3.2 Cells per nest, prey per cell, and families reported as prey of *Gorytes* and *Pseudoplisus*. See text, Evans (1966a), and Bohart and Menke (1976) for references and details.

			Cicadoidea			Fulgoroidea		
Gorytes and *Pseudoplisus* species	Maximum cells per nest	Maximum prey per cell	Cercopidae	Cicadellidae	Membracidae	Cixiidae	Dictyopharidae	Issidae
Gorytes								
G. atricornis			X		X			
G. canaliculatus	4	20		X		X		
G. deceptor					X			
G. laticinctus			X					
G. planifrons	1	4						X
G. pleuripunctatus							X	
G. quadrifasciatus				X				
G. quinquecinctus				X				
G. simillimus				X				
G. sulcifrons			X					
G. tricinctus	1	10	X					
Pseudoplisus								
P. natalensis			X					
P. ranosahae			X					

at a slight downward slope, and cells were 6–10 cm into the bank. Cells in the "apparently multicellular" nests were each stocked with three to four froghoppers, *Ptyelus goudoti* (Cercopidae), captured on nearby *Acacia* trees and carried venter-to-venter. Because the 25–30 mm long prey were longer than the 18–20 mm long females, prey carriage was apparently difficult and prey were often dropped in flight. Although many were recovered by females, "the base of the bank was littered with discarded froghoppers." After landing, the female "straddled the prey and crawled laboriously with it to the nest, the entrances of which were left fully or partially open during foraging." Final closure of the nest was apparently a quick affair, so that the nest entrance remained visible.

Lestiphorus

Lestiphorus contains 18 described species distributed throughout the world, except in the Australasian region (Pulawski 2006a). Very little biological information is available.

Lestiphorus bicinctus (Rossi)—Palearctic (Europe)

This European species, a known predator of the spittlebugs *Philaenus spumarius* and *Aphrophora alni* (Cercopidae) (Bernard 1934; Schmidt 1979), is reported by Benno (1966) to be a possible host in the Netherlands of *Nysson trimaculatus* (Rossi). Benno also notes that when he marked *L. bicinctus* and released them 500 yards from where they were caught (an area dominated by *Cornus* shrubs), they were found back at the place of capture the next morning.

Liogorytes

The biology of this genus of 11 described South American species (Pulawski 2006a) was simply stated as "unknown" in Bohart and Menke (1976). Not much more is available.

Liogorytes joergenseni (Brèthes)—Neotropical (Argentina)

In an aggregation in Argentina, females provisioned nests with adults of a "medium size" cicada of the genus *Chonosia* (Cicadidae; Tettigadinae), carrying them to ground nests with entrances about "one-half inch" (1.3

cm) diameter (Bohart and Stange 1977). No other details of its biology are given, but this is enough to make *Liogorytes* just the third genus of Gorytini known to prey on cicadas.

Hoplisoides

Hoplisoides, the largest genus in the Gorytini, includes 80 species (Pulawski 2006a) and Bohart (1997) lists 17 species from America north of Mexico.

Hoplisoides aglaia (Handlirsch)—Afrotropical (southern Africa)

Gess (1981) noted that this species was associated with sandy soil in the karroid area near Grahamstown, South Africa, and that it provisions with Membracidae.

Hoplisoides ater (Gmelin)—Neotropical (West Indies)

In Cuba, nests were found in an area of bare, friable sandy soil where eight nests excavated by Sánchez and Genaro (1992a)were relatively short (3.5–10.0 cm long burrows) and shallow (cells 0.8–4.0 cm deep). The mound of soil at the entrance was always scattered by the wasp and the entrance was kept closed while the female was hunting. Prey were carried to the nest using the middle legs with the prey held head forward and venter upward. The vast majority of the 318 prey collected were immature and adult Membracidae of four species: 252 *Micrutalis calva* Say, 32 *Monobelus flavidus* Fairmaire, 31 *Monobelus* sp., and 2 *Goniolomus tricorniger* Stål. Among the 169 adult *M. calva* prey, 91% were females. The single anomalous prey record in the sample (and a record anomalous for the entire genus *Hoplisoides*) was one *Lepyronia angulifera robusta* Metcalf and Bruner (Cercopidae). The 3–4 cells per nest were mass provisioned with 10–34 prey per cell, the highest number occurring in cells containing only small *M. calva*. The egg was laid ventrally at the base of a leg on the last prey placed in the cell.

Hoplisoides hamatus (Handlirsch)—Nearctic (western North America)

Evans (1970) added several observations to those in Evans (1966a), where the species was discussed under the name *H. spilographus*. Several dense-nesting aggregations discovered in 1967 in Jackson Hole, Wyoming, contained 20 or more females who sometimes nested within several centimeters of one another, "often in clay loam with many stones." Females constructed shallow one- or two-celled nests (Figure 3.1F). As in the previous report, prey were all immatures of one species of Membracidae (Ceresini, possibly *Stictocephala* sp.). *Nysson rusticus* Cresson was observed digging into a nest entrance.

Hoplisoides iridipennis (F. Smith)—Neotropical (Mexico to Brazil)

In Trinidad, this species nested in sand, provisioning cells with Membracidae: *Darnoides brunneus* (Germar), *Erechtia bicolor* Walker, and *Horiola picta* (Coquebert). The one nest examined had an oblique burrow, 8 cm long, and at least one cell containing three prey (Callan 1976).

Hoplisoides jaumei (Alayo Dalmau)—Neotropical (Cuba)

In Cuba, Sánchez and Genaro (1992a) found 12 nests in a vertical surface of rather friable clay soil adhering to the roots of a fallen tree. Because of roots and stones in the soil, the structure of only three nests was determined. The nests had burrows 9.0–22.0 cm long, with cells 6.0–19.0 cm deep; there were as many as five cells per nest, with cells in close proximity of one another. Prey were carried to the nest in flight with the middle legs, and the nest was kept closed during provisioning. The 149 prey collected were of three species in two families not previously reported for *Hoplisoides:* 137 *Neurotmeta sponsa,* 10 *Remosa spinolae* (Guérin) (Tropiduchidae), and 2 *Melormenis* sp. (Flatidae). The number of prey per cell (5–12) was fewer than for *H. ater,* the other Cuban species studied by Sánchez and Genaro, because *H. jaumei* provisioned with a lower proportion of immatures. Eggs were laid ventrally at the bases of the coxae of prey.

Hoplisoides punctuosus (Eversmann)—Nearctic (Eurasia)

Discussing this species as *Hoplisoides punctatus* Kirschbaum, Evans (1966a) cited several earlier studies by Ferton in southern Europe. Ferton found that females dug nests up to 15 cm deep in sandy soil and provided as many as 60 prey per cell, all prey records being nymphs and adults of *Tettigometra* (Tettigometridae). The nest was always closed by the female before foraging. Since Ferton's work, which is nearly 100 years old, only a brief note by Krombein (1972) has appeared. He observed a copulating pair, in an end-to-end posture, within a group of males and females "flying over low vegetation on the sandy beach at Tolon, Argolis" in Greece. Not much, but it may give a hint as to this species' mating strategy.

Hoplisoides semipunctatus (Taschenberg)—Neotropical (Argentina); Nearctic

In Argentina, in two nests with burrows 15–17 cm long, each had a single terminal cell at depths of 9 and 11 cm on "bare, eroded, sandy soil" on slopes (Evans and Matthews 1974). One nest was found with a burrow that curved "twice to avoid large stones in the soil." When excavated, the nests were still being provisioned. One nest contained one unidentified immature and eight adults of the species *Aconophora* sp., *Enchenopa ferruginea* Walker, and *Publilia* sp. (all Membracidae). The cell in the second nest, probably near completion, contained 21 *E. ferruginea* adults. When approaching the nest, which was left open during foraging, females held the prey with their middle legs and descended "slowly, obliquely downward from a height of about one meter." The form of approach to the nest has been observed in other *Hoplisoides*, but the lack of a temporary closure is unusual (Evans 1966a; Callan 1976).

Hoplisoides splendidulus (Bradley)—Nearctic (southwestern United States)

In Ward County, Texas, Rubink (1977) observed five females nesting in sandy gravel amid sparse vegetation. Each nest had an ~8 cm long oblique burrow leading to 1–3 cells stocked with 2–5 adult and nymphal *Acanalonia similis* Doering (Fulgoridae). After construction of the burrow, the mound was extensively leveled and a temporary nest closure was maintained while the female hunted. Females apparently spent the night away from their nesting area, as do those of *H. spilographicus* (Evans 1966a).

Hoplisoides vespoides (F. Smith)—Neotropical

Callan (1976) published observations on both *H. vespoides* and *Hoplisoides umbonicida* Pate. Bohart (1997) has since synonymized the two, so Callan's observations apparently apply to the same species. Callan observed "*H. vespoides*" nesting in sandy soil in several locations in Trinidad, in association with *Bicyrtes variegatus* (Chapter 6) *and Cerceris dilatata* Spinola at one site, and with *Bembecinus agilis* (Chapter 5) at another. The membracid *Umbonia spinosa* (F.) was the only prey recorded. The wasps reported as *H. umbonicida* nested in coarse sand along a stream and in a sandpit. The single nest excavated had an oblique burrow about 10 cm long and at least four cells provisioned with *U. spinosa*. At 1.5 cm in length, the treehopper was larger than the wasp itself, which carried the prey venter-to-venter in a slow descending flight to the nest; the entrance, which was closed during foraging, was quickly opened with the forelegs, the wasp then entering quickly without dropping the prey.

Overview of *Hoplisoides*

With few exceptions, female *Hoplisoides* nest in flat or slightly sloping sandy soil. One exception, *Hoplisoides jaumei*, was reported nesting in a vertical surface of friable clay. Soil used by those nesting in sand varies from loose sand to sand with a high clay content. Nests of up to 4–5 cells have been reported. Because the prey records reviewed here (Table 3.3) and in Evans (1966a) are often based on samples from multiple populations, they suggest that some species specialize on Membracidae. Females of the population of *H. hamatus* in Jackson Hole, Wyoming, have only been recorded taking membracid nymphs from a single species. On the other hand, prey records of some other species do not include Membracidae. Records of *H. latifrons* and *H. punctuosus* as predators of Tettigometridae are also based on samples from multiple populations. Sánchez and Genaro (1992a) recently added three new prey families, Tropiduchidae, Flatidae, and Cercopidae. However, *H. ater*, should also be included among the treehopper specialists, because all prey were Membracidae outside a single, perhaps anomalous, Cercopidae (among over 300 records). This leaves only *H. denticulatus*, for which we have only a single prey record (Cicadellidae). The number of prey per cell varies widely among species, being related to the relative size of prey.

Sagenista

Pulawski (2006a) lists 10 species of these "medium to small, mostly black wasps" (Bohart and Menke 1976), all from South and Central America.

Table 3.3 Cells per nest, prey per cell, and families of prey reported as prey of *Hoplisoides*. See text and Evans (1966a) for references and details.

			Cicadoidea		Fulgoroidea			
Hoplisoides species	Maximum cells per nest	Maximum prey per cell	Cicadellidae	Membracidae	Flatidae	Fulgoridae	Tettigometridae	Tropiduchidae
H. ater	4	34		X				
H. costalis	2	9		X				
H. denticulatus			X					
H. hamatus	2	14		X				
H. iridipennis				X				
H. jaumei	5	12			X			X
H. latifrons	2	19					X	
H. manjikuli		13		X				
H. nebulosus	3	20		X				
H. punctuosus		60					X	
H. semipunctatus				X				
H. spilopterus				X				
H. splendidulus		5				X		
H. tricolor	4	19		X				
H. vespoides	4			X				

The only report on the behavior of this genus in Evans (1966a) was on *Sagenista brasiliensis,* a species then placed in *Gorytes.* It is still the only member of the genus for which biological information is available.

Sagenista brasiliensis (Shuckard)—Neotropical (Brazil)

In the Maracas Valley of Trinidad, *S. brasiliensis* nested in "friable, sandy soil" in one location and in a "sandy bank" in another (Callan 1976); earlier F. X. Williams (1928) found it nesting "in banks of rich soil along the margin of a jungle" in Brazil. One nest examined by Callan had a burrow 6 cm long and two cells provisioned with Membracidae and Fulgoroidea; another nest contained three cells provisioned with 11 species from several homopteran families. Callan lists all known prey records for this species (including those from an earlier study of Williams (1928)): *Koloptera callosa* Metcalf (Achilidae); *Punama* sp. (Delphacidae), *Dictyophara* sp. (Dictyopharidae); *Epormenis fuliginosa* (Fennah), *Euhyloptera corticalis* Fennah, *Flatormenis* sp. (Flatidae); *Thionia mammifera* Fennah, *Thionia* sp. (Issidae); and *Ceresa vitulis* (Membracidae). Prey were carried venter-to-venter and the egg is attached along the base of one of the legs.

Overview of the Tribe Gorytini

Nesting behavior. By the far the most common general soil type in which the nests of Gorytini are found is that with a high sand content, this having been reported for *Hoplisoides, Gorytes, Tanyoprymnus, Exeirus,* and some *Sphecius. Sphecius hogardii,* in contrast, nests in hard stony soil, *Pseudoplisus ranosahae* in damp clay, and *Clitemnestra plomleyi* in soil with high gravel content. Relatively flat surfaces are most often favored, though some species nest on sloping banks while *Clitemnestra bipunctata, Gorytes simillimus,* and *Hoplisoides jaumei* have been found on vertical surfaces. *Sphecius speciosus'* penchant for nesting in lawns, parks, and sports fields has received attention because the species is often considered a nuisance. Several oddities observed for members of the tribe include nests under leaf litter *(Ammatomus icarioides),* in flowerpots *(Pseudoplisus natalensis, Gorytes laticinctus),* and in preexisting holes in soil *(Clitemnestra plomleyi, Tanyoprymnus moneduloides).* The last observation is surprising, as Evans (1966a) wrote that no Bembicinae nest in preexisting cavities. It is perhaps too early to expand the list of diverse nesting sites among Gorytini to rotten wood *(Argogorytes nipponis),* until the tentative report can be confirmed.

Hunting and provisioning. As far as is known, all Gorytini mass provision cells with Homoptera (Table 3.4). We have records of 16 prey families within two suborders and three superfamilies of Homoptera: Auchenorrhyncha: Cicadoidea (4 families); Auchenorrhyncha: Fulgoroidea (11 families); and Sternorrynchna: Psylloidea (1 family) (Table 3.4). As of yet, no prey from the third suborder, Aphidoidea, have been recorded, though these are preyed on by species of smaller apoid wasps of the Pemphredoninae. The most common prey family, Cicadellidae (9 of 20 genera of Gorytini) are used as prey by many wasps, including *Alysson, Didineis,* and *Bembecinus* (Bembicinae), *Mimesa, Mimumesa,* and *Psen* (Pemphredoninae), and several *Crossocerus* (Crabroninae).

Table 3.4 Families of Homoptera reported as prey of genera of Gorytini. See text and Evans (1966a) for references and details.

Genus (number of species studied)	Cercopidae	Cicadellidae	Cicadidae	Membracidae	Achilidae	Cixiidae	Delphacidae	Dictyopharidae	Eurybrachidae	Eurymelidae	Flatidae	Fulgoridae	Issidae	Tropiduchidae	Tettigometridae	Psyllidae
Ammatomus (1)									X		X					
Argogorytes (5)	X												X			
Austrogorytes (1)										X						
Clitemnestra (3)	X	X		X		X	X	X			X			X		X
Exeirus (1)			X													
Gorytes (11)	X	X		X		X		X					X			
Hapalomellinus (1)		X														
Harpactus (3)	X	X														
Hoplisoides (15)		X		X								X		X	X	
Lestiphorus (1)	X															
Liogorytes (1)			X													
Oryttus (1)		X						X				X				
Psammaecius (1)		X														
Psammaletes (1)											X					
Pseudoplisus (2)	X															
Sagenista (1)				X	X		X	X			X		X			
Sphecius (9)			X													
Tanyoprymnus (1)								X				X	X			
Trichogorytes (1)		X														
Number of wasp genera	6	8	3	4	1	2	2	5	1	1	4	3	4	2	1	1

If we take Bohart and Menke's (1976) and Prentice's (1998) proposed relationships among genera of Gorytini as a working hypothesis, it seems likely that cicada hunting evolved three times independently in the tribe Gorytini, in *Sphecius*, *Exeirus*, and *Liogorytes*. Among the Gorytini discussed here, *Sphecius* is considered both by Bohart and Menke and by Prentice (1998) to be most closely related to such genera as *Ammatomus* and *Tanyoprymnus*, which are predators of various small "hoppers." Although *Exeirus* is in "size, general shape, and prey . . . reminiscent of *Sphecius*, . . . the relationship is not close, and *Exeirus* must have diverged very early" from the main lineage of the Gorytini (Bohart and Menke 1976); Prentice places them in different subtribes of Gorytini, *Sphecius* in the Handlirschiina and *Exeirus* in the Clitemnestrina. Bohart and Menke considered *Liogorytes* (which Prentice places in the subtribe Gorytina) most closely related to *Arigorytes* (whose biology is unknown) and *Hoplisoides* (which prey on Membracidae and other small Homoptera). The behavioral shift to cicadas as prey seems correlated with the evolution of several morphological traits, including larger body size and, at least in the case of *Sphecius*, a "highly developed sting apparatus" (Radović 1985).

Some gorytines in relatively unstudied genera are clearly diet opportunists. Just two *Sagenista brasiliensis* cells examined by Callan (1976) contained six prey families. And three species of *Clitemnestra* are known to take nine prey families in three homopteran superfamilies. In contrast, females of the fairly well studied genera like *Sphecius* are predators only of Cicadidae, with 7 of 10 species being recorded to take prey from a single genus. Nevertheless, we may still have a fairly incomplete record of the range of prey taken by female Gorytini, even at the level of prey family. The genus *Hoplisoides* is a case in point. More species of *Hoplisoides* have been studied than any other genus of Gorytini and most are predators of Membracidae, but it was only recently that Sánchez and Genaro (1992a) doubled the number of known prey families with a study of just two species, *H. jaumei* and *H. ater*.

Natural enemies. Gorytinae, at least the best-studied species of *Clitemnestra*, *Gorytes*, *Hoplisoides*, and *Sphecius*, suffer brood parasitism from miltogrammine flies of the genera *Metopia*, *Phrosinella*, and *Senotainia*. Reported rates of miltogramminae mortality range as high as 45% for *Gorytes canaliculatus* and 38% for *Sphecius speciosus*. As will be discussed in the next chapter, Gorytini (*Argogorytes*, *Gorytes*, *Harpactus*, *Hoplisoides*,

Lestiphorus, and *Oryttus*) are unique among sand wasps in being hosts of brood parasitic Nyssonini.

Male behavior. In 1966, detailed knowledge of gorytine male behavior was restricted to N. Lin's studies of territoriality in emergence/nesting areas by *Sphecius speciosus.* His work has been greatly expanded upon by others working with *S. speciosus* and *S. grandis.* Beyond that we have several brief reports on patrolling by male *Exeirus lateritius* and *Pseudoplisus ranosahae.* Several excellent studies of male *Argogorytes* behavior tell us a lot about their interactions with female-mimicking flowers, but give only a hint about how males find females when they are not being duped by plants.

— 4 —

Brood Parasites of the Nyssonini

The general features of the biology of the Nyssonini were summarized by Evans (1966a), Bohart and Menke (1976), and O'Neill (2001). On the basis of both behavioral and morphological evidence, it is generally agreed that all Nyssonini are brood parasites of other apoid wasps. Evans (1966a) presented arguments that, after the adult female deposits an egg in the host's nest cell, the nyssonine larva destroys the host's egg before feeding on the provisions. As of 1966, knowledge of the host relationships of the tribe was mostly restricted to the genus *Nysson,* though information of varying reliability was also available for a few other genera. Thus it required fewer than 8 of over 500 pages in Evans (1966a) to review known information on the Nyssonini. The intervening years have not improved matters much.

Phylogenetic status. Bohart and Menke (1976) listed 18 genera of Nyssonini, but Pulawski (2006a) just 17 because *Synnervus* has been subsumed into the genus *Nysson*; the 226 species of Nyssonini constitute approximately 13% of the Bembicinae. According to Alexander (1992), the Nyssonini is "apparently paraphyletic with respect to other [Bembicinae]." Prentice (1998), however, considers the Nyssonini to be monophyletic.

Nysson

Nysson, the largest genus in the Nyssonini, now has 102 described species (Pulawski 2006a), 24 species of which have been described for localities in America north of Mexico. Although *Nysson* is the fourth-largest genus of Bembicinae, comparatively little is known of its biology. As with many so-called host records for natural enemies of solitary wasps, associations between particular species of *Nysson* and particular host species are often tentative, because they are not always based on actual rearing of the brood parasites from host cells. We include many tentative records primarily to highlight the scarcity of firmly documented prey records.

Nysson braunsii Handlirsch—Afrotropical (southern Africa)

On the basis of host records for other species, Gess (1981) provided a list of potential hosts for sites near Grahamstown, South Africa: *Harpactus vicarius karooensis* (Brauns), *Hoplisoides thalia* (Handlirsch), *Hoplisoides aglaia*, and *Oryttus kraepelini* (Brauns).

Nysson daeckei Viereck—Nearctic (eastern United States)

In Jackson Hole, Wyoming, Evans (1970) observed females exploring, but not entering, nests of *Philanthus pulcher* Dalla Torre. Because *N. daeckei* is a known brood parasite of Gorytini, which also nested in the area, *Philanthus* may not be the real host.

Nysson interruptus (F.)—Palearctic (Europe)

On the basis of a close correspondence in their seasonal activity periods and locations of flight activity, Benno (1977) suggested that *Argogorytes fargeii* is the host of *N. interruptus* in the Netherlands.

Nysson maculosus (Gmelin) (= *Nysson trimaculatus* Rossi)—Palearctic

In the Netherlands, Benno (1966) monitored the flight periods and abundance of *Lestiphorus bicinctus* and *N. trimaculatus*, finding a tight correspondence in their activities. Benno took this as support for the hypothesis that *L. bicinctus* is a host of *N. maculosus.*

Nysson rugosus Cameron—Oriental (India)

Krombein (1984, 1985) provides circumstantial evidence that *N. rugosus* is a brood parasite of *Bembecinus.* It is abundant in Sri Lanka only where *Bembecinus* is abundant, and at these locations, there are no other abundant wasps that would likely serve as hosts.

Nysson rusticus Cresson—Nearctic (western United States)

Evans (1970) observed a female digging into a nest of *Hoplisoides spilopterus.*

Nysson spinosus (J. Foster)—Palearctic (Europe)

Lomholdt (1975) gives *Argogorytes* as the host of this species. And on the basis of a close correspondence in their seasonal activity periods and locations of flight activity, Benno (1977) also suggested that *Argogorytes mystaceus* is the host of *N. spinosus* in the Netherlands. Correspondence between the chemical composition of mandibular gland secretions of *Argogorytes fargei, A. mystaceus,* and *N. spinosus* (2,5 dimethyl-3-isopentylpyrazine) led Borg-Karlson and Tengö (1980) to hypothesize that the chemical in *N. spinosus* aids the wasp to mimic its host. Radović (1985) examined the sting of *N. spinosus,* finding the stylet "completely straight and devoid of spines on the lancetae," contrasting this with those Bembicinae that take mobile prey and have a well-developed sting apparatus and curved stylet.

Acanthostethus

All 15 species of this genus occur in Australia. Females have been observed within nest aggregations of potential hosts, but these wasps have not been reared from nest cells.

Acanthostethus portlandensis (Rayment)—Australasian (Australia: Victoria)

In a *Sericophorus viridis* Saussure aggregation, *A. portlandensis* was abundant and often investigated, and sometimes entered, open nests (Matthews and Evans 1971). In 1 of 17 cells with *Sericophorus* eggs, a small egg (quite probably of *Acanthostethus*) was concealed on the fourth prey behind the left coxa. The egg of the parasite (1.3 mm long) was smaller than that of the host (2.5 mm), as would be expected.

Zanysson

The 17 species of this genus are restricted to the New World. The closest we have to a firm host record is a reported association of *Zanysson texanus* with nests of *Tachytes* (Cazier in Evans 1966a).

Zanysson armatus (Cresson)—Neotropical (Cuba)

In Cuba, females were observed in the nesting area of *Hoplisoides ater,* but

Table 4.1 Presumptive host relationships of species of Nyssonini, updated from Table 9 in Evans (1966a). A few of the names in this table differ from that earlier version because of taxonomic name changes (names changes for the Nyssonini are given in footnotes).

		Best evidence of host relationship		
Species of Nyssonini	Host or presumptive host	Egg or larva found in host nest cells	Observed entering "host" nests	Observed in "host" nesting areas
Acanthostethus portlandensis (Rayment)	*Sericophorus viridis* Saussure	X		
Brachystegus scalaris (Illiger)	*Tachytes europaeus* Kohl	X		
Epinysson tramosericus (Cresson)[a]	*Hoplisoides placidus* F. Smith	X		
Epinysson bellus (Cresson)[b]	*Hoplisoides tricolor* (Cresson)		X	
Epinysson tuberculatus Handlirsch[c]	*Hoplisoides costalis* (Cresson)	X		
Epinysson moestus (Cresson)[d]	*Hoplisoides hamatus* (Handlirsch)		X	
Metanysson arivaipa Pate	*Cerceris graphica* F. Smith		X	
Metanysson coahuila Pate	*Cerceris conifrons* Mickel		X	
Nysson daeckei Viereck	*Gorytes canaliculatus* Packard	X		
	Hoplisoides placidus	X		
Nysson dimidiatus Jurine	*Harpactus elegans* (Lepeletier)		X	
	Harpactus tumidus (Panzer)		X	
	Harpactus laevis (Latreille)			X
	Hoplisoides latifrons (Spinola)	X		

Nysson fidelis Cresson	*Gorytes canaliculatus*	X		
Nysson interruptus (F.)	*Argogorytes fargeii* (Shuckard)			X[e]
	Argogorytes mystaceus (L.)		X	
	Gorytes quadrifasciatus (F.)		X	
Nysson lateralis Packard	*Gorytes canaliculatus*		X	
Nysson maculosus (Gmelin)[f]	*Gorytes laticinctus* (Lepeletier)		X	
	Gorytes quadrifasciatus			X
	Harpactus tumidus	X		
	Harpactus laevis		X	
	Lestiphorus bicinctus (Rossi)			X
	Oryttus concinnus (Rossi)		X	
Nysson niger Chevrier	*Hoplisoides punctuosus* (Eversmann)			X
Nysson pumilis Cresson	*Hoplisoides hamatus*			X
Nysson rusticus Cresson	*Hoplisoides hamatus*			X
	Hoplisoides splilographus (Handlirsch)		X	
Nysson spinosus Foerster	*Argogorytes mystaceus*		X	
	Gorytes quadrifasciatus		X	
Nysson tridens Gerstaecker	*Harpactus laevis*			X
	Harpactus lunatus (Dahlbom)			X
Zanysson armatus (Cresson)	*Hoplisoides ater* (Gmelin)			X
Zanysson plesius (Rohwer)[g]	*Tachytes distinctus* F. Smith		X	

[a]As *Nysson tuberculatus* in Evans (1966a); [b]as *N. bellus* in Evans (1966a); [c]as *N. hoplisivora* (Rohwer) in Evans (1966a); [d]as *N. moestus* in Evans (1966a); [e]Benno (1977) noted a correspondence between the flight activity of the potential host and parasite; [f]as *N. maculatus* in Evans (1966a); [g]as *Z. tonto* in Evans (1966a).

other other potential hosts were also present in the area (Sánchez and Genaro 1992a).

Zanysson texanus (Cresson)—Nearctic (United States)

In Wyoming, Lavigne and Holland (1969) reported *Z. texanus fuscipes* among prey of the robber fly *Diogmites angustipennis* (Loew).

Overview of the Tribe Nyssonini

Behaviorally, Nyssonini is the anomalous tribe of Bembicinae because, as far as is known, all species are brood parasites of other apoid wasps. The only other brood parasitic bembicines are members of the genus *Stizoides* (Chapter 5). A generic account of the biology of Nyssonini (mainly based on observations of *Nysson* and not much updated from that in Evans 1966a) would go as follows. Females search the host's nesting area for open nests or host females entering their nests. When entering host nests, usually after the host has departed, the parasite is capable of removing temporary nest closures. Upon reaching a nest cell, the female lays a small egg (typically smaller than the host's) in a cryptic location on one of the prey at the bottom of the cell (e.g., beneath the wing of a leafhopper). Because parasite oviposition usually occurs before host oviposition, the adult parasite has no chance to destroy the host's egg. However, when the parasite larva hatches, it uses its sharp mandibles to kill the host's egg and then feeds on prey in the cell. Upon leaving the nest following oviposition, the parasite female closes off the burrow (perhaps to prevent conspecifics from attacking the same nest).

Most host records are for other species of Bembicinae (Table 4.1). As indicated in the table, fewer than one-third of the presumptive host records are based on direct evidence of finding a nyssonine egg or larvae within a host nest. Hosts of *Nysson* and *Epinysson* are species of Gorytini, including *Argogorytes, Bembecinus, Gorytes, Harpactus, Hoplisoides, Lestiphorus,* and *Oryttus.* Host records for other genera are wasps of the subfamily Larrinae (*Sericophorus* and *Tachytes*) and Philanthinae (*Cerceris*) (Evans 1966a; Bohart and Menke 1976). All Nyssonini are obligate brood parasites. Among both solitary bees and apoid wasps, obligate brood parasitism is more common in temperate regions than in the tropics, perhaps because the greater seasonality of host populations in temperate zones provides a more predictable resource (Wcislo 1981).

— 5 —

Stizini:

A Mixed Tribe of Hopper Hunters and Brood Parasites

Despite relatively low morphological diversity at the generic level, the tribe Stizini as a whole exhibits a great diversity of reproductive strategies. No other sand wasp tribe contains mass provisioners *(Stizus)*, progressive provisioners *(Bembecinus)*, and obligate brood parasites *(Stizoides)*. And, across the tribe, prey include grasshoppers, katydids, mantids, and leafhoppers (as well as several odd records of flies). Recent research has uncovered previously unstudied and unknown aspects of nest structure, prey types, and mating strategies. Future studies in Africa and Central Asia, which together include well over half of the world's *Stizus*, should be rewarding and could perhaps uncover further interspecific variation in behavior.

Phylogenetic status. The tribe Stizini consists of three genera that are similar enough structurally to have once been placed in the single genus *Stizus*, originally erected by Latreille in 1802. Although it contains just three genera (4% of Bembicinae), the Stizini has about 20% of the described species because it contains the second- *(Bembecinus)* and third- *(Stizus)* most speciose genera (trailing only *Bembix*). The Stizini "consistently formed a monophyletic assemblage" with Nyssonini, Gorytini, and Bembicini in Alexander's (1992) analysis, but it is "apparently paraphyletic with respect to Bembicini" (i.e., Stizini + Bembicini was "consistently supported as a monophyletic group"). Prentice (1998) also considers the traditional grouping to be paraphyletic with respect to the Bembicini of Bohart and Menke (1976). Therefore, he placed the three component genera into the Bembicini, dividing them between into two subtribes: Stizina (*Stizus* + *Stizoides*) and Bembecinina *(Bembecinus)*. We keep the three genera together here, but add the proviso that this may be a temporary taxonomic grouping.

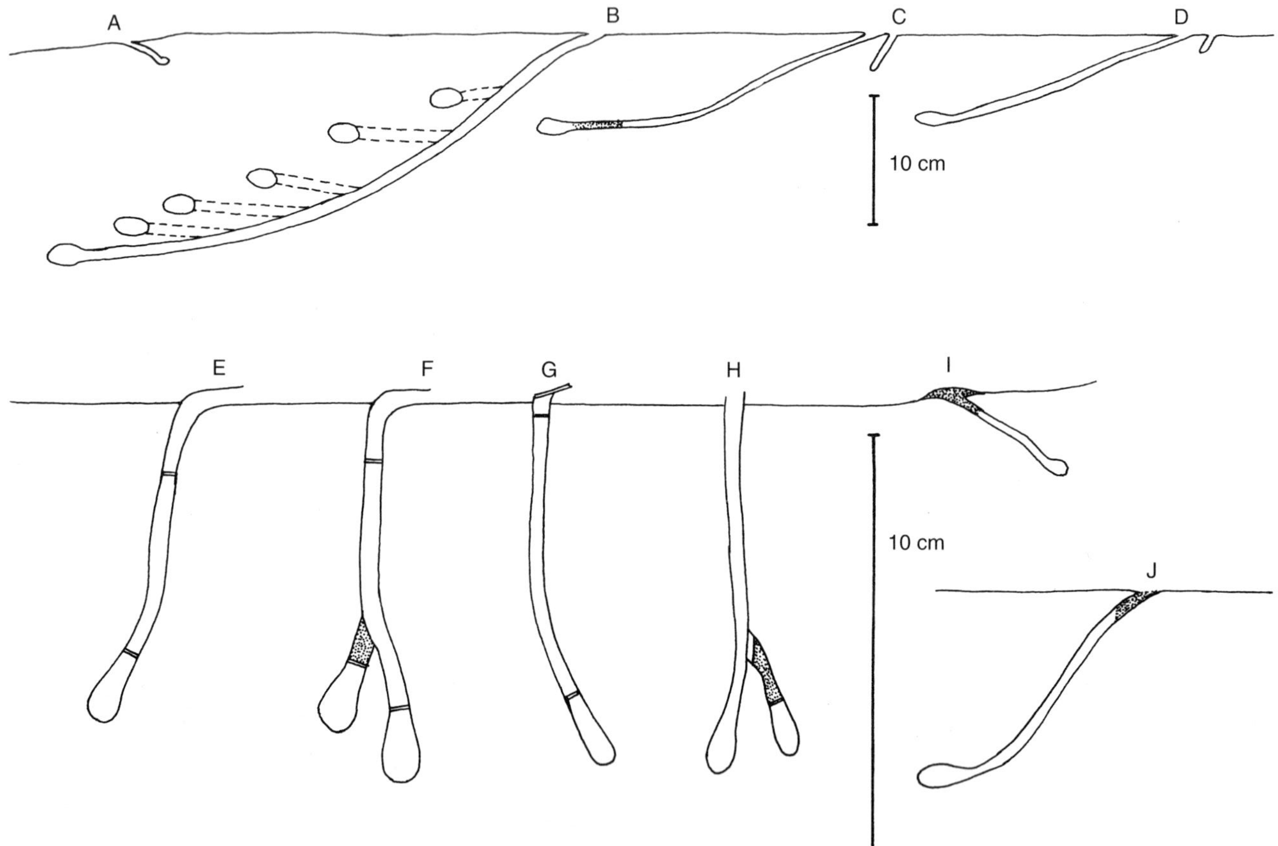
A
B
C
D
10 cm
E
F
G
H
I
10 cm
J

Stizus

Behavioral studies of *Stizus* date back at least to Fabre (1886), who observed *Stizus distinguendus* Handlirsch in Europe. *Stizus* contains 120 species (Pulawski 2006a) of often colorful wasps up to 3.5 cm in length. The genus is widespread in temperate and tropical regions, though apparently absent from South America, Australia, and Southeast Asia. Four species occur in America north of Mexico (Stubblefield 1984). As of 1966, the best-studied species was *Stizus pulcherrimus*, due to several studies by K. Tsuneki (1965b, 1976) in Korea and Mongolia. Recent publications have added information on two European, two South African, and one South Asian species, and increased our knowledge of one North American and one Asian species *(S. pulcherrimus).*

Stizus continuus (Klug)—Palearctic (Europe)

In Spain, Asís et al. (1988) observed females nesting in a salt marsh, where the soil had a texture of "damp sand," and where periods of drought had left a saline crust on the soil surface. The 3–8 cells per nest were 8–18 cm deep and built off the sides of 35–69 cm long burrows (Figure 5.1B). One or two accessory burrows were dug near the nest entrance. Cells were provisioned with 4–8 prey from six species in three families of Orthoptera: 10 nympal *Pezotettix giornae* (Rossi), 3 adult *Heteracris littoralis* (Rambur), 13 nymphal and 2 adult *Tropidopola cylindrica* (Marschall); 5 nymphs of an undetermined species (Acrididae); 2 nymphal and 3 adult *Pyrgomorpha conica* (Oliv.) (Pyrgomorphidae); and 5 adult *Homorocoryphus nitidulus* (Scolopi) (Tettigoniidae). In captivity (see below), hunting female wasps readily stung and provisioned adults of *Acrotylus insubricus* (Acrididae), a species not observed as prey in the field.

As a result of being stung several times beneath the thorax, prey were "almost completely paralyzed" so that only their antennae and palpi moved.

Figure 5.1 *(opposite page).* Structure of nests of (A) *Bembecinus tridens* (Lüps 1969); (B) *Stizus continuus* (Asís et al. 1988); (C) *S. pulcherrimus* (Tsuneki 1976); (D) *S. perrisii* (Asis et al. 1991); (E, F) *Bembecinus cinguliger* (F. W. Gess and S. K. Gess 1975); (G, H) *B. oxydorcus* (F. W. Gess and S. K. Gess 1975); (I) *B. tridens* (Lüps 1969); (J) *Bembecinus quinquespinosus* (Evans 1966a). All redrawn from cited sources. A–D drawn to scale of upper reference line; E–J drawn to scale of lower reference line.

Prey were carried in flight, venter-to-venter and head forward, and grasped with the middle legs. The egg, 3 mm long, was attached laterally to the base of the wing pads of the first prey brought into a cell, such that the head of the newly hatched larva would be in a position to feed near the base of the middle pair of legs of the prey (Figure 5.2). Although most cells were mass provisioned, three females had brought prey into cells in which the egg had already hatched and larva was eating. Evans (1966a) referred to this as "delayed provisioning," whereas Asís and his colleagues referred to it as "slow mass provisioning" following Genise (1982f). Asís and his colleagues were able to induce females to build and provision nests in an observation cage, suggesting that this species may be a good candidate for experimental studies of nesting behavior and prey choice. There were, however, some discrepancies between wasp behavior in the field and in the lab that may limit the value of the technique.

Asís et al. (1988) report that males patrolled the emergence area in "sinuous flights 5–10 cm above the ground," often alighting or pouncing on other insects. Males also "frequently walked, touching the ground with their antennae," a behavior that "seemed to be directed towards finding the females that were emerging from their cocoons" within their subterranean nest cells. This led males to dig for females, with groups of competing males sometimes forming clusters around newly emerged females. Asís et al. hypothesize that males defend areas around emergence holes against other males, referring to these areas as territories. In an extensive follow-up study, Asís et al. (2006) examined the role of body size in determining

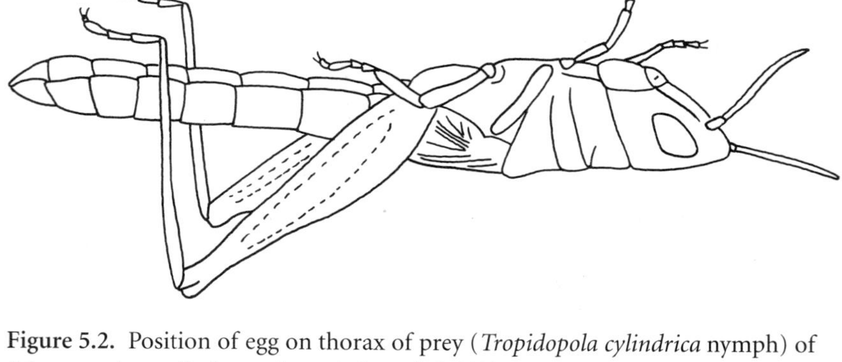

Figure 5.2. Position of egg on thorax of prey (*Tropidopola cylindrica* nymph) of *Stizus continuus*. Redrawn from Asís et al. (1988).

the form and outcome of territorial interactions in the emergence area. Females usually mated with the resident of the territory in which they emerged. Most males seen copulating were larger on average than those never observed to mate, and males that occupied territories; were larger than those never seen on territories. The advantage was attributed to the fact that larger males won 83% of encounters on territories; when removed from territories by the investigators, they were replaced by smaller conspecifics 97% of the time. Fights tended to occur most often between males of similar size, so Asís et al. surmise that males assess each other before engaging in physical combat. However, smaller males were more likely to escalate contests from noncontact "face-to-face hoverings" to grappling of opponents during the peak times of appearance of virgin females (0800–1000 h).

Stizus fuscipennis (F. Smith)—Afrotropical (southern Africa)

In Natal, South Africa, females nested on flat to gently sloping soft, but consolidated, sand atop a coastal dune (Weaving 1989b). The four nests examined were unicellular, although (uniquely) the burrows of two nests dug sequentially by the same female descended from a common entrance. Each of three fully provisioned cells were mass provisioned with 5–13 nymphal mantids of three species: 1 *Phyllocrania paradoxa* Burmeister, 1 *Tarachodes* sp., and 23 of an unidentified species; the largest prey weighed over 11 times that of the smallest. The egg was loosely attached to the dorsum of the prothorax of the first prey placed in the cell. No accessory burrows were found.

Stizus imperialis Handlirsch—Afrotropical (southern Africa, Kenya)

In the eastern Cape of South Africa, nests were found mostly on the vertical north face of a sand pit, although the entrances to some nests were in the roofs of animal burrows in the bank (Weaving 1989b). Few females used the south-facing bank, which received less direct sun than the north-facing bank. The main burrows of the nests entered the bank at an upward angle for 3–5 cm before turning to extend parallel to the face of the bank for another 10–17 cm. The main burrow terminated in a cell, but branches off this burrow led to as many as four further cells. Four fully provisioned cells contained 7–12 prey, the largest prey being about four times heavier than the smallest. All 55 prey were grasshopper nymphs of six species in four families, but most prey were in two families, 30 *Pyrgomorpha* sp. (Pyrgomorphidae) and 13 *Hoplomorpha* sp. (Pamphagidae); the remainder were Acrididae and Lentulidae. As females returned to nests carrying prey in their middle legs, they usually first landed on nearby plants, where they again stung the prey. They then flew to the nest, hovered briefly in front of it, and entered while carrying the prey. Prey were apparently stored within the main burrow before being placed head inward into the cell, where the egg was laid on the side of the thorax of the first prey in. The nest entrance was not closed during foraging and females did not construct accessory burrows.

Stizus iridis Dow—Nearctic (California, Utah)

In Utah, females nested in a sloping terrace above a talus slope and beneath a cliff wall (Dow 1976). Both nests studied had oblique burrows 22 and 35 cm long ending in a single terminal cell. One cell contained three and the other eight adult grasshoppers: 10 *Trimerotropis pallidipennis* (Burmeister) and 1 *Trimerotropis sparsa* (Thomas). In both cases, females were still active at the nests, so they may not have been fully provisioned. An egg was laid on the thorax of one grasshopper in each cell.

Stizus perrisi ibericus de Beaumont—Palearctic (Spain)

In Spain, along with a smaller number of *Bembix sinuata,* 50 females nested in 28 m^2 area of loose sand (Asís et al. 1991). Females used their mandibles and legs to excavate burrows 15–28 cm long that entered the ground with a slight slope, but then angled more directly downward (Figure 5.1D). From 60 to 80 min were required from initiation of digging to the beginning of construction of a temporary closure. All nests had at least one, but usually two, accessory burrows that required 8–12 min for construction, the female alternating between digging of the two burrows. When a nest was completed, females spent 20–30 min leveling the soil near the entrance, but left the accessory burrows open.

Three cells per nest, up to 11 cm deep, were provisioned with 5–12 nymphs and adults of species of Acrididae (Orthoptera): 1 *Calliptamus barbarus* (Costa), 2 *Calliptamus* sp., 1 *Chorthippus binotatus* (Charpentier), 3 *Chorthippus jucundus* (Fischer), 12 *Chorthippus vagans* (Eversman), 2 *Dociostaurus jagoi,* 1 *Dociostaurus* sp., 2 *Euchortippus pulvinatus,* 6 *Euchor-*

tippus sp., 1 *Pezotettix giornae* (Rossi), 5 *Stenobothrus festivus* Bolivar, and 1 *Stenobothrus grammicus* Cazurro. The prey, carried with the middle legs, were not dropped until the temporary closure was removed and most of the prey item was within the burrow. Only then did the female release it, turn around, and pull it in. As with *S. continuus*, the egg was attached laterally near the base of the wing pads of one of the first prey in each cell. Because 2–6 prey were provisioned per day, 2–3 days were required to provision an entire cell ("slow mass provisioning"). Two of the 22 cells examined were parasitized by miltogrammine flies, *Protomiltogramma fasciatum* (Meigen). Asís et al. provide a brief description of cocoon construction which required 20–29 h in the laborartory.

Male *S. perrisi* emerged several days before females and patrolled near the ground in nesting areas, being most active in the morning and at dusk. The form of the flights ("sinuous, just over the ground") is similar to that described for *S. continuus* (Asís et al. 1988).

Stizus pulcherrimus (F. Smith)—Palearctic (China, Japan, Korea, Mongolia)

Evans (1966a) summarized Tsuneki's (1965b) observations in Mongolia and Korea. Here we give a brief update based on Tsuneki's (1976) subsequent study in Japan, where females nested along a sandy road within a 30-year-old pine plantation along with a similar number of *Bembix niponica* F. Smith. Unlike the Korean population, females in Japan built unicellular nests (Figure 5.1C) only and did not provision with Tettigoniidae, though both populations used Acrididae. The nine prey species reported for the Japanese population included *Acrida turrita* L. (nymphs), *Aiolopus japonicus* Shiraki (nymphs and adults), *Atractomorhpa lata* Motchulsky (nymphs), *Chorthippus bicolor* Charpentier (nymphs and adults), *Chorthippus latipennis* Bolivar (nymphs and adults), *Oedaleus infernalis* Saussure (nymphs), *Mongolotettix japonicus* Bolivar (nymphs), *Patanga japonica* Bolivar (nymphs), and *Trilophidia annulata japonica* Saussure (nymphs and adults). All cells were provisioned with nymphs and some contained adults, typically 7–10 per cell in total. Prey carriage, egg placement, and cocoon-spinning behavior were basically the same in Korea and Japan (Tsuneki 1965b). Upon completion of the nest, each burrow was completely filled while the female tamped the soil with the tip of her abdomen.

However, the mound of soil around the nest entrance was left intact, and the females did not fill accessory burrows, of which there were typically two per nest.

Mating was not observed, but males patrolled the nesting area ("sun dance"), chasing and pouncing on other insects. Digging females rejected mating attempts by males, which suggests that mating occurs earlier, soon after they emerge but before they begin nesting. In their attempt to find mates, males frequently pounced upon female *Bembix.*

Stizus vespiformis (F.)—Oriental (India, Mauritius, Sri Lanka)

In Bangladesh, Begum et al. (1989) observed four female *S. vespiformes* [*sic*] constructing multicellular nests "under soil in the floor of houses." One nest examined in detail had an oblique burrow 11 cm long with four 2 × 3 cm cells. The burrow ended in a cluster of cells that, judging from the nest pictured, were 4–11 cm below the soil surface. Females loosened soil with their mandibles, then used their legs to carry clumps of soil several centimeters from the entrance, where it was dropped. The authors indicate that the "total time to complete a nest was about 2.5 hours," but do not indicate whether this included provisioning of all of the cells (presumably it did not). Cells were provisioned with "grasshoppers" brought to the nest every three minutes on average, an interval that included 20–25 s to place the prey into a cell. In flight, prey were held venter up and head forward, and were carried using both the mandibles and the middle legs, a technique also reported for *Stizus pulcherrimus* by Tsuneki (1965b). Also as in *S. pulcherrimus,* the egg was laid "ventrolaterally between the abdomen and thorax of the prey."

Overview of *Stizus*

Biological information is now available for about a dozen species, the level of detail varying greatly among studies. *Stizus* females tend to nest in sandy soil, in some cases near water. At least three species nest in saline soil and one seems to prefer vertical banks. Nests are usually multicellular, and accessory burrows are common though not universal. Females of some species *(S. pulcherrimus, S. perrisi ibericus),* but not others *(S. distinguendus, S. fasciatus),* maintain temporary closures while hunting. All *Stizus* mass provision cells (sometimes over several days—delayed or slow mass provisioning) with grasshoppers, katydids, and mantids (Table 5.1). *Stizus*

Table 5.1 Cells per nest, prey per cell, and families reported as prey of *Stizus*. See text and Evans (1966a) for references and details. Note that the sample sizes for prey vary from a single record for *S. brevipennis* to >100 for *S. pulcherrimus*.

			Orthoptera					Mantodea
Stizus species	Maximum cells per nest	Maximum prey per cell	Acrididae	Lentulidae	Pamphagidae	Pyrgomorphidae	Tettigoniidae	Mantidae
S. brevipennis							X	
S. continuus	8	8	X			X	X	
S. distinguendus	1	12						X
S. fasciatus	3	8	X					
S. fuscipennis	1	13						X
S. imperialis	5	12	X	X	X	X		
S. iridis	1	8	X					
S. marshalli								X
S. perrisi ibericus	3	12	X					
S. pulcherrimus	9	8	X				X	
S. ruficornis		5						X

continuus and *S. imperialis* use both grasshoppers and katydids, but so far as is known, the three species known to prey on Mantodea do not also take Orthoptera. Prentice (1998) notes that *Stizus,* like "most apoid taxa that utilize particularly large orthopteroid prey[,] possess a correspondingly long stylet" on their stings. The egg is laid on the first prey in the cell or on one of the first prey (on the thorax or at the juncture of the thorax and abdomen). Natural enemies reported for *Stizus* include miltogrammine flies (for *S. perrisi* and *S. pulcherrimus*), Mutillidae, and Rhipiphoridae (for *S. imperialis*).

In three species, *S. continuus, S. perrisi,* and *S. pulcherrimus,* males patrol the emergence/nesting areas in search of mates. For *S. continuus,* the description of males digging for females and forming clusters around those that emerge, though brief, is generally similar to behavior described later for *Bembecinus neglectus, Bembecinus quinquespinosus, Glenostictia satan,* and *Bembix rostrata.*

Stizoides

Of the 29 species of *Stizoides,* only *Stizoides renicinctus* is widespread in North America (Gillaspy 1963). The other North American species, *Stizoides foxi* Gillaspy, has been collected in southern Arizona and Baja California. The biology of *Stizoides* has been reviewed by Gillaspy (1963), Evans (1966a), and Ohl (1999); Ohl (1999) recently revised the genus. From a behavioral standpoint, the name *Stizoides* (= "like *Stizus*") is inaccurate. Whereas all *Stizus* are nest provisioners, available information indicates that all *Stizoides* are brood parasites of digger wasps. Our best picture of *Stizoides* behavior comes from studies of *S. renicinctus,* discussed in Evans (1966a) as *S. unicinctus.*

Stizoides renicintus (Say)—Nearctic

Female *S. renicinctus* search for a closed nest of their host species and burrow into it. They then destroy the host egg and lay one of their own. The female returns to the soil surface and plugs the entrance and main burrow with debris (Evans 1966a). Several recent papers address characteristics of this species related to its brood parasitic lifestyle. For this and three species of Nyssonini, Wcislo (1998) concluded that "females are not conspicuously larger than conspecific males, and have similarly sized scapes and flagella" on their antennae. This contrasts with most nest-provisioning species, in which females are larger on average than males (see O'Neill (2001) for discussion of the exceptions); in nonparasitic species, males also have shorter antennal scapes. The convergence between the stizine and nysonnine species is intriguing, but the significance of Wcislo's observations is unclear. The morphological differences may relate to differences in reproductive strategies between nest-provisioning and brood parasitic species. For example, in brood parasites, female search strategies may be more male-like than in nest-provisioning species. Furthermore, because female *Stizoides* do not need to carry prey, selection on body size of females may be somewhat relaxed, resulting in a lack of sexual size dimorphism.

Females of nest-provisioning Stizini have three ovarioles (strands of developing eggs) in each of their two ovaries and, when dissected, are found to carry no more than two mature oocytes (Iwata 1955, 1965; O'Neill 1985). However, Ohl and Linde (2003) dissected a single female *S. renicintus* and found that the four ovarioles in each of its two ovaries carried a total of five mature oocytes.

Recently, we had the opportunity to examine a somewhat larger sample of females of *S. renicinctus* collected at two locations in Montana (KMO and A. Pearce, unpublished). The number of mature oocytes in 30 females examined ranged from 1 to 6, the average being 3.6. Each female had 8–10 ovarioles, this apparently being the only known case of apoid wasp females having more than eight ovarioles. The number of mature oocytes carried was higher in larger females, but lower in older females, age being judged from the degree of wear and tear on the wings. Thus, compared with smaller conspecifics, larger females may be better able to exploit a series of host nests in rapid succession, and their ability to do so may decline with age. Larger females also tended to carry larger mature oocytes. Nevertheless, mature oocytes of this species are similar in size to those for several species of nest-provisioning digger wasps with smaller abdomens (*Bembecinus quinquespinosus* and *Philanthus pulcher*); and *S. renicinctus* oocytes are smaller in size than those of *Philanthus zebratus*, which has abdomens comparable in size to those of *S. renicinctus* (O'Neill 1985). The mean number of mature oocytes per ovariole (0.40) was within the range of values observed for other bembicine brood parasites (0.25–0.84), and higher than the vast majority of values reported for bembicine nest provisioners (Ohl and Linde 2003). We will discuss the significance of these observations more fully in Chapter 8.

Mixed-species sleeping clusters of *Stizoides* with potential hosts were observed by Rau (1938). Kazenas and Tobias (1993, in Ohl 1999) recorded a single individual of *Stizoides tridentatus* (Fabricius) in a sleeping cluster with large numbers of *Prionyx* and *Sphex*. In Montana, we (KMO and A. Pearce, unpublished) have observed sleeping clusters of *S. renicinctus* and *Prionyx* sp. that sometimes also included *Ammophila* sp., eumenine wasps, and halictid bees.

Overview of *Stizoides*

Evans's (1966a) review of *Stizoides* noted that *Prionyx* and *Palmodes* were hosts of *S. renicinctus*, that *Prionyx kirbyi* Linden hosted *Stizoides crassicornis* (F.) (based on Deleurance 1944), and that *Sphex flavipennis* F. (in Evans 1966a as *S. maxillosus* (F.) (in Ohl 1999 as *Sphex rufocinctus*) hosted *Stizoides tridentatus* (based on Arens and Arens 1953). In addition, Myartseva (1965, cited in Ohl 1999) gave host records not cited in Evans (1966a). In Turkmenistan *Prionyx crudelis* (F. Smith) are also hosts of *S. crassicornis*, whereas *Stizus transcaspicus* Radoszkowski is a host, not only of both *S. crassicornis*, but of *Stizoides assimilis* (F.), *Stizoides cyanopterus* (Gussakovskij), and *S. tridentatus*. Even this short list of "host records" needs to be viewed with caution. For example, Arens and Arens (1953) found a consistent association of *Stizoides crassicornis* with *Sphex flavipennis* nest aggregations, and even observed *Stizoides* digging into "host" nests. However, no eggs of the presumed brood parasite were found, and no *Stizoides* offspring were reared from nests. Thus while the information is strongly suggestive of the true host–parasite relationship, definitive proof remains to be found. As we noted in Chapter 4, this cautionary statement can be extended of all parasite–host relationships for sand wasps, where spatial and temporal associations are often used to infer (we believe too confidently) true host records. The record of a *Stizoides* parasitizing *Bembix ugandensis* Turner discussed in Gillaspy (1963) is about as tentative as one can get; the "*Stizoides*" is identified only from a brief verbal description of its appearance in the field and no evidence is given that the supposed parasite was reared from nests.

Although *Prionyx, Palmodes,* and *Sphex* are within a different family (Sphecidae) than *Stizus* (Melo 1999), they share the trait of preying on grasshoppers. This led Evans (1955, 1966a) to predict that *Stizus* was the original host of *Stizoides*. The observations of Myartseva brought to light by Ohl (1999) are consistent with this hypothesis. *Stizoides* may then have exploited other predators of Orthoptera in their habitats. Ohl (1999) expands on this theme and its implications for speciation in *Stizoides*.

Bembecinus

Bembecinus contains 186 described species, the greatest number occurring in the Afrotropical region (Bohart and Menke 1976; Pulawski 2006a). Bohart (1996a) provides keys to North and Central American species. Certain species exhibit intrapopulational color variation among males that sometimes makes taxonomic determination problematic (Arnold 1945; Krombein and Willink 1951; Evans and Matthews 1971; Bohart and Menke 1976), but which represents an adaptive polymorphism linked to mating tactics in at least one species (O'Neill and Evans 1983; O'Neill et al. 1989).

Evans (1955, 1966a) discussed biological information on more than 15 species. Here we review work of varying detail on three U.S. species, along with 19 others. In his prefatory remarks to a discussion of this genus, Evans (1966a) remarked that, while *Bembecinus* has some unique behavioral features relative to other sand wasps, there are few interspecific differences of note within the genus. More recent work shows that the latter is not strictly true, particularly with regard to observations on several South African species. In addition, several aspects of the behavior of *Bembecinus* have been worked out in greater detail since 1966, and prey and natural enemy records have been expanded.

Bembecinus agilis (F. Smith) (= *B. cingulatus*)—Neotropical

In Trinidad, females nested in a sandpit and provisioned nests with prey (carried prey venter-to-venter) of five genera of Cicadellidae: *Hortensia, Oncometopia, Parathona, Phera,* and *Ponama* (Callan 1991b). Several of these genera were also among the prey of *B. agilis* in British Guiana, where Richards (1937) reported it to be a progressive provisioner of unicellular nests dug in sand.

Bembecinus antipodum (Handlirsch)—Australasian (Australia)

North of Sydney, Evans and Matthews (1971) found females nesting in hard-packed sand along a road. They examined a single nest that had just been completed by a female who had closed the nest by filling in the top 4 cm of the 5 cm long burrow with sand scraped from the side of the tunnel. The single cell at a depth of 3.5 cm contained a large larva, with prey fragments, and four adult homopterans, one *Rhotidoides* sp., two *Tartessus* sp. (Cicadellidae), and one *Ipoella* sp. (Eurymelidae).

Bembecinus argentifrons (F. Smith) (= *Bembecinus braunsii* (Handlirsch))—Afrotropical (eastern and southern Africa)

Near Grahamstown, South Africa, females dug one- or two-celled nests with oblique burrows in "loose dry fine sand," provisioning them with *Macropsis octopunctata* China and *M. chinai* Metcalf (F. W. Gess 1981), and on occasion one to two Fulgoroidea or Membracidae (F. W. Gess 1981). Females maintained a temporary closure while hunting.

Bembecinus asiaticus Gussakovskij—Palearctic (central Asia, Middle East)

Myartseva 1976 (in Kazenas and Esenbekova 1995) gives Cicadellidae as prey.

Bembecinus bicinctus (Taschenberg)—Neotropical (Argentina)

In Argentina, near Yacochuya, Salta, Evans and Matthews (1974) found a single nest in "bare, eroded sandy soil" near nests of *Hoplisoides semipunctatus.* The burrow entered the ground at 30° then angled downward at about 85° to a single spherical cell containing two immature Cicadellidae.

Bembecinus bimaculatus (Matsumura and Uchida)—Palearctic (Ryukyu Islands)

Kifune and Yamane (1985) and Kifune (1988) found that 3 of 12 individuals of this wasp collected in the Ryukyu Islands of Japan were stylopized by the strepsiteran *Paraxenos nagatomii* Kifune.

Bembecinus bolivari (Handlirsch)—Neotropical

In Trinidad, females nested in both inland and coastal areas, sometimes at the same sites as *B. agilis,* and taking prey that "included some of the same genera" (Callan 1991b). In Costa Rica, Cane and Miyamoto (1979) found *B. bolivari* nests mixed in with those of *Stictia heros* and *Bembix multipicta* in "open areas of coarse old beach sand dissected by patches of scrubby herbaceous second growth." Although they provide no details on nesting behavior, they did note that aggregations of this species were regularly plundered by *Solenopsis* ants that entered nests and removed their contents.

Bembecinus cinguliger (F. Smith)—Afrotropical (southern Africa)

F. W. Gess and S. K. Gess (1975) studied *B. cinguliger* near Grahamstown, South Africa. The biology of this species, and that of *Bembecinus oxydorcus* studied at the same site, is the most atypical of any *Bembecinus* yet studied, in terms of nest-site selection, nest structure, and prey use. Females nested in aggregations in sparsely vegetated clay soil. Burrows entered the ground more vertically than most *Bembecinus* and extended downward 6–9 cm to the bottom of a terminal cell that had an oblique to near vertical orientation (Figure 5.1E, F). Five of 34 nests excavated had a second cell, at the end of a short side burrow arising above the first cell, which had been completely provisioned and sealed.

The most unique features of the nests, indeed features unique for the entire subfamily Bembicinae (with the exception of those of *B. oxydorcus*), were the elaborate mud turrets constructed by females above the nest entrance. The digging of the main nest burrow and the construction of the turret were part of an integrated process that began with a female carrying load of water in her crop and regurgitating it on the soil surface. Mud formed as she worked the water into the clay with her mandibles. As the shaft was excavated, the resulting mud was used to construct the turret, which took about 45 minutes and 15 loads of water carried in flight from 18 m away. The 2.6–4.3 cm long turret arose from the nest entrance, but lay prostrate and mostly in contact with the ground. After completion of the turret, soil brought up from the ever-lengthening burrow was carried in flight 30–40 cm away and dropped. However, when the side shaft was excavated by the wasp, it probably used that soil to fill the shaft above the first cell. Before nightfall each day, females placed thin mud seals in the nest, one just below the entrance and one just above the cell currently being provisioned.

Bembecinus cinguliger differs from most other *Bembecinus* in the structure of foretarsal spines. In those species nesting in friable soil, the female's rake spines (forming the "sand rake" or pecten) are elongated to facilitate sweeping of loose sand beneath the body. However, *B. cinguliger* and the other turret-building species, *B. oxydorcus,* nest in nonfriable soils and have "dense row of short spines" (Figure 5.3, left), perhaps used for scraping harder soils or for constructing the turret. The structure of the pecten of these two species differs from that of *Bembecinus* (such as B. *haemorrhoidalis;* Figure 5.3, right) that nest in loose, sandy soil.

The egg was laid on the floor of the cells ("on a small cone of earth"), which were progressively provisioned much faster than prey were consumed. The 223 prey collected from 31 cells included 179 nymphs and adults from about 10 species of Cicadellidae, 30 adults from two Fulgoroidea, and 14 from two species of Tephritidae (Diptera). The latter is astonishing, given that all other prey records for *Bembecinus* (again with the exception of the sympatric *B. oxydorcus*) are for Homoptera. Gess and Gess speculate that the flies inhabit the same vegetation as homopteran prey of this species. An examination of one sealed cell in which no wasp larva developed suggests that as many as 41 prey were provided to a single larva; this cell also contained a single fly puparium that may have been the cause of the offspring's demise and which was the only parasite found in 34 nests.

Males began emerging 7–11 days before females during the two years of study, a degree of protandry greater than that observed for *Bembecinus*

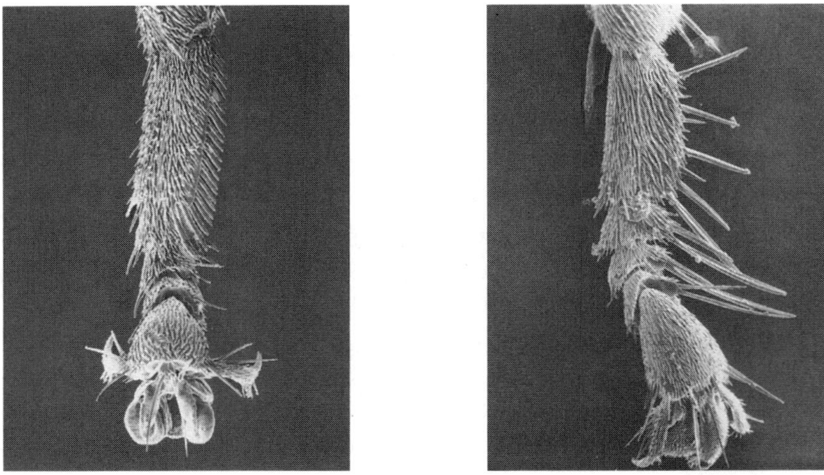

Figure 5.3. Foretarsi of females: (left) *Bembecinus cinguliger;* and (right) *Bembecinus haemorrhoidalis.* Photos by Fred Gess, from F. W. Gess (1981); used with permission.

quinquespinosus (see below). Within the "bare patches of earth utilized for nesting by this wasp season after season," males "patrolled . . . on an irregular path . . . at a height of 2–8 cm." They constantly investigated other flying insects, which generally turned out to be other patrolling males. When a receptive female was intercepted, the copulating pair flew to "low bushes fringing the bare areas." By the description, patrolling behavior in *B. cinguliger* seems similar to that of *B. strenuus,* whereas the tendency for copulating pairs to leave the emergence area is reminiscent of behaviors observed for *B. neglectus* and *B. quinquespinosus* (both described below), but with the exception that copulation does not occur in those species until the pair is outside of the emergence area.

Confirming earlier observations by Brauns (1911) and Jacot-Guillarmod (pers. comm. in F. W. Gess and S. K. Gess 1975), the Gesses found a large *B. cinguliger* sleeping cluster on vegetation. The cluster consisted of thousands of individuals, mirroring Brauns's description of a sleeping cluster the size of a baby's head. The cluster observed by Gess and Gess included both sexes early in the season, but only females later. It was situated at some distance from nesting areas in a sheltered location on a grass tussock and was apparently present in the same location over 46 nights (though it was not checked every day). Wasps leaving the cluster in the morning first basked in the sun before going to nesting areas. However, on overcast days they remained in the cluster or returned to it as weather turned unfavorable.

Because the cocoons of sand wasps are lined with soil taken from within the cell, they are typically made with sand grains. The walls of the hard, smooth 14.5–16.0 mm long cocoons of *B. cinguliger*, however, are impregnated with particles of clay typical of the soil in which these wasps nest. Each cocoon contains 1–3 pores (usually two).

Bembecinus comberi (R. Turner)—Oriental (Sri Lanka)

In Sri Lanka, females nested in hard, dry soil along a vehicle trail through a jungle and in damp sand in open areas with sparse vegetation (Krombein 1984). Shallow nests of up to at least three cells (an unusual feature in the genus) had burrows 6.0–9.5 cm in length, entering the soil typically at angles of 30–45°. Some evidence suggests that females spent the night in the nest and left nests open while foraging.

Cells were provisioned with a wide variety of prey from six families of Homoptera. The diversity of prey could be seen even within nests. One three-celled nest contained 12 species from 4 families, and one cell was provisioned with six species from three families. The maximum number of prey documented for a single cell was 17. The 49 prey removed from six cells included *Acostemma prasina* Walker, *Kutara brunnescens* Distant, *Petalocephala* sp. (Cicadellidae); *Ketumala thea* Ghauri (Flatidae); *Forculus viridis* Distant, *Gergithus complicatus* Distant, *Gergithus cribratus* Melichar, *Narayana fryeri* Distant, *Narayana pundaluoyana* Distant, *Sarima creata* Distant (Issidae); *Centrotus indicatus* Melichar, *Coccosterphus tuberculatus* Motschulsky, *Cryptaspidia piceola* Melichar, *Gargara* spp., *Leptocentrus* sp. (Membracidae); *Ricania fenestrata* F. (Ricaniidae); *Eporiella* sp. (listed as a tentative identification), and *Stacota breviceps* Walker (Tropiduchidae). Other records gathered from cells and from females returning to nests with prey included seven of the above taxa, as well as *Coccosterphus obscurus* Distant. Only *Bembecinus tridens* is recorded taking a greater array of prey families than *B. comberi,* and the former has been the subject of multiple studies over many years.

Bembecinus egens (Handlirsch)—Australasian (Australia)

At two localities along the coast of Queensland in Australia, females nested in a "fine-grained, pale beach sand" and "in patches of white, very pale friable sand" along a river (Evans and Matthews 1971); *Bembix* also nested at both sites. The five nests excavated had burrows 10–15 cm long and a single cell at depths of 7–9 cm. Evans and Matthews note that, although they found only unicellular nests, they suggest that further study might reveal multicellular nests.

After completion of the burrow, the female completely leveled the mound and constructed 0.5–2.0 cm long and 0.1–1.0 cm deep accessory burrows ("quarries") that often disappeared quickly due to action of the wind. Two cells contained only an egg, sitting obliquely erect on a sand pedestal. The other three cells held prey and were provisioned progressively. One had three large adult *Tartessus* sp. (Cicadellidae). A second contained two adult *Aprivesa exuta* Melichar (Ricaniidae), one adult *Parasalurnis roseicincta* Walker (Flatidae), and many prey fragments. The third contained a wasp larva, along with 19 small fulgaroid and cicadellid prey, including *Perkinsiella saccharicida* Kirk. Burrows were closed while females hunted.

Bembecinus haemorrhoidalis (Handlirsch)—Afrotropical

Near Grahamstown, South Africa, F. W. Gess (1981) reported *B. haemorrhoidalis* nesting in "loose, dry sand" in a sandpit, using foretarsal pectens with long, flattened spines (in constrast to the structure of the spines in the sympatric, clay-soil–nesting *B. cinguliger* and *B. oxydorcus*). The one- and two-celled nests had oblique burrows, and temporary closures when females were absent. Cells were provisioned primarily with Cicadellidae that included *Coloborrhis corticina* Germar, *Exitianus nanus* (Distant), *Macropsis octopunctata* China, *Macropsis chinai* Metcalf, *Macropsis* sp., *Idioscopus* sp., and *Batracomorphus subolivaceous* (Stål); more rarely, cells contained Fulgoroidea or Membracidae.

Bembecinus hirtulus (F. Smith)—Australasian (Australia)

Evans and Matthews (1971) present observations on this species from three sites in Australia. Detailed records of nests are presented for two of the locations. As Evans and Matthews found "no conspicuous behavioral differences between the various populations studied," we will combine some of the information they present for different populations. Along the Murrumbidgee River near Canberra (Pine Island), females nested in sand or sandy loam, often several centimeters from one another, whereas at Peak Hill, New South Wales, nests were dug within "somewhat coarser, more compact soil . . . in piles and banks of more or less sandy soil left from mining operations." Among 16 nests excavated, burrows were 7–12 cm long, cells 5–8 cm deep. At both sites, females laid eggs in empty cells and erected temporary closures before hunting.

The progressively provisioned cells contained as many as 27 prey when examined. The frequency distribution of prey at two sites differed, probably because of differences in prey availability. The nine species of Cicadellidae (and the number taken at the two sites—Pine Island/Peak Hill) were: *Batracomorphus* sp. (1/80), *Deltocephalus* sp. (31/0), *Exitianus* sp. (0/4), *Idiocerus* sp. (1/0), *Limotettix* sp. (1/1), *Macropsis* sp. (16/0), *Nesoclutha pallida* (Ev.) (2/0), *Orosius argentatus* (Ev.) (1/14), and unidentified nymphs (5/20). In addition, five specimens of Psyllidae (*Psylla* sp.) were recorded from a nest at Pine Island, whereas one Delphacidae (*Toya lazulis* (Kirk.) was found among prey at Peak Hill.

Bembecinus hungaricus (Frivaldsky)—Palearctic

This widespread species has several described subspecies that have been studied in Austria, Taiwan, and Japan. Evans (1966a) summarized earlier accounts of this species by Kunio Iwata who reported Cicadellidae, Psyllidae, and Fulgoroidea as prey.

In Austria, Zolda et al. (2001) and Zolda and Holzinger (2002) studied *B. hungaricus* where it nested in level sand and where adjacent nests were separated by 30–120 cm. Depending on the weather, females took 1–6 hours to build their unicellular nests, and from 3 to 7 days from initiation of nests to final closure. Nests had 5.0–12.5 cm long burrows that entered the soil at ~25° and ended in terminal cell at a depth of 3–10 cm. Cells, which were apparently expanded as provisioning progressed, were stocked with 13 species of Cicadellidae from 12 genera, plus 1 species of Delphacidae and 1 unidentified Psyllidae. The most common prey (41% of the sample) were the cicadellid *Idiocerus stigmaticalis* Lewis; no other species made up more than 9% of the prey. The plant hosts of over 90% of the prey are deciduous trees, particularly willows, poplars, and alders. Prey, which were "as large as the wasps," were carried in flight venter-to-venter and shifted to the hind legs as females opened the temporary closure with their forelegs. Upon completion of provisioning, females filled burrows with sand and leveled the tumuli, which typically measured 2 × 3 cm. A female often went on to start a new nest within a half meter of her previous nest.

In Taiwan, females of *Bembecinus hungaricus formosanus* (Sonan) nested in a newly made road and in a "seaside sand bank" (Tsuneki and Iida 1969). Burrows entered the soil at various angles from vertical to oblique, with 5–7 cm long burrows and 3–9 cm deep cells; the burrow length of the deepest nest was not provided. The one- or two-celled nests were provisioned with nymphal and adult Cicadellidae after the egg was laid erect on sand pedestal in the empty cell. In several nests, females continued foraging after the egg hatched (progressive provisioning), but one nest with 18 prey was closed before the egg hatched (mass provisioning). However, the possibility that the egg in the latter nest failed to hatch cannot be excluded.

In Japan, female *Bembecinus hungaricus japonicus* (Sonan) nested in sand or sandy gravel (Tsuneki 1969). Females dug with their legs and mandibles, sometimes carrying small pebbles in the mandibles up from the depths of the burrow. As in most *Bembecinus*, the nest was simple, with a 7–12 cm long oblique and usually straight tunnel leading to a single terminal cell 3–10 cm below the surface. Tsuneki notes that this is a depth sufficient to place the cell within the moist sand, which began at a depth of 3–5 cm. Before provisioning, the females constructed a large pedestal consisting of a "block of sand grains, about half the size of the wasp's head" in the center of the cell. The egg was then laid in an erect position on the pedestal.

Following oviposition, the females leveled the soil around the nest entrance "quite elaborately" and then constructed a temporary closure that was maintained whenever the female was away from the nest. Tsuneki (1969) describes what he says is "no doubt the orientation flight" that followed the construction of the temporary closure:

> The wasp then flys [*sic*] up and begins to fly around the site of her nest. The flight is not circular or spiral with the nest at the centre, but is to come hither and thither, always passing above the nest at each flight. From time to time she stops above and in front of the nest entrance like a helicopter, directing her head toward the entrance as if to gaze at it, and after several moments flies off. During the course, the wasp often sits on sand near the entrance . . . After 2–3 minutes of such a flight the wasp goes out of sight. After a moment, however, she turns back, alight [*sic*] on sand in front of the nest entrance, but at once flies away. At an interval of 30–100 seconds the wasp used to return twice or thrice . . . to the site of her nest before she finally goes away. (Tsuneki 1969, 9)

Females did not bring prey to nests until the day after digging, though why this should be so is not clear. Cells were then provisioned progressively, but rapidly over 1–2 days while larvae were present in nests. The number of prey provisioned per cell is unknown, but as many as 32 were found in one cell, along with larvae that had already consumed some prey. When provisioning ended, larvae were less than half grown. A large number of prey are required because females took nymphs and adults of "comparatively small species" that included one species of Psyllidae (*Anonomoneura mori* Schwarz). Most prey species, however, were Cicadellidae: *Deltocephalus dorsalis* Motschulsky, *Drabescus ogumae* Matsumura, *Eutettix disciguttus* (Walker), *Jassus praesul* Horváth, *Nephotettix bipunctatus cincticeps* Esaki and Hashimoto, *Parabolocratus prasinus* Matsumura, *Penthimia* spp., and *Platymetopius cinctus* Matsumura. Prey were carried in flight venter-to-venter, head-to-head with the wasp's middle legs.

Tsuneki reared *B. hungaricus* in the lab to examine larval growth and cocoon formation. Despite their strict use of Homoptera as prey, he was able to rear larvae in the lab feeding them partially on "cabbage caterpillars." Tsuneki provides a detailed account of cocoon structure and cocoon building, the first of a species of *Bembecinus,* which we quote here (because the original report appears in a relatively obscure journal). The cocoons are "elongated egg-shaped," though one end is more broadly rounded (the "outer zone"). As with *B. cinguliger,* the outer silk shroud of the cocoon is mingled with prey fragments. The larva

> . . . began to stretch the silk thread between the walls of the artificial cell that was made of wet sand in a dish by impressing with my finger. It was in the evening, about 17:00 . . . The next day at 9:00 the form of the cocoon was completed. It was an acorn-shaped pouch of silk thread, about twice as long as wide, still semitransparent and . . . one of the ends was left open and at the end there was a funnel-like stretch of silk membrane, as in the case of *Bembix* and *Stizus* . . . At 10:30, the equatorial zone of the pouch was narrowly covered with the layer of sand grains. At 11:00 the larva protruded its anterior body up to the 5th segment from the opening and was collecting the sand grains from the floor. At 14:20 the . . . sand had covered already the broad central zone and the outer zone, leaving a narrow portion at the outer opening area and about one fourth of the cocoon length at the inner portion free from the sand covering. At 18:00 the larvae was lining the inner portion of the silk pouch with sand grains. The result showed that the order of the sand inlaying work was first com-

menced in the central zone, then shifted to the outer zone (leaving the entrance opened) and to the inner zone and finally closed the entrance with the sand lid . . . On the third day . . . at 8:00, the silk pouch was completely lined with sand grains and the portion of the entrance funnel was also completely closed with the sand layer. (Tsuneki 1969, 14)

Bembecinus luteolus Krombein—Oriental (Sri Lanka)

Krombein (1984) made brief observations of females nesting on the flat or gentle slopes of sand dunes on a beach in Sri Lanka. He also notes that females spent the night in their burrows rather than in sleeping clusters, suggesting that this was due to the cool oceanic winds prevalent at the site. He was unable to determine nest structure (other than that the main burrow was 15.5 cm long) or prey types. He does provide a detailed description of nest digging, which, though based on observations of a single female, are of sufficient detail to be worth quoting, as they provide a good narrative of the sequence of burrow excavation, mound leveling, and temporary closure:

At 1420 on the same date [20 January 1979] I found another female . . . just beginning to burrow on slightly sloping bare sand. The slope was about 20° from the horizontal and the burrow went in at an angle of 30° to the slope. As she got deeper into the sand she brought out two or three loads of sand, scattered them backward beneath her, and then made short flights low to the ground for three or four seconds. She was in the burrow about 20 seconds between each emergence to scatter sand from the entrance. As the burrow deepened she was excavating below for 30 to 40 seconds before bringing up sand. The spoil heap was flat, oval, about 40 mm wide and 60 mm long. The burrow entrance was about 4 mm in diameter. The wasp came to the burrow entrance five times and scattered sand. On the sixth exit she backed up 40 mm from the burrow to scatter the spoil heap over a wide area. By 1530 the wasp in scattering sand from the spoil heap had made a slightly depressed path about 2 mm deep near the entrance that widened gradually to 10 mm and about 40 mm from the burrow. By 1600 the wasp was spending as much as 1¾, 2, and 2⅔ minutes underground before backing to the surface with a single load. At 1618, after being below for 10¾ minutes, she came to the surface head first, scraping sand backward beneath her to close the burrow entrance. Presumably she deposited an egg in the cell during the protracted period below the surface. She then proceeded to scrape more sand over the burrow entrance though not filling the depression entirely. About 1½ minutes after appearing at the entrance she flew away without making an orientation flight.

Bembecinus neglectus (Cresson)—Nearctic (central, southcentral United States)

Evans (1955) first studied *B. neglectus* in dune blowout in Kansas, where females nested in hard-packed, pebbly sand, often within 5 cm of one another. Females progressively provisioned their shallow one- and two-celled nests with Cicadellidae and Dictyopharidae after first ovipositing in an empty cell.

Evans's brief report of male behavior in this species, suggesting that they patrolled the emergence area for females, is presently being further investigated by Alan Hook. The following is a paraphrase of a description that he sent us (and which we gratefully use with his permission). Each season, males emerge on average before females, then search for mates in the emergence area, sometimes digging through the surface crust of the soil to reach females digging their own way up from their natal cells. Clusters of competing males form around newly eclosed females. In our study of *B. quinquespinosus,* we referred to these clusters as "mating balls," and Hook's video recordings of the events, which we have seen, suggest that the two species have a similar form of mate competition. Upon grasping a female, the successful male then attempts to carry her in flight out of the emergence area where they mate, presumably without harassment from other males. Larger males of the population frequent the emergence area, whereas smaller males patrol nearby vegetation. So far, all of this matches what we observed in *B. quinquepsinosus* (see below).

Perhaps the most unique morphological feature of *Bembecinus* is the color polymorphism exhibited by certain species, though the exact form of the polymorphism is variable (Arnold 1945; Bohart 1996a). In *B. neglectus,* females are mostly black with pale white stripes (the standard for *Bembecinus*), whereas males seem to come in two color morphs, pale yellow and yellow (the latter being mostly yellow with very little black). Bohart suggests that yellow males tend to be larger than dark-colored males, a trend confirmed by Hook. Hook also confirms earlier observations that *B. neglectus* forms sleeping aggregations on branches, usually out in the open.

Figure 5.4. Female *Bembecinus oxydorcus* using mud to seal turret above nest entrance. Photos by Fred Gess, from F. W. and S. K. Gess (1975); used with permission.

Some sleeping aggregations were primarily of one sex, though many are mixed.

Alan Hook and Greg Palmer (pers. comm.) also observed several types of interactions between *B. neglectus* and fire ants *(Solenopsis invicta)* in Pedernales Falls State Park, Blanco County, Texas. They recorded four instances of fire ants stealing prey from inside nests, and others of the ants scavenging abandoned prey and dead adult *B. neglectus* in the nesting area. The cause of the wasps' death was not directly observed, but Hook and Palmer also found evidence that fire ants attack adults and, by attaching themselves to the wings, prevent them from taking flight. Finally, these authors also found a fire ant queen in a nest cell, an instance, perhaps, of supersedure by a nest-founding female ant.

Bembecinus oxydorcus (Handlirsch)—Afrotropical (southern Africa)

F. W. Gess and S. K. Gess (1975) studied *B. oxydorcus* in the same location in South Africa as *B. cinguliger,* where the two species share several remarkable morphological and biological features: the lack of a sand rake of elongate spines on the foretarsi, the use of Tephritidae as prey, construction of a mud turret above the nest entrance, nests with one or two cells, and clay-lined cocoons (in this case with four pores each). Like *B. cinguliger,* females nested in clay soil, digging 7.2–15.5 cm deep vertical burrows that lead to one terminal cell and, in 10% of 20 nests, a second cell off of a short side shaft (Figure 5.1G, H). Prey, which were progressively provisioned, included Cicadellidae and Tephritidae, but no Fulgoroidea. Relative to that of *B. cinguliger,* the turret on *B. oxydorcus* nests is shorter and more erect than that of *B. cinguliger* and has a tapered lip on one side (Figure 5.4). Also, *B. oxydorcus* constructs three (rather than two) temporary thin mud seals in the nest at night, one above the cell, one just below the soil surface, and one within the turret.

Bembecinus posterus (Sonan)—Palearctic (Taiwan)

In Taiwan, females nested in a dirt lane of "loamy soil" and hard clay (Tsuneki 1969). The oblique burrows, 3–7 cm long, ended in a single brood cell that was as close as 3 cm to the soil surface. The eggs were laid in empty cells on pedestals of moistened sand. The female continued provisioning the cell, always with Cicadellidae, until the larvae were almost full grown. Perhaps because of the soil type in the nesting area, a female making a temporary closure "packs and pounds" the soil into place.

Bembecinus proximus (Handlirsch)—Oriental (India, Sri Lanka)

In Sri Lanka, females nested in a "newly cleaned, flat, dirt road with scattered gravel on the surface," in a "flat, sandy loam road," and "in a sand pile" (Krombein 1984). Burrows entered the ground at 30–45°, and there was apparently one progressively provisioned cell per nest. Prey, as many as eight per cell when they were examined, were *Acostemma prasina, Batracomorphus* sp., *Selenocephalus* sp. (Cicadellidae), *Orthophagus* sp. (Dictyopharidae), *Paruzelia salome* Fennah, and *Stacota breviceps* (Tropiduchidae). Prey were carried venter-to-venter and not dropped while the female removed the temporary closure. The egg was laid on a pedestal. One cell contained prey, wasp larva, and "six acalyptrate dipterous maggots," which could not be further identified.

Bembecinus pusillus (Handlirsch)—Oriental

In Sri Lanka, females nested in sand, where burrows entering the soil at an angle of 20–35° ended in a single terminal cell (Krombein 1984). An egg was found lying on its side in one cell, not on a pedestal. Prey were carried venter-to-venter between the legs and held while female removed the temporary closure. Provisions were mostly Cicadellidae of the species *Batracomorphus* sp., *Nephotettix nigropictus* (Stål), *Nephotettix virescens* (Distant), *Nephotettix* sp., *Platytretus marginatus* Melichar. Krombein's prey list also includes *Hecalus apicalis* (Matsumura), but note that *Hecalus apicalis* Van Duzee is a North American species (McKamey 2006), so the name in the list may be a mistake. Other prey were identified as *Kirbyana* sp. (Cixiidae) and *Matutinus* sp. (Delphacidae). Cells, which were probably provisioned progressively, contained up to 14 prey when examined, the cell with 14 also housing a half-grown wasp larva and a maggot.

Bembecinus quinquespinosus (Say) (= *B. godmani godmani* (Cameron))—Nearctic (western United States), Neotropical (Central America)

In flat, coarse, sandy gravel along a wash on the Pawnee National Grasslands in northern Colorado, active nests were typically 5–10 cm apart, and perhaps much denser overall (Evans et al. 1986). Eight unicellular nests had oblique burrows 8–10 cm long with a terminal cell (Figure 5.1J) stocked with (Cicadellidae). No accessory burrows were found at nest entrances. Among the 757 prey were 121 adult *Cuerna striata* (Walker), 13 adult *Amphigonalia* sp., and 623 immatures, likely of the same two species. Forty-three other prey records previously recorded (under *Bembecinus g. godmani*) from a site in Mexico as unidentified Cicadellidae (Evans 1966a) were since found to be 42 *Carneocephala sagittifera* (Uhler) and 1 *Acinopterus angulatus* Lawson (Evans 1968). In addition, Bohart (1996a) excavated a single nest of this species near Lake Texoma in Oklahoma and found nine nymphal and adult cicadellids of the genera *Carneocephala* and *Graphocephala*.

At the Pawnee grasslands in 1982, we excavated 19 *B. quinquespinosus* nests to extract cell contents. Females, because of their use of large numbers of immature leafhoppers, need to provide a large number of prey in each cell, larger than any previously recorded for *Bembecinus*. We found as many as 45, 59, 62, and 74 prey in cells (KMO and HEE—previously unpublished), but because this is a progressive provisioner, the actual number of prey provided may be higher. For example, the cell with 45 prey, 42 of which were small nymphs, also contained a large wasp larva, indicating that many prey had already been consumed. It was clear that many prey were brought into the cell before the egg hatched. Two cells with eggs, in each of which an unhatched larvae was visible, contained 23 and 24 leafhoppers.

The female *B. quinquespinosus* in the Pawnee Grassland population ranged in size from 2.0 to 3.1 mm in head width, a measure that is strongly correlated with body mass in this species (O'Neill 1985). Larger females brought larger prey to their nests on average (O'Neill 1985) and a greater proportion of adult leafhoppers (KMO and HEE—previously unpublished). The 357 prey taken from nests of ten females of 2.4–2.6 mm head width included just 2.5% adult leafhoppers. In contrast, seventeen 2.7–2.8 mm females provisioned with 24.1% adults ($N = 187$ prey), and six 2.9 mm females used 48.8% adults ($N = 80$). The ability of larger females to take larger prey provides them with two possible advantages: they provide each offspring with a greater mass of prey or they can provide a given level of investment in fewer foraging trips. Larger females also have an advantage in egg production. The size of largest oocyte carried by females (measured either as its length or its volume) increased with female head width. This may reflect a greater investment (in yolk) at the beginning of an offspring's development. In addition, the total length of terminal oocytes in the six ovarioles increased with female head width. This may indicate that females can produce more mature oocytes in a given period of time, and thus be less likely to be limited by egg production capacity when provisioning a series of nests during good weather.

Interactions among males and females resulted in temporal shifts in the use of potential nest sites (Evans et al. 1986). We began our studies of this species in the early 1980s, often finding it difficult to locate nesting aggregations. One major reason is that, rather than nest where their mothers did (as is usual for many bembicines), females shifted to new sites as much as 100–150 m away every year (Figure 5.5). Uniquely, it appears that the entire group of newly emerged females (or at least a large proportion of them) moves to the new location, though how they end up at the same site is unknown. The most plausible benefit of moving is that those females nesting where they emerged suffered from continual harassment by mate-seeking males, harassment that is actually life-threatening (see below). The alternative hypotheses that females move to escape parasites or that the high density of emerging wasps disrupts the nesting substrate seem unlikely. We saw few parasites and rains quickly reconstituted the integrity of

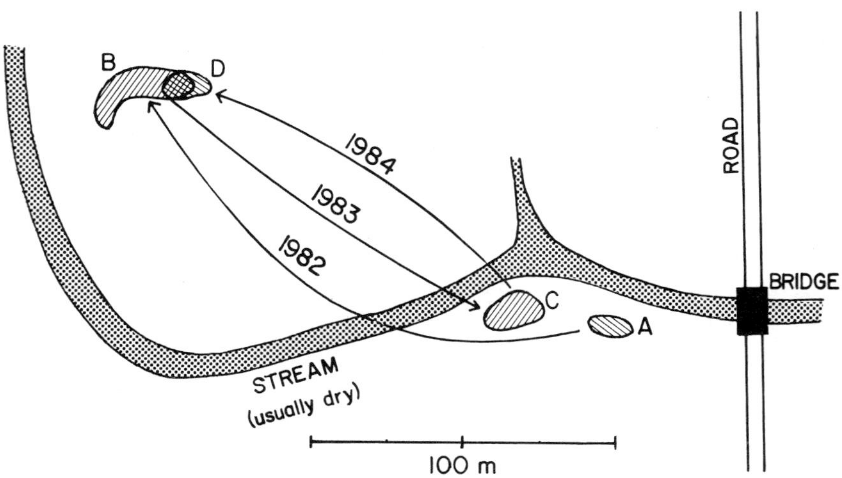

Figure 5.5. The location of nest aggregations of *Bembecinus quinquespinosus* from 1981–1984 (A through D, respectively). Arrows with dates show the movement patterns of females each year as they emerged in one area and nested in another. From Evans et al. (1986); used with permission.

the surface crust on the soil after the emergence period. The aggregations are also dynamic within seasons. Once a nest site was chosen, rather than gradually fill in the available space, the first females nested close together within several square meters. A week later, all females nested in an adjacent patch of sand, and similar shifts occurred over the next several weeks, with the aggregation sometimes fragmenting into multiple clusters. Perhaps as a result of the high density of nests, patches of ground eventually became saturated with nest cells, so females shifted to nearby soil.

The male mating strategies of this species were studied by O'Neill and Evans (1983); Evans and O'Neill (1986); and O'Neill et al. (1989). Because of high nest density, the density of emerging wasps the following year and the potential for intense male–male competition were also high. Assuming a 50:50 sex ratio, the density of female *B. quinquespinosus* cocoons was 134/m^2 and 144/m^2 in two 0.25 m^2 samples, whereas the density of emerging females was 42/m^2 and 94/m^2 in two 1 m^2 emergence traps (KMO, previously unpublished). Males began emerging several days before females and immediately started patrolling the emergence area. They detected females about to emerge from the soil, and often congregated at these sites, digging downward through the surface crust. Males of *B. quinquespinosus* have more developed foretarsal rake spines than males of other *Bembecinus,* though they are less elaborate than those of females. When a female emerged, males (as many as ~50, but usually fewer) clustered around her in a tight, but constantly shifting "mating ball" (Figure 5.6). A male succeeded in mating only if he was able to mount the female dorsally and carry her in flight to the surrounding vegetation, where mating lasted several seconds. Males larger than their mates were more likely to succeed at this (Figure 5.7), apparently due to a greater load-carrying capacity in flight. At least among those in the emergence area, large males also tended to mate with larger females than did small males (and therefore with females that take larger prey and lay larger eggs). In addition, we could identify no corresponding advantages accruing to being small. Mortality in the pupal stage was not correlated with body size, small males did not emerge earlier in the year than large males, and they were not more active at peak periods of females emergence.

The size advantage enjoyed by larger males in the emergence area explains several morphological and behavioral features of this species. First, unlike most apoid wasps, there is no sexual size dimorphism in *B. quinquespinosus,* although there is a great deal of size variation within each sex.

Figure 5.6. A cluster ("mating ball") of males of *Bembecinus quinquespinosus;* female in center of cluster is not visible. Photo by H. E. Evans.

Second, males below average size (and therefore below the average size of females as well) tended to avoid emergence areas (where they are unlikely to succeed), opting to patrol nonaggressively just outside the emergence area. Females mated just once upon emergence, which we confirmed by sequestering both mated and unmated females with males soon after their emergence. However, at least 5% apparently emerged without being detected by the larger males in the emergence area and were available to the smaller males. This may actually benefit females, because some are injured or killed while in the "mating balls" in the emergence area.

A third male morphological feature associated with the mating strategy is a color polymorphism. In most *Bembecinus* (e.g., 7 of 11 Central and North American species; Bohart 1996a), males and females have the same basic coloration, mostly black. The *B. quinquespinosus* population we studied exhibits both an allometric color polymorphism among males and a sexual color dimorphism. In *B. quinquespinosus,* males below average size are mostly black, but as size increases above the mean, males take on increasing yellow coloration, so that the largest males are nearly completely

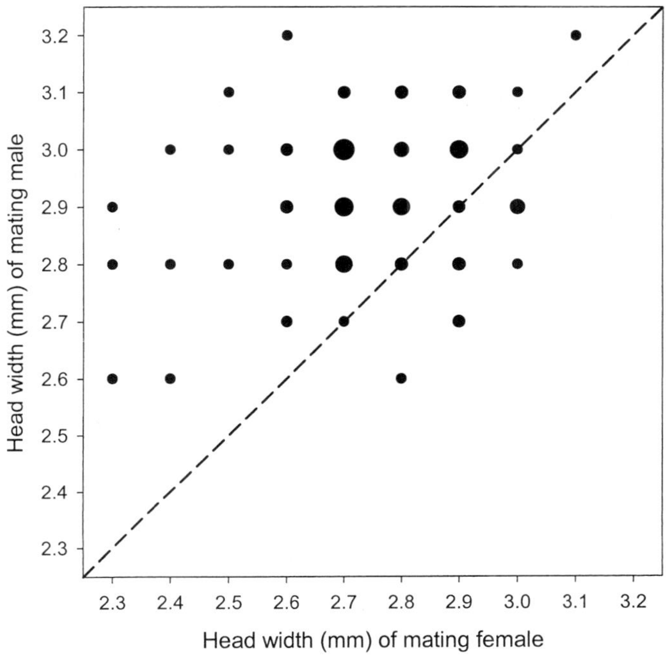

Figure 5.7. Size of males and females in 110 mating pairs of *Bembecinus quinquespinosus.* Size of each circle is proportional to the sample size for each combination of head widths; sample size ranged from 1 to 12. Redrawn from O'Neill et al. (1989).

yellow (Figure 5.8). Female *B. quinquespinosus* are primarily black, the same color as small conspecific males and as both sexes of the nonyellow species. The correlation between size and color was also noted for several *Bembecinus* in Madagascar, including *Bembecinus mirus* (Arnold) and *Bembecinus assentator* (Arnold) (Arnold 1945).

The intraspecific correlation between color and size led Krombein and Willink (1951) to suggest that nutrition may be responsible. O'Neill et al. (1989) argued that it may not be simply that small individuals cannot produce yellow coloration due to some nutritional deficiency. Although small male solitary wasps undoubtedly receive a lower quantity of provisions than large males (O'Neill 2001), it seems unlikely that they do not have access to food species that contain some critical precursor of the yellow pigment. Rather, the link to nutrition may be indirect. Large males search for females in the emergence areas, often at times when the fully insolated soil

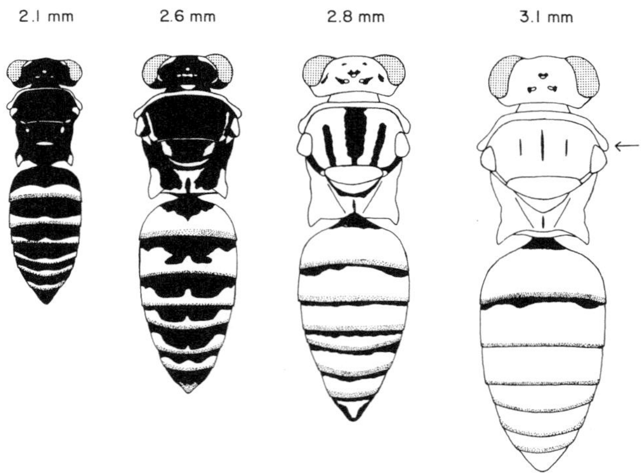

Figure 5.8. Variation in color patterns of male *Bembecinus quinquespinosus* (from O'Neill and Evans 1983, © Springer-Verlag, 1983; used with kind permission of Springer Science and Business Media). Male head width is given for each specimen; head-width range corresponds to body dry mass range of 2.3–15.4 mg (O'Neill et al. 1989).

surface reaches temperatures exceeding 45°C. The yellow pigmentation increases body surface reflectance from 9% for black integument to 29% for yellow integument in the visible and ultraviolet range of wavelengths, likely reducing body temperature or at least rates of heating. Thus large males directly benefit from yellow coloration. Small males, however, tend to avoid the emergence area because they cannot compete with large males due to their inability to carry females. In the areas where small males patrol for females, they may in fact benefit from darker coloration, as it may help them maintain high body temperature. Thus color variation in *B. quinquespinosus* may represent an adaptive color polymorphism, rather than resulting from a nutritional deficiency in smaller males.

Three predators of adults have been reported: robber flies (*Megaphorus willistoni* (Williston)) and *Proctocanthus micans* (Schiner)) (Dennis and Lavigne 1975), and horned lizards (*Phrynosoma douglasi*) (O'Neill et al. 1989). Although the lizards were seen feeding on males just 11 times, further evidence of their taste for male *B. quinquespinosus* was found in lizard feces that were clearly the result of a diet based mainly on the wasps (when males were present). While feeding on the males, lizards apparently ignored passing harvester ants (*Pogonomyrmex* spp.) (KMO—previously unpublished), their typical prey (Hölldobler and Wilson 1990). We also observed predation attempts by tiger beetles, but these were always unsuccessful, apparently because the beetles could not maintain a grasp on the smooth exoskeleton of the wasps (KMO, personal observation).

Of 134 cocoons excavated from the emergence area in 1983, one contained a *Dasymutilla cruesa* Cresson female, six had dead *B. quinquespinosus*, and one had a live male with vestigial wings. Although 80% of the cocoons were empty, it is possible that some had produced parasitoids rather than *B. quinquespinosus.* In 1985, one of 28 cocoons examined had an unidentified mutillid female, but 36% were filled with fungal myceliae, perhaps as a result of a long period of heavy rains that saturated the soil (KMO and HEE, previously unpublished). Low emergence following rains ended our studies of this species in 1985, and during periodic visits by HEE up through the late 1990s no nesting or emergence areas were found.

Both sexes of *B. quinquespinosus* were reported to spend the night in clusters on vegetation (Evans 1955). At the Pawnee site, however, males and females clustered under rocks near the emergence area, whereas later in the season, females did the same under rocks near the nesting area (Evans et al. 1986). The number of wasps in 17 clusters averaged 16 and ranged as high as 85, but cluster size probably exceeded 100 in several that we did not capture and count.

Bembecinus strenuus (Mickel) (= *B. nanus strenuus*)—Nearctic (eastern, central United States)

In early to late July, at Roggen in northern Colorado, females nested on the slope of a sparsely vegetated sand dune blowout (Evans and O'Neill 1986), along with *Philanthus albopilosus* Cresson and *Philanthus psyche* (Evans and O'Neill 1988). The unicellular nests, which were usually as least 30 cm apart, had 9.0–16.5 cm long straight, oblique burrows; no accessory burrows were found. After the egg was laid on top of several grains of sand in the empty cell, the cell was provisioned progressively over several days with adults and immatures of one species of Cixiidae (*Oecleus excavatus* Ball), two of Dictyopharidae (*Scolops maculosus* Ball and *Scolops* sp.), and nine of Cicadellidae from three subfamilies (*Acinopterus viridis* Ball, *Athysanella wilburi* Ball and Beamer, *Carneocephala* sp., *Ceratagallia* sp., *Cuerna striata* (Walker), *Dicyphonia ornata* (Baker), *Draeculacephala* sp., *Flexamia inflata*, Osborn and Ball, and *Mesamia* sp.).

Recently, R. Matthews and J. Matthews (2005) studied a population of *B. strenuus* in Keith County, Nebraska, where females also nested in sparsely vegetated sand along with *Philanthus psyche*. Nests, which were unicellular and similar in dimension to those in Colorado, were dug in one day and provisioned the next. Unlike those in Colorado, most nests were fully provisioned in a single day. Prey from two fully provisioned nests were all of a single unidentified species of Cicadellidae; one cell contained 2 adults and 18 late-instar nymphs, the other 3 adults and 12 nymphs, all "profoundly paralyzed and unresponsive to touch." Matthews and Matthews report that females undertook extensive leveling and final closures almost two hours in duration. During initial leveling, the female moved toward the nest entrance, while scraping sand backward; later, during final closure, she repeatedly moved away from the entrance, each time along a different path, forming a "faintly visible series of radiating lines."

In both Colorado and Nebraska, males patrolled widely for females in the emergence/nesting area; unlike male *B. neglectus* and *B. quinquespinosus*, they did not dig for females. Rather, they pounced upon small dark objects on the sand surface that in some cases turned out to be receptive females with whom they then briefly copulated without leaving the emergence area. One of these at each site was a female working at her nest entrance. Mating usually proceeded without interference from other males, so there was none of the frenzied competition among males that was so prevalent in our study *of B. quinquespinosus*. A comparison of the mating strategy of *B. strenuus* to that of *B. quinquespinosus* is instructive. In the former, females nest in lower density, with the result that the density of emerging females is lower the following year. Assuming a 50:50 sex ratio, the density of female *B. strenuus* cocoons ranged from 0 to 16/m^2 (in four 0.125 m^2 samples; KMO—previously unpublished), compared with estimates of 42–144 female *B. quinquespinosus*/m^2. Because of the lower density, mate-seeking *B. strenuus* males needed to search more widely for females. Thus females are less likely to be detected prior to emergence, and males are less likely to be disturbed when they find virgin females. Several traits of *B. strenuus* seem correlated with these differences: (1) the smaller foretarsal spines of males (males do not dig for females), (2) a lack of a color polymorphism (males have no need to spend much time on the soil surface when it is hot), and (3) a typical direction of sexual size dimorphism (males do not need to carry females). In addition, nesting *B. strenuus* females do not suffer the same degree of harassment by males, and nest in the same site each year.

Bembecinus tridens (Fabricius)—Palearctic

Earlier work on this Eurasian species, among the first *Bembecinus* to be studied, was reviewed by Evans (1955, 1966a). The first behavioral studies of this species since the mid-1960s were those of Lüps (1969, 1973), who found nests in hard-trodden, partly loose sand on and near a little-used path. Females nested in the same place each year, even though vegetation changed with time. Nests were often within a few cm of each other, so that females interfered with one another and sometimes abandoned nests. After females oviposited in empty cells, at least 43 h passed before they brought in prey, though they sometimes visited cells in the interim during inspection visits ("Kontrollbesuche"). Provisioning seems to be truly progressive; Lüps presents a few observations indicating that females made inspection visits in the morning to gauge whether prey were needed. Cells were provisioned with leafhoppers ("Zikaden") that were carried in flight, venter-to-venter, head-to-head, with the middle and hind pairs of legs. As a female landed at a nest to open the temporary closure, she shifted prey to the midlegs, supported herself with her hindlegs, and removed the closure with her forelegs. However, when a female could not immediately find her nest, she laid the prey down to search for it, but this usually resulted in ants stealing the prey.

At sites near Vienna, Austria, Zolda and Holzinger (2002) collected a sample of 25 prey that included eight species of Cicadellidae in seven genera, plus two species of Cercopidae and one each of Cixiidae and Tropiduchidae. About one-quarter of the prey were the cercopid *Philaenus spumarius.* The majority of prey use grasses and herbs as hosts, particularly the grass *Calamagrostis epigejos.* Zolda and Holzinger compared the prey of this species with that of *B. hungaricus* nesting at the same site. Of the 24 species of prey taken, 15 by *B. hungaricus* and 12 by *B. tridens,* only three species were used by both, probably because females hunted in different habitats.

Although copulation has been not described, some aspects of the basic male mating tactic in the nesting areas seem similar to that of *B.*

quinquespinosus: males chase and pounce on insects and gather to dig (i.e., scrape weakly) at places where females are about to emerge (Lüps 1973). Within the nesting area, females are active at higher temperatures than are males. This difference could be due to different physiological temperature optima, as Lüps suggests. However, two untested alternative hypotheses are that (1) sexual differences in activity times are constrained by other factors such as parasite activity (Karsai 1989) or prey availability or (2) males are active elsewhere where temperatures are the same as in the nesting area.

Overview of *Bembecinus*

Habitat and nest density. Of the two dozen or so *Bembecinus* for which general reports exist regarding soil at nest sites, most nest in sand, often described as being "loose." *Bembecinus hungaricus japonicus* and *B. quinquespinosus* seem to prefer coarse sandy gravel rather than fine sand, whereas *B. antipodum* and *B. tridens errans* nest in hard-packed sand; *B. neglectus* was found nesting in hard-packed sand mixed with many pebbles of various sizes. The major exceptions to the preference for sand among those that dig their own nest burrows are *B. posterus, B. cinguliger,* and *B. oxydorcus,* which nest in clay soils. Ferton (1908; in Krombein 1984) found *B. fertoni* and *B. gazagnairei* nesting in preexisting cavities in fine-grained limestone that is hard enough to have been used by Romans as a source of stone for building monuments. Females plugged burrows with sand after cells were provisioned.

Density of nests is variable between and within species, though some of the variation may be due to the time of year at which aggregations were observed; brief observations, early or late in a season, may miss the peak of nesting activity. However, some of the interspecific variation may be real. In Colorado, at sites separated by about 80 km, we observed females *B. strenuus* nesting a minimum of 30 cm apart in a dune blowout and *B. quinquespinosus* nesting commonly 5–10 cm apart. *Bembecinus hirtulus* females nest within several centimeters of one another. Long-term observations of *B. quinquespinosus* provide some evidence that females are attracted to one another, and not just to a specific substrate. Females nested close to one another early in the season and, as the substrate initially occupied became saturated with nests, they abandoned it en masse and moved to adjacent patches of soil while continuing to nest close together. And after females emerged, nearly all of them moved to new locations, maintaining close contact with other females. Despite their apparent tolerance of conspecifics, two females occasionally butted heads or grappled if they happened to be present at the same time at adjacent nest entrances.

Nesting behavior. Nest burrows are commonly less than 10 cm long, the longest recorded being those of *B. strenuus* (25 cm); all recorded cell depths were 3–20 cm. All but one species have 1–2 cells per nest; three cells have been reported only for *B. comberi,* the longer-term nests perhaps being enabled by its nesting in hard dry substrates. But even those species nesting in clay make no more than two cells. The most unique nests constructed by members of this genus are those of *B. cinguliger* and *B. oxydorcus,* which construct mud turrets at the entrance of their nests. No other bembicines are known to build turrets, which are more common in solitary vespid wasps (O'Neill 2001). Although accessory burrows have been observed for *B. neglectus* (Evans 1966a) and *B. egens,* it appears that they are uncommon in *Bembecinus.*

Soil and small pebbles are cleared from the burrow in several ways. Pebbles and pellets of moist sand may be carried using the mandibles, and pellets of moist sand may be held beneath the head; both methods are used by *B. neglectus* (Evans 1966a). Dry sand is loosened and swept away beneath the body, using the forelegs. Like other ground-nesting sphecids, most female *Bembecinus* sport rake spines on their foretarsi, but the spines are much reduced in males. Several recent studies indicate that selection has molded the form of tarsal combs to specific uses in several species of *Bembecinus.* First, male *B. quinquespinosus* have unusually long rake spines that aid in their attempts to dig for preemergent females. Although the spines are not as long as those of females, they are longer than those of male *B. strenuus,* who do not dig for females. Second, the tarsal combs of the South African *B. cinguliger* and *B. oxydorcus* females have evolved into a set of short, closely spaced spines that presumably do a better job of scraping away the clay soil. The sympatric *B. argentifrons* and *B. haemorrhoidalis,* which nest in loose, dry sand, have the typical longer and more widely spaced rake spines.

As a rule, females maintain temporary outer closures when they are away from nests. Evans noted that this was the case for *B. agilis, B. fertoni, B. hungaricus, B. mexicanus, B. neglectus, B. prismaticus, B. quinquespinosus,* and *B. tridens.* To this list, subsequent reports add *B. argentifrons, B. cinguliger, B. hirtulus, B. oxydorcus, B. posterus, B. proximus,* and *B. pusillus.* No reports, other than an earlier study of *B. quinquespinosus* by Rodeck

(1931), explicitly state that temporary closures are not used. Some species seem not to level the soil mound at the nest entrance *(B. mexicanus)*, whereas others do so partially *(B. fertoni)* or completely *(B. neglectus)*. Various degrees of leveling have been recently described for *B. egens*, *B. luteolus*, and *B. hungaricus* (the latter contradicting a report by Iwata 1936).

Provisioning. Evans (1966a) suggested that the standard sequence of oviposition and provisioning began with the female laying an egg in an erect or oblique orientation atop a 1–2 mm pedestal of sand grains near the middle of the floor of the cell. Observations on other species since then report no strong evidence to contradict the assertion that *Bembecinus* female oviposit in empty cells, with the egg deposited in an erect position. Occasionally, as in the case of our study of *B. quinquespinosus*, prey are found without eggs. However, given the difficulty of excavating cells in loose sand and locating a tiny egg or larva, we are not ready to conclude that the initiation of provisioning precedes oviposition in this species.

Provisioning is known to be progressive in all 16 species for which observations are detailed enough to draw any conclusion. The larva is generally less than half grown when a cell is closed by the female, so provisioning may occur rapidly, and is best classified as truncated progressive provisioning. Female *B. hungaricus formosanus* provision so rapidly that Tsuneki categorized some cells as mass provisioned; so the distinction between slow mass provisioning and rapid progressive provisioning may not be discrete; R. Matthews and J. Matthews also reported rapid truncated progressive provisioning in one population of *B. strenuus*.

Among nearly two dozen species for which some prey records are available (Table 5.2), all but one include Cicadellidae in their diets, and the only exception *(B. mexicanus)* is based on prey records from just three nests at one site (Evans 1966a). Prey records also include an array of other families of Cicadoidea and Fulgoroidea, as well as records of three species taking Psyllidae (also Homoptera) and tephritid flies; *B. tridens* alone uses at least eight families. As with other progressive provisioners, it is difficult to estimate the number of prey provided to individual larvae, because many prey may be consumed before the cell is completely provisioned. We do know that the number of prey provided to each offspring often exceeds the maximum values known for *Stizus* (typically 8–12 prey), which are mass provisioners. Some maximum values of prey per cell for *Bembecinus*, which may be underestimates, include: 27 by *B. hirtulus;* 32 by *B. hungaricus japonicus;*

41 by *B. cinguliger;* and 74 *by B. quinquespinosus.* High values for the last species are likely due to the preponderance of nymphs among prey.

Natural enemies. In 1966, Evans noted that it was odd that no fly parasites had been recorded attacking species of *Bembecinus.* Since then, there have been several reports of Diptera larvae or puparia in *Bembecinus* cells. A single puparium of an unidentified fly was found in one sealed cell (out of 34 nests examined) of *B. cinguliger,* along with a large number of prey, but no evidence of a wasp egg or larva (F. W. Gess and S. K. Gess 1975). Krombein (1984) observed miltogrammine flies following prey-laden *Bembecinus,* but the only fly larvae found in cells were judged to be commensals rather than parasites. As far as we know, earlier studies revealed just one mutillid parasitoid of *Bembecinus,* that being a record of *Smicromyrme viduata* attacking *B. tridens* (Grandi 1961). To that, we added records of a low level of parasitism of *B. quinquespinosus* by mutillids: just 2 of 162 cocoons examined during two years housed mutillids (one of which was *Dasymutilla cruesa*). Much more devastating to this species was mortality directly or indirectly caused by heavy rains that saturated the soil one year. Recent reports of natural enemies of adults include stylopids, robber flies, and horned lizards. Ants have been reported to steal prey of *B. bolivari, B. neglectus,* and *B. tridens.*

Male behavior. Thus far, the range of mate-seeking behaviors reported for *Bembecinus* (and the *Stizini* as a whole) seems to fit along a rather narrow continuum. Males of *B. cinguliger* and *B. strenuus* (as well as of *Stizus perrisi* and *S. pulcherrimus*) patrol widely in emergence areas, seeking post-emergent virgins. Males of *B. neglectus, B. quinquespinosus,* and *B. tridens* (as well as *S. continuus*) take a more proactive approach, attempting to reach females just as they break through the soil surface and often digging to reach them before their competitors. In these species, competition among males becomes very intense, particularly after the female emerges into the waiting crowd of males. It may be that the difference among species is due to the high density of emerging females, which concentrates large numbers of mate-seeking males in a small areas. Perhaps males of a species dig for females depending on the prevailing level of mate competition. Studies of multiple populations or long-term studies of one population with varying population densities could shed light on this problem.

Sleeping and adult feeding. Evans (1966a) reviewed the information on sleeping behavior in *Bembecinus,* citing six species in which the wasps are known to sleep in clusters on vegetation that sometimes included hun-

Table 5.2 Cells per nest, prey per cell, and families reported as prey of *Bembecinus*. Includes records from citations in this text and those in Evans (1966a).

			Cicadoidea			Fulgoroidea								Other	
Bembecinus species	Maximum cells per nest	Maximum prey per cell	Cercopidae	Cicadellidae	Membracidae	Cixiidae	Delphacidae	Dictyopharidae	Flatidae	Issidae	Ricaniidae	Tropiduchidae	Tettigometridae	Unspecified Fulgoroidea	Homoptera: Psyllidae	Diptera: Tephritidae
B. agilis				X												
B. antipodum				X												
B. argentifrons	2			X	X									X		
B. bicinctus				X												
B. bolivari				X												
B. cinguliger	2	41		X												X
B. comberi	3	17		X	X				X	X	X	X				
B. egens	1	19		X					X		X					
B. fertoni				X										X		
B. haemorrhoidalis	2			X	X									X		
B. hirtulus	1	27		X			X								X	
B. hungaricus				X			X								X	
B. hungaricus formosanus				X												

B. hungaricus japonicus		32		X											X	
B. luteolus																
B. mexicanus[a]					X											
B. neglectus		15		X				X								
B. oxydorcus	2			X												X
B. posterus				X												
B. prismaticus				X												
B. proximus				X				X				X				
B. pusillus		14		X		X	X									
B. quinquespinosus	1	74		X												
B. strenuus	1		X	X	X	X		X						X		
B. tridens			X	X	X	X		X		X		X	X		X	

[a]Based on three nests at one site.

dreds of individuals. Since then, *B. neglectus* and *B. cinguliger* have been found sleeping on plants. Female *B. comberi* and *B. luteolus,* on the other hand, sleep alone in their burrows. Sites chosen for roosts may vary among populations. Evans's (1955) earlier report of *B. quinquespinosus* sleeping on vegetation in Arizona differs from our observations of these wasps sleeping in clusters under rocks in northern Colorado. Evans (1966a) noted that adult *Bembecinus* are not commonly found on flowers. Since then, Evans and Matthews (1971) reported seeing *B. egens* feeding on blossoms of mangrove and herbs, F. W. Gess and S. K. Gess (1975) collected *B. cinguliger* and *B. oxydorcus* of both sexes visiting various flowers to obtain nectar or glandular exudates (in South Africa), and Krombein(1984) found *B. luteolus* visiting "prostrate creepers" for nectar (in Sri Lanka). In addition, we observed large numbers of *B. quinquespinosus* feeding on honeydew on sunflowers (*Helianthus* sp.) in Colorado.

— 6 —

Bembicini:

The Diverse New World Genera

The strictly New World genera of the tribe Bembicini considered in this chapter are treated in the same sequence as in Bohart and Menke (1976). The only worldwide genus of Bembicini and the largest genus of sand wasps, *Bembix,* is covered in the next chapter. Two New World genera, *Carlobembix* Willink (1947) and *Chilostictia* Gillaspy (1983), are known from a single species each and nothing is known of their ethology. At least some biological information is available for all 15 other genera.

Phylogenetic status of the Tribe Bembicini. The 17 genera of Bembicini include over 500 species (30% of Bembicinae) (Pulawski 2006a). If the phylogenetic relationships among genera of Bembicini suggested by Bohart and Menke are correct, the genera treated in Chapter 6 are paraphyletic with respect to *Bembix* (e.g., *Stictia* and *Rubrica* may be more closely related to *Bembix* than they are to such genera as *Steniolia* or *Bicyrtes*). *Stictiella, Microstictia, Glenostictia, Xerostictia,* and *Steniolia* have been considered by some authors to represent a separate clade within the Bembicini, sometimes referred to as the subtribe Stictiellina (Bohart and Menke 1976; Bohart and Gillaspy 1985). Prentice (1998), however, considers the subtribal relationships among the genera of the Bembicini to be "currently unresolved," and tentatively places all species in the subtribe Bembicina of the Bembicini, along with what we consider the Stizini and Gorytini here (Table 1.1).

Bicyrtes

Pulawski (2006a), following the revision by Bohart (1996b), lists 27 species of *Bicyrtes,* five in America north of Mexico, all but seven of which have a basically neotropical distribution. Biological information is available for 13

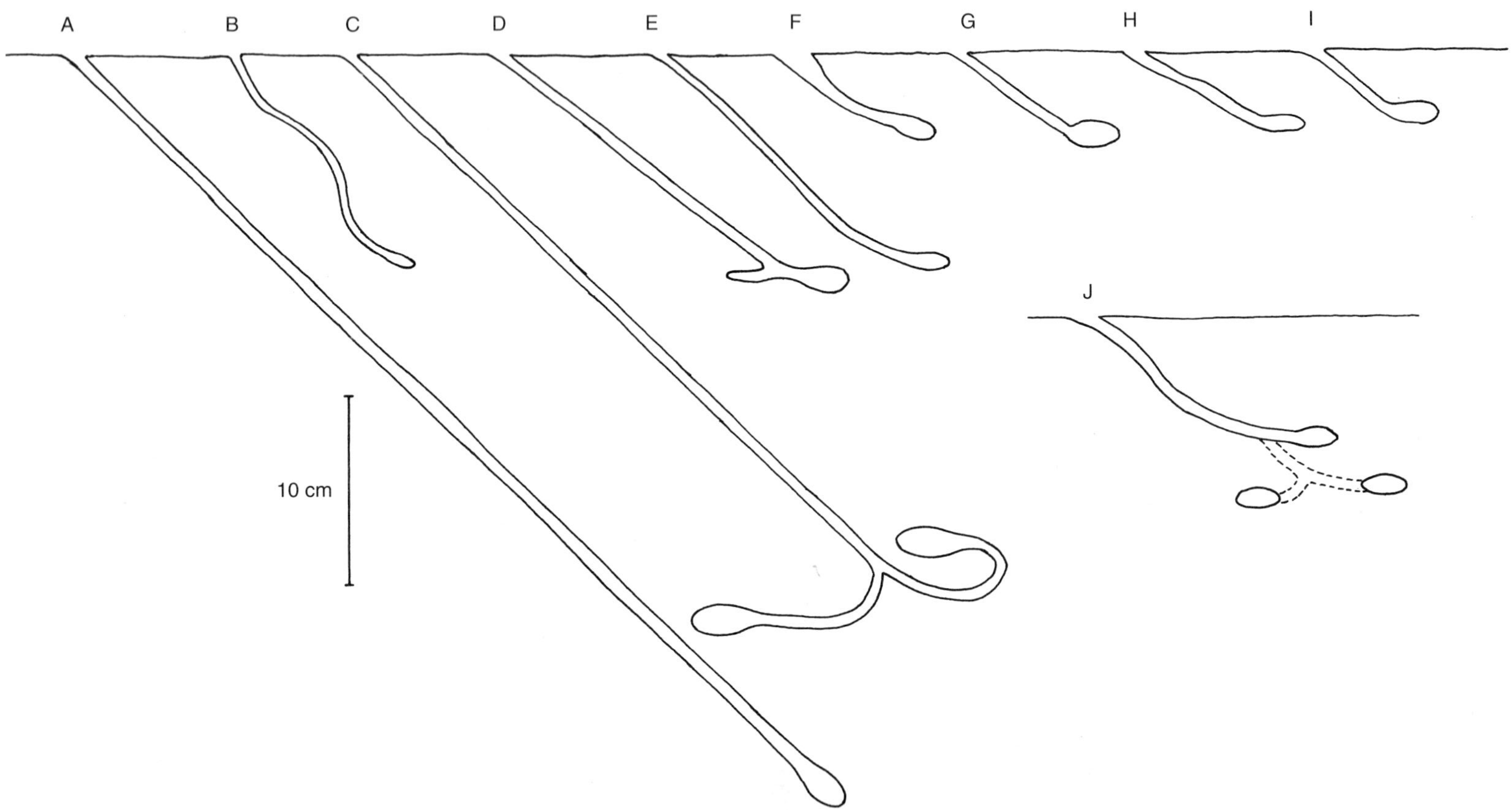

Figure 6.1. Structure of nests of (A) *Stictia flexuosa* (Genise 1982c); (B) *Microbembex argentifrons* (Genaro and Sanchez Alonso 1990); (C) *Xerostictia longilabris* (Alcock 1975c); (D) *Zyzzyx chilensis* (Genise 1982e); (E) *Hemidula singularis* (Genise 1982a); (F) *Bicyrtes angulatus* (Martins et al. 1998); (G) *Rubrica nasuta* (Pimenta and Martins 1999); (H) *Glenostictia pictifrons* (Alexander et al. 1993); (I) *Stenolia obliqua* (Evans 1970); (J) *Bicyrtes ventralis* (Evans 1966a). All redrawn to same scale from cited sources.

species. Note that Bohart's list includes the species names with feminine endings (i.e., "-a"), whereas Pulawski lists the same species with masculine endings (i.e., "-us"); we have adopted the names in Pulawski's catalog.

Bicyrtes angulatus (F. Smith)—Neotropical (South America)

In Brazil, Martins et al. (1998) studied 227 nests over a two-year period in compacted, well-drained sand along a dirt road on the campus of Universidade Federal de Minas Gerais. The nesting wasps here benefited from high densities of the weed *Waltheria indica* (L.) (= *W. americana* L.) (Sterculicaceae), whose flowers provided nectar for *B. angulatus* (Macedo and Martins 1998). In 1993, the overall density of *B. angulatus* nests was 6.5/m^2, and its nests occurred within a mixed aggregation that included five other species of wasps and a few solitary ground-nesting bees. Other wasps included *Ammophila gracilis* Lepeletier (54 nests), *Bicyrtes discisus* (4 nests), and *Rubrica nasuta* (193 nests).

Of the 227 nests monitored, 147 were provisioned by 103 different females that had been individually marked by the investigators. Nests were unicellular and averaged less than 10 cm in depth (Figure 6.1F). Digging females first loosened soil with their mandibles before scraping it back beneath their bodies, "lifting the abdomen each time the soil [was] thrown, and simultaneously tilting the head down over soil," which formed a mound near the nest entrance. Digging by *B. angulatus* is similar to that of *Bicyrtes quadrifasciatus* Evans (1966a), with the notable exception that, in the latter species, "several spurts of sand are thrown out each time the head goes down." Nests were excavated by females in less than 5 h, but provisioned progressively over ~6 days, most often beginning on the day following nest excavation. During final closure, the burrow was filled with soil from the tumulus and finally compacted by the female using the tip of her abdomen.

The 7–24 prey (mean = 11.9) provided per cell included 302 Alydidae of the genera *Apidaurus* and *Megalotomus* (mostly nymphs) and 5 undetermined Pentatomidae (all nymphs and all from a single nest). The partially paralyzed prey were carried venter-up by the wasp using its middle legs, which continued to grasp prey as the wasp opened the nest entrance with her front legs. The egg was laid erect between the middle or hind coxae on the first prey in the cell.

About 90% of completed nests failed to produce adult wasps, largely as a result of predation by ants (primarily *Solenopsis* sp.) and termites (100 nests); in fact, one entire aggregation of 25 nests was destroyed in 1991. Other natural enemies included Bombyliidae (*Ligyra morio* (F.); 6 nests) and Sarcophagidae (*Metopia* sp; 1 nest—listed as a probable parasite). The duration of time from egg to adult varied from 44 to 375 days, the high values resulting from dormancy within the cocoon.

Males patrolled the nesting aggregation close to the soil surface, in a manner similar to that described for *Bicyrtes quadrifasciatus* by Evans (1966a). They were active as early as 0845 h and as late as 1600 h. While patrolling, they "occasionally clashed briefly in mid-air with other males, females, or other insects." Both male and female *B. angulatus* were also seen sleeping together on *Panicum maximum* (Poaceae) along with *R. nasuta.*

Bicyrtes cingulatus (Burmeister)—Neotropical (Argentina)

At Cafayete, Salta, Argentina, *B. cingulatus* nested in bare sandy soil (Evans and Matthews 1974). The three nests studied had 12–14 cm long burrows that began at a relatively shallow angle (20–45°), then became steeper before leveling off near a single cell at a depth of 8.5–10.5 cm. Cells were progressively provisioned with immature and adult Rhopalidae of the genera *Arhyssus, Liorhyssus,* and *Xenogenus.*

Bicyrtes discisus (Taschenberg)—Neotropical (Mexico to South America)

In Argentina, females nested within an aggregation that also included *Bicyrtes variegatus* (Olivier). Females sometimes made as many as three cells per nest, mass provisioning with an average of 16 prey that included 72 immature *Nezara viridula* (L.) and 24 immature *Euschistus* sp. (Pentatomidae) (Genise 1979a, 1982f). Digging was similar in form to that of *B. variegatus.* In Trinidad, Callan (1991b) found shallow, unicellular nests in sandy or stony soil that were provisioned with adult Pentatomidae (*Edessa* sp.) and Coreidae (*Hypselonotus fulvus* [deGeer] and *Cebrenus* sp.). The single nest excavated was similar to those reported elsewhere: a single cell, 10 cm deep at the end of an oblique burrow.

Bicyrtes quadrifasciatus (Say)—Nearctic (North America, east of Rocky Mountains)

This well-studied species has been the subject of studies in New York, Mississippi, Kansas, and Texas, reported in publications going back to

1923 (Evans 1966a). Pentatomidae and Coreidae made up the vast majority of previously known prey, which also included Cydnidae, Lygaeidae, Reduviidae, and Scutelleridae. Supplementing these records, Kurczewski and Kurczewski (1971) reported six nymphs of Pentatomidae as prey from Presque Isle, Pennsylvania: four *Acrosternum hilare* (Say) and two *Banasa* sp., the latter being a new prey record. The only other recent mention of the species that we are aware of is from Waldbauer et al. (1977), who listed it, as well as *B. ventralis*, as part of a complex of the models mimicked by Conopidae and Syrphidae.

Bicyrtes simillimus (F. Smith)—Neotropical (Brazil to Argentina)

In Argentina, this species nested in relatively firm soil, where it produced an appreciable tumulus at nest entrances and, sometimes, accessory burrows (Genise 1982f). One nest excavated had a tunnel 24 cm long and 0.8 cm in diameter that led to a 1.7 × 2.0 cm cell 20 cm deep. Prey consisted of immature *Nezara viridula* (Pentatomidae) that were often brought in over a second day. Such delayed provisioning (called "slow provisioning" by Genise) is evidently similar to that reported for *B. quadrifasciatus* by Evans (1966a).

Bicyrtes spinosus (Fabricius)—Neotropical (South America, West Indies)

In Cuba, where this species has several generations per year, unicellular nests had burrows averaging 9.9 cm long, including the terminal cell, which sat at depths of 4.5–8.5 cm (Sánchez and Genaro 1992b). (Alayo (1969) notes that the species is often found in the company of *Microbembex* and *Stictia* in Cuba.) Cells were provisioned progressively with Hemiptera; all but two of the 268 prey were immatures. Identified prey included: 56 *Acanthocerus lobatus* (Burm.), 5 *Anasa scorbutica* (F.), 3 *Chariesterus gracilicornis* Stål, 1 *Harmostes harmatus* (F.), 6 *Leptoglossus balteatus* (L.), 2 *Leptoglossus gonagra* (F.), 5 *Leptoglossus* sp., 5 *Phthia picta* (Drury) (Coreidae); 19 *Acrosternum marginatum* (Beauv.), 2 *Banasa subrufescens* (Walker), 8 *Edessa* sp., 3 *Euschistus bifibulus* (Beauv.), 24 *Modicia* sp., 17 *Mormidea palma* (Stål), 24 *Oebalus* sp., 4 *Proxis punctulatus* (Beauv.), 14 *Thyanta perditor* (F.) (Pentatomidae); 3 *Dioclus* sp., 40 *Sphyrocoris obliquus* (Germar) (Scutellaridae). Other prey specimens (Lygaeidae and Pyrrhocoridae) were not further identified. Several of the prey species are known pests of beans, squashes, and tomatoes.

Bicyrtes variegatus (Olivier)—Nearctic and Neotropical (Texas, Mexico, South America)

This widely distributed species was studied in Argentina (Genise 1979a), in Mexico (Martin and Martin 1990), and in Trinidad (Callan 1991b). In Argentina, nests were dug in pure sand, and sometimes occurred in aggregations. Females spent about an hour constructing a 12 cm long burrow down to the first cell. The form of digging is characteristic of the genus: females dug with their first pair of legs and digging movements were accompanied by rapid inclinations of the body. The sand was dispersed backward, as far as 30 cm, while the abdomen was elevated and the head was near the soil surface. Nests contained 3–5 cells, mass provisioned with 3–6 fifth instar *Nezara viridula* (Pentatomidae) after the egg was laid on the first prey (attached in a semi-erect position to the coxae of the first pair of legs and inclined toward the head).

In Mexico, along the coast of Quintana Roo, *B. variegatus* nests were isolated and scattered among those of *Stictia signata.* Burrow length averaged 13 cm, cell depth 9 cm. Of 24 nests excavated, 23 had one cell, 1 had two cells. Prey identified consisted of 10 Alydidae (including 9 *Hyalimenus* sp.); 14 Coreidae (including 1 *Leptoglossus* sp.); 40 Pentatomidae (including 13 *Arvelius* sp.); and 10 Scutellaridae. Most provisioning was completed on the first day, but it sometimes extended over a second day and perhaps longer, because large larvae were found in two nests still being provisioned. On Trinidad, *B. variegatus* is a coastal species that preys on immature Pentatomidae (*Edessa affinis* Dallas) and Coreidae (*Leptoglossus balteatus, Phthia picta*). Evans (1976b) observed several females digging nests on a beach in Baja California Sur ("well below the high water mark"), but made no further observations.

Bicyrtes ventralis (Say)—Nearctic

Previously studied by Evans (1966a) in New York and Kansas, *B. ventralis* is a delayed mass provisioner that stocks multicellular nests (Figure 6.1J) with stinkbugs (Pentatomidae; 80 specimens in 10 species) and Coreidae (4 specimens of one species). Spofford and Kurczewski (1990, 1992) recently reported that 5% of 25 *B. ventralis* nest cells were parasitized by the miltogrammine *Senotainia trilineata* in a study in New York. Females made diversionary flights to avoid larviposition on prey and sometimes abandoned prey after being trailed by a fly. In several cases, Spofford and Kurczewski observed a *B. ventralis* larva devouring several *S. trilineata* maggots that had been brought into the cell attached to stinkbug prey. There is not much else to add since Evans's earlier studies, except that one female *B. ventralis* was found as prey of *Philanthus basilaris* Cresson at the Great Sand Dunes National Monument, Colorado (Evans and O'Neill 1988).

Bicyrtes viduatus (Handlirsch)—Nearctic (southwestern United States, Mexico)

In Arizona, Alcock and Gamboa (1975) found two nests with burrows 14 and 15 cm long and cells 7.5–11 cm deep in coarse, rocky gravel in the middle of a dry wash. One of the nests "had been constructed in extraordinarily rocky material" so that a trowel had to be "used as a pick to chip through the crumbling rock." Cells were provisioned with immature Coreidae and the egg was laid on the first prey in each cell, attached ventrally between the midlegs. Alcock and Gamboa also note that the "orientation flight of [*B. viduatus*] was similar to that of [*B. quadrifasciatus*], consisting of a slow vertical flight to a height of about 2 m before circling still higher and then flying off."

Overview of *Bicyrtes*

Studies over the past several decades confirm many of the basic generic features of *Bicyrtes,* though several new details have been discovered (see also review in Martins et al. 1998). Some *Bicyrtes* nest in rather dense aggregations, generally in sandy soil (with the exception of *B. simillimus* and, perhaps *B. viduatus*). Some occupy sand near water, others inland sands or gravels, in some cases quite compacted. Females dig shallow nests with cells typically at depths of 5–20 cm, depending on the species studied. Nests often have just a single cell, but up to three per nest have been found in three species and up to five per nest in two others (Table 6.1).

Oviposition occurs on the first prey brought into the cell. *Bicyrtes discisus, B. fodiens,* and *B. variegatus* are mass provisioners; *B. quadrifasciatus, B. simillimus,* and some *B. ventralis* are delayed mass provisioners; *B. angulatus, B. cingulatus,* and *B. spinosus* are progressive provisioners. Updated prey records include nine families of Hemiptera: Alydidae, Coreidae, Cydnidae, Lygaeidae, Pentatomidae, Pyrrhocoridae, Reduviidae, Rhopalidae,

Table 6.1 Form of provisioning, cells per nest, prey per cell, and families of Hemiptera reported as prey of *Bicyrtes*. See text and Evans (1966a) for references and details.

Bicyrtes species	Form of provisioning[a]	Cells per nest	Prey per cell	Alydidae	Coreidae	Cydnidae	Lygaeidae	Pentatomidae	Pyrrhocoridae	Reduviidae	Rhopalidae	Scutellaridae
B. angulatus	P	1–2	7–24	X				X				
B. capnopterus								X				
B. cingulatus	P	1									X	
B. discisus	M	1–3	16		X			X				
B. fodiens	M	2–5	10–23					X				X
B. quadrifasciatus	D	1–3	4–14		X	X	X	X		X		X
B. simillimus	D	1	11					X				
B. spinosus	P	1			X		X	X	X			X
B. variegatus	M	1–5	3–6	X	X			X	X			X
B. ventralis	M or D	1–3	3–11		X			X				
B. viduatus					X							

[a]M = mass provisioning; D = delayed mass provisioning; P = progressive provisioning.

and Scutelleridae; Alydidae and Rhopalidae were not known as prey in 1966. Prey are mostly immatures or a mixture of nymphs and adults. Alydidae, Coreidae, and Pentatomidae often have defensive secretions that produce odors that are noxious, at least to humans. Perhaps *Bicyrtes* females and their larvae have means of tolerating these chemicals.

Microbembex

Microbembex are in one major way the most idiosyncratic of solitary apoid wasps, because they scavenge dead or disabled arthropods, a trait unique among solitary wasps. Pulawski (2006a) lists 34 species, nine in America north of Mexico. There are several recent reports on North American and Neotropical species.

Microbembex argentifrons (Cresson)—Neotropical (West Indies)

In Cuba, females nested in areas of fine sand, in spaces amid vegetation, building shallow, unicellular nests with oblique burrows 16–32 cm long and cells 10.2–13.5 cm deep (Figure 6.1B) (Genaro and Sánchez Alonso 1990). Cells were provisioned with dead insects of at least 18 families in eight orders, more than half of the provisions being made up of four families: Formicidae (34%), Scarabaeidae (7%), Lygaeidae (5%), and Tenebrionidae (4%). Males and females spent nights and other periods of inactivity within sleeping burrows.

Microbembex argentina Brèthes—Neotropical (Argentina)

In Salta and Catamarca, Argentina, this relatively large species was common in sand dunes within intermontane valleys, where it nested on flat to gently sloping surfaces (Matthews and Evans 1974). At Salta, one nest with a 15 cm long burrow entered at a 60° angle and led to a 1.2 × 3.0 cm cell 13 cm beneath the surface. A second nest, excavated after the female made the final closure, had a 12 cm deep cell at the end of a 18 cm long burrow, the upper 10 cm of which had been "filled solidly" during closure. The cell in the first nest contained a large wasp larvae, but the only trace of prey remaining was the leg of a spider; the cell in the second nest contained a nearly full-grown larvae along with two adult pentatomids, two adult mirids, two adult cicadellids, and parts of a beetle. Other females carrying prey had an adult antlion, a tachinid fly, and a tenebrionid beetle larva.

Near Santa Maria, Catamarca, one nest had a burrow 19 cm long with a cell 12.5 cm deep, whereas a second nest had a 21 cm long burrow with a 12.0 cm deep cell. The cell in the latter nest contained the typical odd assortment of detritus consisting of parts or complete bodies of arthropods that were undoubtedly dead when "captured" by the provisioning female: 1 tenebrionid beetle larva, 1 rhipiphorid beetle, 1 dermestid beetle, 2 pyralid caterpillars, and 1 spider. Nests were clearly provisioned progressively. When the researchers captured flies, crushed them, and placed them on the sand near the nests, female wasps readily picked them up and carried them into their nests. Males were observed patrolling "close to the sand, chiefly in blowouts" in irregular flight patterns, but no sexual interactions were reported.

Microbembex argyropleura Bohart—Nearctic (Arizona, California, Nevada)

In Dateland, Arizona, an aggregation of 1000–2000 wasps occupied an area about 10 × 20 m in extent along a sandy roadside (Evans 1976a). One nest excavated had a cell at 29 cm vertical depth. Alcock (1975c) had previously studied this species in Arizona, finding cells 45–67 cm in depth, burrows 70–95 cm long. Nests were dug into very dry dunes, so the females may have been seeking moister soil. When Alcock placed fragments of insects on the soil, they were quickly retrieved by hunting females, as has been observed in other studies of *Microbembex* (e.g., Evans 1966a; Matthews and Evans 1974).

Microbembex californica Bohart—Nearctic (United States and north Mexico west of Continental Divide)

In Humboldt County, California, Goodman (1970) found *M. californica* nesting in vegetation-free blowouts of dunes, describing it as a "general scavenger." The unicellular nests each had an oblique burrow that entered the soil at 30° and extended about 25 cm to a cell 5–17 cm deep. One nest contained remains of a spider, along with those of six families (Cercopidae, Carabidae, Curculionidae, Asilidae, Therevidae, and Formicidae) in four insect orders. Earwigs were also common in provisions. In general, there were no important differences from *M. monodonta.* The egg was laid in the empty cell before provisioning began. Beyond a bit of sand scraped into the entrance, no temporary closure was maintained during foraging, so that a female returning with prey entered the nest quickly without releasing her provision.

Microbembex ciliata (F.)—Neotropical (South America)

In Venezuela, over 200 nests occurred in a 5 m^2 area of crushed "coral sand having considerable loam content" about 80 m behind a beach (Matthews and Evans 1974). The shallow, 8.5–10.0 cm deep cells were at the end of 14.0–15.0 cm long burrows in the unicellular nests; a previous study reported as many as eight cells per nest (Janvier 1928). Cells within two of the nests were provisioned with diverse dead or disabled arthropods, particularly ants of the genus *Azteca.* In the third nest, an egg found was glued upright in the innermost portion of the cell. Females maintained outer nest closures while foraging. In coastal Trinidad, Callan (1991b) found *M. ciliata* nesting and collecting dead arthropods from the sand surface.

Microbembex cubana Bohart—Nearctic and Neotropical (Bahamas, Cuba)

On Staniel Cay, in the Bahamas, Toft (1987a,b) found *M. cubana* and *Microbembex monodonta* on sandy soil. The overall population of *M. cubana* was subdivided into small aggregations ranging in size from about 10 to 100 individuals; the sex ratio in aggregations averaged about 2:1 in favor of males. Individuals tended to focus reproductive activities (nesting and mating) within their "home" aggregations, although they did intermingle during other activities (feeding at flowers, hunting, and within clusters of sleeping burrows). Male *M. cubana* were also larger on average than the females of their species (11.1 mm), a reversal of the direction of sexual dimorphism typical among solitary aculeate wasps (O'Neill 2001). The significance of this reversal is unclear, because the mating system of the species is unknown. Toft does state that males of both *M. cubana* and *M. monodonta* on Staniel Cay are "territorial," but she does not provide details (i.e., do individuals defend specific plots of ground, aggressively excluding conspecific males?). In a further study of these same populations, Fraizer (1997) developed models that explore the relationships between sexual and trophic competition and thermal constraints.

Microbembex evansi Bohart—Nearctic (Texas)

At Monahans Sand Dunes in Texas, Evans (1976a) noted a "strongly maculated form" of *M. monodonta* which "showed no unusual features" in behavior, and which has since been described as a distinct species (Bohart 1993). Two nests were excavated and originally reported as belonging to *M. monodonta;* they had single cells 15 and 21 cm deep, and burrows 21–30 cm long. One female was observed carrying the body of a queen *Camponotus* (Formicidae) with several missing legs.

Microbembex hirsuta Parker—Nearctic (Colorado, New Mexico, Texas)

In Colorado and New Mexico, "nests were scattered widely over the available sand, chiefly in blowouts and windward slopes" (Evans 1976a). Provisions in the Colorado nests contained pieces or bodies of a camel cricket, robber fly, scoliid wasp, and ants. Three unicellular nests, whose cells were 15–23 cm deep (Figure 6.2), were similar to those of the *M. monodonta,* beside which the species often nested. At LaJoya, New Mexico, Rubink (1978) found that, when female *M. hirsuta* made consecutive nests, they tended "begin a new nest in the immediate vicinity of the old one, and within minutes of completing its final closure." Like other *Microbembex* (Evans 1966a), female *M. hirsuta* may test the soil in several places before finally settling down to dig a nest. Rubink presented preliminary evidence that compared with those sites abandoned after the female dug no more than 1 cm deep, microsites accepted by females always had higher soil shear strengths, shear strength being a measure of soil cohesiveness; thus females appeared to reject sites where the surface soil was too loose.

Microbembex monodonta (Say)—Nearctic (North America east of Continental Divide, Bahamas, Central America)

This widespread, well-studied species (Evans 1966a) has received further study in the United States (Evans 1976a) and the Bahamas (Toft 1987a,b). Evans's studies at two sites in Colorado and one in Texas were the first made west of the 100th meridian; he observed no major behavioral differences between western and eastern populations, or between western *M. monodonta* and other western species of *Microbembex.* Four nests, all unicellular, examined at the Great Sand Dunes National Monument in Colorado were somewhat shallow for this species (10–15 cm, with 15–26 cm long burrows), but contained a typical array of provisions: "femur of a grasshopper, wing of a true bug, a complete membracid, a small beetle of the family Scarabaeidae, 2 Chrysomelidae, 2 unrecognizable beetles, an incomplete ichneumon wasp, wasps of the genera *Dryudella* and *Anci-*

Figure 6.2. Unicellular nest of *Microbembex hirsuta* from Hasty, Colorado; ruler is 15 cm long. Photo by H. E. Evans.

strocerus (in poor condition), and several worker ants"; if you added "eye of newt" and "wing of bat," the list would sound something like a recipe for a witch's brew!

Several reports of natural enemies of *M. monodonta* have also appeared. Spofford and Kurczewski (1990, 1992) reported that 17% of 24 *M. monodonta* nest cells were parasitized by the miltogrammine flies *S. trilineata* and *S. vigilans* Allen, as well as undetermined species of the same genus. When trailed by a fly, some females made diversionary flights in attempts to avoid pursuit. At the Great Sand Dunes National Monument, 27 females, but no males, were found as prey of *Philanthus basilaris* (Evans and O'Neill 1988). In Alberta, Canada, one male and one female were reported as prey of *Philanthus albopilosus* (Hilchie 1982). Waldbauer et al. (1977) listed *M. monodonta* as part of a complex of models mimicked by several syrphid flies.

In the study by Toft (1987a,b), *Microbembex monodonta* occurred in fewer and larger aggregations than those of the sympatric *M. cubana*. Female *M. monodonta* were active longer each day, beginning earlier and retiring later than females of *M. cubana.* Females of *M. monodonta* were larger (length averaging 11.0 mm) than males (10.5 mm) and than females of *M. cubana* (10.4 mm). Differences, though slight, suggest that larger females *M. monodonta* may be capable of longer periods of activity, since they may be able to attain body temperatures that enable them to become active earlier in the day, as Willmer (1985) showed to be the case for *Cerceris arenaria* (L.).

Microbembex nigrifrons (Provancher)—Nearctic (eastern United States and Mexico)

In central Washington, this species nested in dunes, making unicellular nests provisioned with dead arthropods (Alcock and Ryan 1973). Females were seen hovering over seeds and pieces of dried vegetation, apparently inspecting them as possible food items. Nests dug in drier soil were appreciably deeper than those in moister soil (average cell depth 32 cm as compared with 13–19 cm). In Arizona, Alcock and Gamboa (1975) found that, in a dry period, cells averaged 29 cm in depth, but following a period of rain they averaged 10 cm deep. Their observations suggest that outer nest closures served to protect nests from conspecific females, as well as from bee flies and cuckoo wasps.

Microbembex uruguayensis (Holmberg)—Neotropical (Argentina, Uruguay, Paraguay)

Llano (1959) reported that *M. uruguayensis* used living carabid beetles as prey, stinging them before taking them to the nest. Two recent reports on *M. uruguayensis* suggest Llano was mistaken. Matthews and Evans (1974) studied the species (under the name *M. schrottkyi* Willink, a synonym) in Santiago del Estero, Argentina. They found two unicellular nests with cells at depths of 11.0 and 12.5 cm. The combined provisions in the cells included 39 ants (of two *Camponotus* species), two cicadellid leafhoppers, and two pentatomid bugs. In Entre Rios, Argentina, Alcock (1975b) excavated four nests in sand dunes; two nests contained eggs in an upright position along with several ants that were missing some of their legs. The presence of several food items with eggs suggests that females may have been mass provisioning. Alcock noted that females, after picking up an object from the soil, often curved their abdomen forward, as if stinging their prey, possibly "a vestigial behavior pattern." This behavior has been reported for several species of the genus, and may have deceived Llano into believing that the wasps were taking living prey.

Overview of *Microbembex*

In summary, there appears to be little behavioral variation among *Microbembex* studied. All species occur in broad areas of sand, and it is not unusual to find two or more species nesting in the same area, sometimes along with *Bembix*. Most *Microbembex* make unicellular nests, laying the egg in an empty cell. As noted, Janvier (1928) reported up to eight cells per nest for *M. ciliata*, but Matthews and Evans (1974) found only unicellular nests for this species; so Janvier's report requires confirmation (as do many of his observations—see Evans 1966a). Dead or disabled arthropods are brought in progressively over a variable length of time, probably reflecting the females' success in finding food items nearby.

The bulk of the behavioral evidence argues that females of this genus are scavengers; there is no solid evidence to the contrary. All major orders of insects have been found in nest cells as well as Dermaptera, Ephemeroptera, Psocoptera, Trichoptera, and four orders of Arachnida (Araneida, Phalangida, Scorpionida, and Solpugida). Provisions are often rich in ants, likely a result of the abundance of these insects. No other genera of Bembicini scavenge to obtain provisions, with the exception of those species that steal prey from other females. Despite the relative behavioral homogeneity of the genus, *Microbembex* provides opportunities for studies of diet selection, habitat sharing, and interspecific differences in mating behavior.

Hemidula

Pulawski (2006a) lists two species, both confined to southern South America. Until the 1980s, there was but a single, brief published report of the behavior of *Hemidula* (Willink 1947), and the first prey record did not appear until 1982 (Genise 1982a).

Hemidula burmeisteri Willink—Neotropical (Argentina)

In Argentina, this species excavated unicellular nests in saline soil, first bringing in a very small prey that is apparently used only as a pedestal for the egg (Genise 1989); the prey pictured in the paper is a bombyliid fly.

The nest excavated by Genise had an 8 cm long oblique burrow, which entered the soil at an angle of 45° and ended in an ovoid 10 × 15 mm cell.

Hemidula singularis Taschenberg—Neotropical (Argentina)

Willink (1947) had reported finding a large aggregation of *H. singularis* nesting in a field in Mendoza, Argentina. Nests were often in small depressions, 10–12 cm deep (Figure 6.1E), but the wasps were evidently not active at that time. More recently, Genise (1982a) studied *H. singularis* in Mendoza and found it to be a predator on tabanid flies, the egg being laid erect on the side of the first fly placed in the cell. Nests were dug in soil of high salt content, and were unicellular; the burrows were about 17 cm long, the cells 10 cm deep.

Rubrica

This is a neotropical genus of four species (Pulawski 2006a) of large and colorful wasps, one of which, *Rubrica nasuta,* has recently been the subject of extensive studies of both adult and larval behavior. Evans (1966a) and Pimenta and Martins (1999) both reviewed information on *Rubrica.*

Rubrica nasuta (Christ) (= *Rubrica surinamensis* (DeGeer))—Neotropical (Colombia to Argentina; Trinidad)

In recent years, *R. nasuta* has been the subject of at least seven reports. Evans et al. (1974) reported observations made in Trinidad, Colombia, and Argentina, where females nested in firm clay-sand that contained many pebbles. At Cali, Colombia, about 150 nests occurred along a little-used road over a distance of 25 m. Here, one- to three-celled nests were often clumped and separated by as little as 3–15 cm (Figure 6.3); one nest possibly had five cells, but there was some uncertainty in attributing all of the cells to a single nest. Burrow lengths at the Colombia and Argentina sites were 9–22 cm, cell depths 5–13 cm. The egg was laid on the side of the first fly placed in the cell, and provisioning was progressive. Observations at one site in Colombia confirmed earlier publications noting that, before provisioning each morning, females often clean the cells of uneaten prey. Here females were seen flying from the entrance and dropping uneaten prey, or pieces thereof. "Cell cleaning" required 5–8 min, during which 3–6 loads were removed from the nest. In Argentina, Genise (1982b) also ob-

Figure 6.3. Unicellular nest of *Rubrica nasuta* from Cafayate, Argentina; ruler is 15 cm long. Photo by H. E. Evans.

served *R. nasuta* females removing uneaten prey from the nest, especially when the larva was young, but he called it "prey discarding behavior" rather than "cell cleaning."

In Brazil, on the campus of the Universidade Federal de Minas Gerais, philopatry seemed more important than soil hardness in choice of a nesting site, though the many other aspects of soil quality were not investigated (Pimenta and Martins 1999). Females making an outer closure used the tip of the abdomen to compact the soil. From the third day, females made short flights to discard prey fragments from the nest. Most nests the authors excavated were unicellular (Figure 6.1G), but two nests had a second cell. Burrow length averaged 10.6 cm, cell depth 5.3 cm. Provisioning was progressive and took from 8–10 days to complete.

Prey records were provided in all three of the studies cited above. The 206 records of Evans et al. (1974) from multiple sites included 32 species of flies of eight families: 4 *Villa* sp. (Bombyliidae); 4 *Cochliomyia macellaria (F.),* 2 *Phaenicia* sp. (Calliphoridae); 1 *Morellia scapulata* Bigot, 9 *Musca domestica* L., 8 *Stomoxys calcitrans* L. (Muscidae); 1 *Oxysarcodexia* sp., 2

Sarcophaga sp. (Sarcophagidae); 3 *Hedriodiscus chloraspis* Wiedemann, 65 *Hedriodiscus pulcher* Wiedemann, 8 *Hermetia illucens* L., 12 *Hoplitimyia fasciata* F., 5 *Hoplitimyia subalba* Walker, 15 *Labostigmina inermis* Wiedemann, 4 *Stratiomys connexa* van der Wulp (Stratiomyidae); 3 *Eristalis erraticus* Curran, 1 *Eristalis tenax* L., 1 *Eristalis testaceicornis* Macquart, 1 *Eristalis vinetorum* F., 12 *Eristalis* spp., 3 *Quichuana aurata* Walker, 16 *Volucella obesa* F. (Syrphidae); 1 *Esenbeckia prasiniventris* Macquart, 1 *Leucotabanus exaestuans* L., 2 *Tabanus nebulosus* DeGeer, 4 *Tabanus claripennis* Bigot, 2 *Tabanus lineola* Bellardi, 4 *Tabanus colombensis* Macquart (Tabanidae); 1 *Cylindromyia* sp., 2 *Protogoniops* sp., 9 *Ptilodexia* sp. (Tachinidae). There were also several records of the use of nondipterous prey. On Trinidad, one female was seen carrying a dragonfly (Libellulidae: *Perithemis moona* Kirby), another a small skipper (Hesperiidae: *Panoquina* sp.); in Argentina, one nest contained the wings of a skipper (*Monca* sp.) (along with flies and a wasp larva), whereas another nest had the wings of two moths (Pyralidae).

In Argentina, Genise (1979b, 1980) noted that prey used varied among nests and across seasons. In one aggregation in 1976, nine nests were stocked solely with stratiomyid flies (*Hermetia* sp.), whereas one contained a variety of Diptera, mostly Syrphidae and Stratiomyidae (*Lobostigmina* sp., but no *Hermetia*). In another aggregation in 1979–1980, prey were principally Syrphidae (*Eristalis* sp.) during January and February, but Tabanidae during the end of February and March. Soares and Martins (1995) described cocoon-spinning behavior in *R. nasuta* in some detail, finding the behavior consistent with that described for species of *Bembix*. In Brazil, each larva was provided with about 50 prey, mostly Syrphidae and Stratiomyidae (Pimenta and Martins 1999). Flies of nine families were recorded as prey, including those from the following genera: *Cerotainia, Ommata* (Asilidae); *Anthrax, Exoprosopa, Phthiria, Villa* (Bombyliidae); *Chrysomyia, Lucilia, Phormia* (Calliphoridae); *Hedriodiscus, Hermetia, Hoplitimyia* (Stratiomyidae); *Copestylum, Ornidia, Palpada, Meromarcus, Spilomyia, Volucella* (Syrphidae); *Chrysops, Tabanus* (Tabanidae); *Cylindromyia*, and *Trichopoda* (Tachinidae). In another study, females either searched for prey during continuous flights along pathways where prey density was low or made more restricted searches within areas of higher prey density, such as flowers and feces (Fontenelle and Martins 2002). Females usually attacked prey that were perched on vertical stems, approaching them from about 10 cm below and 25 cm away. Captured prey were stung twice ventrally.

Pimenta and Martins (1999) found males patrolling with rapid zigzag flights about 20 cm above the ground. Marked males patrolled a particular group of nests and struck at other males that intruded, which sometimes resulted in two males wrestling on the ground. When a male was removed from a territory, he was quickly replaced by another (up to six males in 5 h when this was repeated). Males also struck other insects that approached their territories.

Overview of *Rubrica*

Pimenta and Martins (1999) reviewed the observations of Bodkin (1917) on *R. denticornis* and Llano (1959) on *R. gravida* (Handlirsch). *Rubrica denticornis* and *R. gravida* construct unicellular nests, but females in several populations of *R. nasuta* often make multicellular nests. *Rubrica nasuta* and *R. denticornis* nest in aggregations in relatively compact soil, whereas *R. gravida* occupy well-drained sand. In all three species, provisioning is progressive and females prey primarily on flies. However, observations by Evans et al. (1974) that *R. nasuta* may take a variety of other insects hint at a high degree of behavioral flexibility in this species. Observations at multiple locations confirm that *Rubrica nasuta* females clean cells, removing leftover prey fragments, during the process of progressive provisioning.

Selman

Selman notatus (Taschenberg)—Neotropical (Brazil to Argentina)

The single species in this genus, whose females produce a high-pitched whine in flight, has been the subject of two short reports, both since 1966. In the province of Salta, Argentina, Evans and Matthews (1974) found *S. notatus* nesting in coarse, sandy soil. One completed nest had an open accessory burrow beside the entrance. The burrow was 15 cm long, leading to a 9 cm deep cell that contained a partially grown larva with several flies, all one species of Stratiomyidae (*Hedriodiscus pulcher* Wiedemann). Genise (1981a) studied this species in the province of Entre Rios, Argentina, where nests had burrows 11–15 cm long and cells 7–10 cm deep, and the egg was laid on the first fly placed in the cell. The nests were provisioned progressively with flies (Asilidae, Tabanidae, and Therevidae).

Stictia

Stictia is a genus of 28 species, all but one of which are Neotropical. These large and conspicuous wasps have attracted the attention of many observers, so that more than 20 reports have been published since Evans (1966a).

Stictia arcuata (Taschenberg)—Neotropical (Brazil to Argentina)

Genise (1981b) describes a nest in Argentina that had a 30 cm long burrow entering the soil at a 30° angle and terminating in a cell 15 cm deep. During digging of the nest, sand accumulated in the lower tunnel, so that the wasp periodically moved back to push it outside. The unicellular nests were provisioned progressively with Diptera, the principal type being *Eristalis* sp. (Syrphidae), the most abundant fly on local flowers. The egg was laid on the first prey in the cell. Genise provides circumstantial evidence that *S. arcuata* may sometimes construct two-celled nests.

Stictia carolina (Fabricius)—Nearctic (southeastern United States to New Mexico)

Stictia carolina, commonly known as the "horse guard" for its habit of hawking flies around livestock, received extensive coverage in Evans (1966a). The reports reviewed there identified *S. carolina* strictly as a predator of flies, 80% of which were Tabanidae. Recent reports suggest otherwise. C. S. Lin (1971) studied three populous aggregations of the Nearctic *S. carolina* (F.) in Oklahoma. Several hundred males flew in irregular patterns over the sand and from time to time oriented to spots where a female was about to emerge. Males sometimes formed a ball surrounding a female; when a pair made contact, they would fly off to a tree, where mating lasted 20–30 s. Nests were unicellular, burrows 30–78 cm long, cells 17–30 cm deep. The egg was laid flat in the center of the empty cell and provisioning was progressive, lasting 3–5 days. Although diverse Diptera were the usual prey, Lin found that some females preyed on small, brown cicadas, *Melampsalta calliope* (Walker) (Cicadidae), or brown skippers, *Atalopedes campestris* (Boisduval) (Hesperiidae). He hypothesized that the departure from the usual prey was the result of the high population density and an inadequate supply of flies. Most cells he excavated contained only flies, but four contained cicadas and one skippers. Ten years later, Hook (1981b) also found *S. carolina* taking similar nondipteran prey in New Jersey. Diptera were present in all nine nests he excavated. But four nests also contained skippers or their remains, one of which was *Epargyreus clavus* Cramer.

Interested in the impact of *S. carolina* on horse flies around cattle in Louisiana, Roberts and Wilson (1967) attempted to determine wasp densities that would effect significant control. When counting the number of wasps, both *S. carolina* and *Bembix texana*, around the cattle, they observed as many as five per animal during censuses. Roberts and Wilson estimated that densities of ≥2 wasps per animal "noticeably reduced the tabanid population," though there was marked seasonal and spatial variation in wasp densities. During a three-day study, nine marked *S. carolina* females brought an average of 3.1 tabanids per day to their burrows, a total of 85 flies; no other prey were used. *Bembix texana* brought a greater number of flies (3.9 per day), but only 7% were tabanids, so most horse fly control was due to *S. carolina.*

Stictia flexuosa (Taschenberg)—Neotropical (Argentina, Brazil)

In Argentina, females formed restricted groups of several nests (Genise 1982c). Genise describes a nest with an oblique main burrow 55 cm long leading to a 1.8 × 3.5 cm cell that was 23 cm below the surface (Figure 6.1A). Nests were provisioned progressively with a variety of Diptera, principally Tabanidae.

Stictia heros (Fabricius)—Neotropical (Central America to Brazil)

In a coastal area of Guanacoste, Costa Rica, more than 1500 females nested on a sand flat near the ocean (Sheehan 1984). Nests were unicellular, burrows 49–105 cm long, cells 32–67 cm deep. The egg was laid on the side of the first fly brought in and provisioning was progressive. Cell cleaning usually followed the initial inspection trip in the morning and occurred late in provisioning, usually close to the last day. In Sheehan's study, flies of six families served as prey, Calliphoridae, Sarcophagidae, Stratiomyidae, Syrphidae, Tabanidae, and Tachinidae. Females sometimes ventured more than 1 km inland in their search for tabanids around horses, but flies were also taken at a nearby garbage pit. At a short distance from the site of Sheehan's study, Larsson and Larsson (1989) marked and measured many females and found that those with lower wing-load (high wing length/head width ratio) tended to be active early in the morning. Also, females with high ratios tended to nest closer to their neighbors than those with low ratios, suggesting that individuals choose nest sites in response to the size of their neighbors.

Also in Guanacoste Province, Costa Rica, Villalobos and Shelly (1996) observed several instances of predation on house flies in a village, but within aggregations, prey were also obtained through theft. Females carrying prey were often pounced upon by other females, who sometimes stole the prey and carried it to their own nest. Female "marauders" also entered the nests of other females, removing fresh flies for use by their own offspring. Some females persistently robbed other nests; one marked female made four raids into two nests within 13 min, and later made seven raids into three different nests in 10 min. The probability that returning *S. heros* females would be attacked by prey thief was related to the size of the prey being delivered. Prey that elicited attack averaged 12.1 mm in length, while prey that were delivered safely averaged 9 mm. Competition was so intense that a female returning with prey was sometimes attacked by two or three females simultaneously. Marauders usually raided nests near their own nest and tended to concentrate on nests that had previously yielded prey. Chance meetings between nest owners and marauders were rare, but when they occurred the owners chased the marauders away. Outer closure of nests appeared to function to deter marauding. Considering both the large population size and the fact that two other species of sand wasps nested here (*S. signata* and *Bembix multipicta*), there may have been severe competition for flies of suitable size. *Stictia signata* females were seen on eight occasions to steal prey from *S. heros.*

Larsson (1989a, 1991) reported that although males perform a typical "sun dance" in the morning hours, most males disappeared after about 1045 h. The few remaining males then hovered in fixed positions for another hour or so, as described for other species such as *S. signata* and *S. vivida* (Evans 1966a). Larsson interpreted this as a conditional alternative mating strategy related to the increasing midday temperatures. Males that assumed territorial hovering had larger body size and longer wings compared with males patrolling earlier in the morning. The larger, hovering males were probably able to reduce their thoracic heat and could continue mate-seeking for a longer period.

Stictia maccus (Handlirsch) (= *lineata* Fabricius)—Neotropical (French Guiana to Argentina)

In the province of Salta, Argentina, females nested solitarily or in small groups in a variety of places where the soil was sandy (Evans and Matthews 1974). Burrows were quite long, with cells at depths of about 30 cm, and nests were provisioned with stratiomyid flies (including *Hedriodiscus pulcher*). A nest in Tucuman Province was shallower (17 cm) and provisioned with Bombyliidae, Tabanidae, and Tachinidae. At both sites, females maintained nest closures and had thrown up a large mound of soil at the entrance. One such mound measured 12 × 15 cm and was 9 cm high in the center. At one end were three accessory burrows, the deepest 7 cm long, and on the sides were broad depressions where the wasp had dug sand and scraped it onto the mound. Another mound was still larger.

Stictia maculata (F.)—Neotropical (Mexico to Brazil, Peru)

On the Osa Peninsula of Costa Rica, a group of about 10 females nested in sandy loam in a forested area about 20 m from the Pacific Ocean (Matthews et al. 1981). Nests were unicellular and widely spaced, mostly in shady locations. Burrows were 52–54 cm long, with cells at depths of 30–34 cm. Mounds at nest entrances were leveled completely and, in some cases, camouflaged by the wasps' picking up twigs and bits of leaf matter and moving them over the closed nest entrance. Cells were provisioned progressively with a variety of large flies, but no intact specimens were present in excavated cells. At least some flies were taken near an open-air privy, where flies were captured and stung in midair.

Stictia pantherina (Handlirsch)—Neotropical (Colombia to Brazil)

In Trinidad, *S. pantherina* tended to have dispersed nests, though groups of 2–3 nests were found (Callan 1991b). At one site, females nested "in the compacted sand of a small stream-side beach," whereas at another they occupied "hard-packed, sandy soil" in a grassland. At the latter site, nests were often close beside those of *Rubrica nasuta*. The one nest that was excavated had one cell at a depth of 20 cm that contained a large larva and fragments of syrphid flies. Other females were seen with Calliphoridae and Syrphidae, including *Quichuana aurata* Walker.

Stictia punctata (F.)—Neotropical (Mexico to Argentina)

Evans and Matthews (1974) studied this species briefly in Colombia, finding about 10 females occupying a steep slope, the nests 1–2 m apart. The soil was a mixture of clay and soft rock; holes made by the females were left open at all times. Females bringing in prey produced a loud, low-pitched hum and plunged directly into the open burrows. Cells were quite deep, one at a depth of 35 cm, two others at depths of 45 and 60 cm. Prey consisted Tabanidae, Tachinidae, and Syrphidae, the last including flies of the genera *Crepidomyia, Eristalis,* and *Volucella.*

Amarante (1996) published a brief account describing "an interesting hunting tactic" of *Stictia punctata* in which he described females hunting flies near garbage, particularly bits of food dropped from picnic tables along the seashore. Here they were joined by hunting females of *Rubrica nasuta, S. signata,* and *Microbembex.* His description bears repeating in detail, partly because it appears in a difficult-to-find source. Whether it represents a learned behavior is not clear at this time.

> On one sunny afternoon, . . . I was lazily tasting some fried shrimp [at a picnic table] when . . . [a female *S. punctata*] . . . flew facing the table and moved laterally, left and right, always hovering below the upper surface of the table. At intervals, she would quickly rise above the table surface and promptly return below it. When she detected a fly on the table, she immediately backed down and flew to a point closest to her prospective prey, doing so below the upper surface and out of view of the fly. Reaching the point closest to the fly, she quickly rose above the table, dove straight at the fly, grasped it with her legs and flew away carrying her prey. After that, I started to watch these creatures with increasing interest, and I noted that this was a frequent behavior, but only by females of *Stictia* [*punctata*].

Stictia signata (L.)—Nearctic, Neotropical (southern United States and West Indies to Argentina)

One of the most widespread of *Stictia, S. signata* (Figure 6.4) occurs on sea beaches and sandy stream-sides. Evans (1966a) reviewed information from a variety of locations where females dug unicellular nests 14–35 cm deep and stocked them with various Diptera. Recent work has been similarly

Figure 6.4. Female *Stictia signata* at nest entrance from St. John, U.S. Virgin Islands. Photo by H. E. Evans.

widespread, having been done on St. John (U.S. Virgin Islands) Puerto Rico, Cuba, and in the Bahamas, Mexico, Costa Rica, Trinidad, and Brazil.

On San Salvador Island, Bahamas, females nested at the edge of a softball field in sand, and cells varied in depth from 13 to 27 cm; the completeness of temporary outer closures by provisioning females varied (Elliott et al. 1979). On St. John, in February 1969, HEE (previously unpublished) found *S. signata* nesting in small numbers behind a beach in hard-packed sand, where some nests were just 20 cm apart. One female making a final closure over a 10 min period scraped soil from a quarry opposite the burrow and from a ring around the entrance, then packed it in place with vigorous blows of the deflected tip of the abdomen. Two nests excavated here had burrows 21 and 26 cm long, and single cells 13 and 16 cm deep. Another nest excavated by HEE (previously unpublished), at Bahia las Cabezas, Puerto Rico, had a burrow 23 cm long and a cell 15 cm deep. On Trinidad, *S. signata* was an inland species that was more or less a solitary nester at the time of Callan's (1991b) obsevations. However, on the sandy shores of Quintana Roo, Mexico, there were a large number of nests, with burrows averaging 33 cm in length and cells 21 cm in depth (Martin and Martin 1990). Similarly, near Santarem, Brazil about 50 females aggregated along a gully (Post 1981). Among 23 nests excavated, 3 had two cells, but 20 had just one. Cell depth varied from 13 to 38 cm, and were provisioned progressively with 3–9 flies per day; a total of 16–44 fly thoraces were found in cells containing cocoons.

On San Salvador Island, Bahamas, where females hunted at a garbage dump and near vertebrate carrion, prey were Calliphoridae, Muscidae, Sarcophagidae, and Syrphidae. On St. John, prey in both nests excavated consisted of the calliphorid *Cochliomyia macellaria,* whereas in Puerto Rico nests contained both *C. macellaria* and the syrphid *Eristalis vinetorum* F. (Syrphidae). In Mexico, prey consisted of *Chrysomyia megacephala* (F.), *C. macellaria* (Calliphoridae); *Helicobia* sp., *Muscopteryx* sp. (Tachinidae); and three unidentified Sarcophagidae. In Brazil, five families of flies served as prey, with Syrphidae being most abundant. Identified prey species included: *Cochliomyia macellaria* (F.), *Hemilucilia segmentaria* (F.) (Calliphoridae); *Oxysarcodexia fringidea* (Curran and Walley), *Paraphrissopoda chrysostoma* (Wiedemann), *Paraphrissopoda retrocita* (Hall), *Xarcophaga favorabilis* (Lopes) (Sarcophagidae); *Copestylum pallens* (Wiedemann), *Palpada* sp., *Ocyptamus climidiatus* (F.), *Ocyptamus flavidipennis* (Wiedemann), *Ornidia obesa* (F.) (Syrphidae); *Acanthocera* sp., *Tabanus consequa* Walker, *Tabanus guyensis* Macquart (Tabanidae); *Acroglossa* sp., *Prophasiopsis* sp., *Uruleskia* sp. (Tachinidae). Syrphids were the most common prey and *C. pallens* was the most common species in the provisions (30% of 30 intact specimens and 64% of 636 prey fragments). Elliott et al. (1979) watched females capturing and stinging prey in midair, and then flying off with the prey "still inserted on the sting."

Philippi and Eberhard (1986) made an experimental study of *S. signata* hunting behavior in a pasture and landing strip in Costa Rica, where Tabanidae were being captured from horses and humans. Visual cues were clearly involved in locating hunting sites and prey, as evidenced by the fact that wasps were more attracted to persons wearing dark clothes than light. Further, when responses to model prey were tested, females responded more readily to 7 mm long black cylinders than to 14 mm long black cylinders or to tan cylinders; the short, black cylinders more closely approximated the size of the usual tabanid prey. Cane and Miyamoto (1979) had worked at this same site in Costa Rica a few years earlier, noting that females searched up and down the legs of horses and would also search up and down other dark, vertical objects, such as fence posts. They identified the hunting wasps they observed as *Bembix multipicta,* which had been identified at nest sites, but Philippi and Eberhard suggest that the hunting females may have been *S. signata.* Cane and Miyamoto noted that as many as 12 wasps foraged around a singe horse at any one time, and that up to 21 wasps were active in a group of ten horses.

Evans (1966a) reviewed the scant information on nest associates of *S. signata,* citing a few records of maggots inhabiting cells but noting that identification of the flies was inconclusive. On Caimito Beach, South Havana Province, Cuba, Genaro (1999) observed that the choropid fly *Liohippelates* sp. occurred in all nests and fed on prey detritus, but did not harm larvae of the wasp; a eucoilid wasp *Hexacola* sp. was also present, but was parasitizing the chloropid larvae.

In Brazil, male *S. signata* hovered in restricted locations about 1 m in diameter and 0.5 m above the ground in the nesting area, during the morning hours. Interactions between males were rare, but sometimes involved head butting and grappling. Post (1981) also found evidence that males are territorial. After removing nine males from their hovering stations, they were replaced within 5–70 s by other males that then adopted the location as their own hovering station. However, when the original resident was released, he returned to his station and expelled the intruder. The experimental removal experiments were natural equivalents of a phenomenon that Post frequently observed: males often left their stations for short periods, only to be replaced while gone by other males. No mating was observed, which is not unusual in solitary wasp studies, but males often chased and pounced on females, often interfering with nesting activities. It seems likely that the nesting females were unreceptive at the time of Post's observations, having mated earlier in their lives.

Post monitored activities of individually marked males over a period of 16 consecutive days. When he analyzed his census data for the 0900–1000 h period, he found that one marked male ("YR") occupied the same station on 13 days, and at one point for eight consecutive days. Similarly, "W" was present at the same station for 14 days, including two stretches of seven straight days. This contrasts with "BW," who showed up on three different territories on six different days, and was never present at the same station for more than three straight days. There was similar variation in fluidity of movement within days. On 11 November 1978, "W" spent 28 of 31 five-minute census periods on territory no.4, and was away from the nesting area for the other three, perhaps feeding on flowers. During the same period, male "RM" was less site-specific; he divided his time between two territories, being present on one during 12 census intervals and another for seven. Similar forms of both site specificity and movement among territories by males have been observed in the territorial systems of other male Hymenoptera, including cicada-killers (N. Lin 1963a), tarantula hawk wasps (Alcock 1979, 1981), beewolves (Gwynne 1978, 1980; O'Neill 1979, 1983a,b), and bumble bees (O'Neill et al. 1991). Other observations of males were provided by Martin and Martin, who observed males making sinuous patrolling flights from 0615 h to late morning, at which time they began stationery hovering flights. Hovering males were spaced about 4 m apart and interacted with neighboring males.

Overview of *Stictia*

All species of *Stictia* build unicellular nests, except *S. signata*, for which both unicellular and bicellular nests have been reported. Most species commonly nest in bare sand or sandy loam, often on beaches, placing nest cells at depths of <40 cm. The major exception to this rule is *S. punctata*, which was found nesting in clay and soft rock and provisioning cells at depths of up to 60 cm. Progressive provisioning has been reported in eight species. The number of prey per cell, often difficult to judge for progressive provisioners, is largely unknown, though a range of 15–63 has been reported for *S. carolina*, and up to 44 prey have been found in *S. signata* cells. All species use Diptera as prey (Table 6.2), with the exception of *S. carolina*, which have also been reported using skippers (two reports) and cicadas. The addition of cicadas and skippers may have been a response to shortages of normal prey. It is informative that females of two different populations of *S. carolina* added skippers to their diets, as it suggests that there is something "fly-like" about skippers that facilitates their addition to the wasps' diet.

A number of interesting points emerge from recent studies of the behavior of *Stictia*. While there is much uniformity in nest structure, it is clear that in at least *S. carolina* the egg is laid in the empty cell, though in other species, the egg is laid on the initial prey. A number of unusual features are reported for individual species: prey stealing and nest marauding in *S. heros* and *S. signata*, cell cleaning in *S. heros*, construction of large mounds at nest entrances *in S. maccus*, the occasional use of nondipterous prey in *S. carolina*, bicellular nests in *S. signata*, and a novel hunting technique by females of *S. punctata*.

Table 6.2 Families of Diptera reported as prey of *Stictia*. See text and Evans (1966a) for references and details.

	Diptera								Lepidoptera	Homoptera
Stictia species	Calliphoridae	Culicidae	Muscidae	Sarcophagidae	Stratiomyiidae	Syrphidae	Tabanidae	Tachinidae	Hesperiidae	Cicadidae
S. arcuata						X				
S. carolina	X	X	X	X	X	X	X	X	X	X
S. flexuosa[a]							X			
S. heros	X				X		X			
S. maccus					X					
S. maculata										
S. pantherina	X					X	X			
S. punctata						X	X	X		
S. signata	X		X	X	X	X	X	X		
S. vivida	X			X			X			

[a]Genise (1982c) notes that prey are principally Tabanidae, implying that other prey are also used.

Editha

Editha contains seven species restricted to South America (Pulawski 2006a), and is one of three genera of Bembicini that prey solely on adult Lepidoptera, though it is the only one that specializes on butterflies (Table 6.3).

Editha fuscipennis (Lepeletier)—Neotropical (Brazil, Paraguay)

Martins (1993) mentions unpublished observations that this species preys on adult butterflies (Lepidoptera).

Editha integra (Burmeister)—Neotropical (Argentina, Brazil)

In the province of Entre Rios, Argentina, *E. integra* nests were dug in sandy soil covered with grass. Females took Lepidoptera of the families Erycinidae, Hesperiidae, Nymphalidae, Pieridae, and Satyridae (Genise

Table 6.3 Families of Lepidoptera reported as prey of species of *Editha, Microstictia,* and *Stictiella.* See text and Evans (1966a) for references and details.

	Moths										Butterflies							
Species	Crambidae	Epipaschiidae	Gelechiidae	Geometridae	Noctuidae	Olethreutidae	Phycitidae	Pterophoridae	Pyralidae	Pyraustidae	Erycinidae	Hesperiidae	Libytheidae	Lycaenidae	Nymphalidae	Papilionidae	Pieridae	Satyridae
Editha																		
E. adonis												X						
E. integra											X	X			X		X	X
E. magnifica												X	X	X	X	X	X	
Microstictia																		
M. minutula			X	X		X		X	X									
Stictiella																		
S. callista												X		X				
S. emarginata					X													
S. evansi					X					X								
S. formosa												X	X	X	X			
S. pulchella					X					X				X				
S. serrata	X	X				X	X			X								

1981a); Genise also cited one record of predation on Papilionidae. His photographs show females taking prey into the nest, an egg laid on the first prey taken to a cell, and a larva surrounded by the wings of butterflies. The nest was bicellular, with a burrow 20–25 cm long.

Editha magnifica (Perty)—Neotropical (Brazil)

Editha magnifica is the largest bembicine wasp, with some females reaching 4.5 cm long. In Sao Paulo Province, Brazil, females nested in vertical earthen banks along a dirt road (Martins 1993). The mean distance between neighboring nests for 133 nests was 0.6 m. Although females dug nests as close as 2 cm apart, no aggressive encounters were observed between neighbors. Nests were unicellular, burrows averaging 32.8 cm in length, and were dug at about a 45° angle into the bank. Nest entrances were closed between provisioning trips, and at final closure, soil was scraped into the hole and pounded in place with strokes of the tip of the abdomen. On average, females spent about 2 weeks progressively provisioning a nest with prey belonging to 6 families: Hesperiidae, Libytheidae, Lycaenidae, Nymphalidae, Papilionidae, and Pieridae. Skippers made up slightly more than half of the prey. Lepidoptera were taken primarily on shrubs (*Vernonia*), and both males and females took nectar there. In Mato Grosso, Brazil, Richards (1971) observed this species hunting Pieridae. An earlier record for *E. magnifica* preying on Homoptera (Bondar 1930) is likely an error.

Early in the reproductive period, males far outnumbered the females and patrolled the nesting area (Martins 1993). Later, the sex ratio was one male to two females; during this period some males became territorial, defending clusters of nesting females. Resident males were aggressive toward conspecifics and other insects that flew over the site. During mating, the male held his middle legs around the female, the notch on the mid femora holding down the female's wings and preventing her from flying.

Trichostictia

The three species of this genus are confined to the southern parts of South America (Pulawski 2006a). No biological information was available for *Trichostictia* in 1966, but several papers have since rectified this.

Trichostictia guttata (Taschenberg)—Neotropical (Brazil to Argentina)

In the province of Entre Rios, Argentina, an aggregation in sandy soil with a density of about one nest per square meter was reported (Genise 1981a). Nesting took place in late summer (March), when temperatures were cool and other Bembicini had finished nesting; females were active mostly in the morning and late afternoon. Nests, which were in shadows of bushes, were unicellular; the egg was laid on the first prey in the cell. Prey were flies of the families Bombyliidae, Syrphidae (*Eristalis* sp.), and Tabanidae.

Trichostictia vulpina Handlirsch—Neotropical (Bolivia, Chile, Peru)

At a site 1500 m in altitude in the province of Cordoba, Argentina, nests were unicellular and the egg was laid on the first prey in the cell (Genise 1985). Prey consisted of flies of the families Bombyliidae and Syrphidae (*Volucella* sp.). Like *T. guttata* and *Zyzzyx chilensis,* this species appears adapted to cool temperatures.

Zyzzyx

Zyzzyx chilensis (Eschscholz)—Neotropical (Argentina, Chile, Peru)

Zyzzyx chilensis, the sole member of its genus, was studied first by Janvier (1928) in Chile. Genise (1982e) has clarified some of the more questionable aspects of Janvier's report and added several details from studies in the province of Mendoza, Argentina. Nests were dug in sandy soil adjacent to bushes; like *Trichostictia*, this species appears adapted to activity at low temperatures. Burrows were about 21 cm long and the single cell about 17 cm deep (Figure 6.1D). The egg was laid in an erect position on the first prey in the cell, the prey consisting of flies. These observations contrast with those of Janvier, who had reported multicellular nests, the egg laid in the empty cell, and skippers as prey from one nest. As noted in Evans (1966a), reports by Janvier on this and other wasps need confirmation.

Stictiella

This is a genus of 14 species that are well known to prey on adult Lepidoptera (Gillaspy et al. 1962) and that collectively range from southern Canada to Mexico (Pulawski 2006a). Evans (1966a) reviewed the substantial data on *Stictiella* known at the time, but relatively little data have appeared since.

Stictiella callista Parker—Nearctic (southwestern United States)

On sandy floodplains of the Gila River in Arizona, a female was seen making a fairly elaborate closure by repeatedly landing on the sand at the nest entrance and moving away a distance while kicking sand behind her (Alcock and Gamboa 1975). The burrow was 68 cm long, reaching a 40 cm deep cell that contained a recently hatched larva attached to the side of a skipper (Hesperiidae).

Stictiella emarginata (Cresson)—Nearctic (southern Canada, United States)

Stictiella emarginata was one of two species of *Stictiella* in Evans (1966a) for which scant prey records were available, in this case a single record of *Euxoa incallida* (Smith) (Noctuidae) (Gillaspy et al. 1962). We found confirming evidence of noctuid use in Larimer County, Colorado, in the form of two female *S. emarginata* seen flying with adult noctuids or "miller moths" *(Euxoa auxiliaris)* held in an inverted position beneath their bodies (HEE, previously unpublished). Neither female could be followed to a nest. During early summer, *E. auxiliaris* migrate in numbers from the plains of North America into the Rocky Mountains.

Kurczewski and Boyle (2006) recently provided many more details on the biology of *S. emarginata,* which they studied in Ontario, Canada, and upstate New York at sites with soil described as "loamy sand" or "coarse loamy sand." Before digging burrows, females often made multiple "false starts," while walking on the sand, tapping the soil surface with their antennae, and breaking into it with their mandibles. Once settled down to a single location, females spent 50–69 min digging burrows, before constructing a temporary closure (0.25–0.75 min) and leveling the mound (5.5–11.0 min). Nests had one or two cells 4–7 cm deep and burrows 6–14 cm long that entered the surface at angles of 45 to 60°.

Females provisioned nests from mid-morning to early evening, spending from 9 to 77 min on foraging flights. Prey were carried venter up using the middle legs. Prey records, all adults, included 8 butterflies (Hesperiidae: *Polites thermistocles* (Latreille)) and 124 noctuid moths of 7 species: 8 *Apamea ampulatrix* Fitch, 7 *Apamea ophiogramma* (Esper), 25 *Calophasia lunula* (Hufnagel), 46 *Chytonix sensilis* Grote, 3 *Euxoa scandens* (Riley), 14 *Lacinipolia vicina* (Grote), and 21 *Nedra ramosula* (Guenee). The egg was laid laterally near the base of the abdomen of the first prey placed in cells. The 5–9 prey provided per cell weighed on average 87 g, whereas as the total mass of provisions placed in a cell averaged 582 g; two adult female wasps, in contrast, weighed 147 and 151 g. Five of 26 cells were parasitized by 2–4 miltogrammine flies each, two of which were reared to adulthood and proved to be *Senotainia* (= *Sphixapata*) *vigilans* Allen.

Stictiella formosa (Cresson)—Nearctic (central United States)

At Riviera Beach, Kleburg County, Texas, in May 1979, a female was seen arriving at a nest with a butterfly whose wings were folded back (HEE, previously unpublished). The entrance had been closed, so she scraped it open, entered, and closed it behind her. The mound of soil at the entrance was 6 × 10 × 1.5 cm deep. The nest had two cells that appeared to have been made in tandem, one behind the other. The burrow was 25 cm long, the cell 11 cm deep. Cells measured 12 × 35 mm and were both stuffed with butterflies. Both cells contained an egg on one of the bottom prey. The deepest cell contained two snout butterflies (Libytheidae, *Libytheana* sp.), one Lycaenidae, and five skippers (Hesperiidae) of three species. The other cell contained one lycaenid and three skippers of three species. Previous prey records included Hesperiidae, Libytheidae, Lycaenidae, and Nymphalidae (Gillaspy et al. 1962).

Stictiella pulchella plana (W. Fox)—Nearctic (California to Colorado)

Evans (1966a) was able to give some data on the nests and prey of this species from several sites in Texas, where it constructed the longest and deepest burrows of any Stictiella. No other information has appeared since then, with the exception of the fact that we found four females of *S. pulchella* as prey of the beewolf *Philanthus basilaris* Cresson at the Great Sand Dunes National Monument, Colorado (Evans and O'Neill 1988).

Overview of *Stictiella*

It appears that mass provisioning is usual among species of *Stictiella*, although provisioning may extend over one day in some species. Multicellular nests have been reported in *S. emarginata* (up to 2 cells), *S. formosa* (as many as 17 cells), and *S. pulchella* (1 two-celled nest among the four examined). Six families of moths and four families of butterflies have been reported as prey, and none have been added to the list since 1966, though prey records for *S. emarginata* have expanded to Hesperiidae (Table 6.3). Prey are placed in cells with their wings intact. As in other predators on Lepidoptera, the wings are not consumed but are left to surround the cocoon. As reported in 1966, fully provisioned cells, which were no deeper than 20 cm in any of the four species discussed, contained as many as 11 prey in *S. formosa*, 15 in *S. evansi*, 19 in *S. pulchella*, and 21 in *S. serrata*. No higher values have been reported since.

Microstictia

This is a genus of 12 species confined to the southern and western United States and northern Mexico (Ohl 2005; Pulawski 2006a). No behavioral studies of *Microstictia* appeared until 1983.

Microstictia minutula (Handlirsch) (= *M. texensis* Gillaspy)—
Nearctic (Texas)

Near Kingsville, Texas, females nested on the banks of an intermittent stream (Gillaspy 1983; see also Bohart and Gillaspy 1985). The flight of females, both to and from the nest was distinctive: "Whining, hovering, and occasionally even backing flights were characteristic. Provisioning wasps approaching their nests with prey threaded their way noisily and rather slowly through vegetation, but departed abruptly with a burst of sound towards open areas" (Bohart and Gillaspy 1985). The authors go on to state that "were it not for this sound, it would be difficult for the observer to spot them in the field." Nest burrows were relatively short and the 1–7 cells per nest sat at depths of 6–17 cm. Females did not level nest mounds, and they often omitted construction of temporary nest closures during the peak of daily foraging.

Fully provisioned cells contained 10–11 prey, the egg being glued to the hind wing of the first prey placed in the cell. The 31 prey identified were small moths of the families Gelechiidae, Geometridae (*Idaea* sp., *Narraga fimetaria* (Grote and Robinson)), Olethreutidae, Pterophoridae (*Oidaematophorus* sp.), and Pyralidae, including Crambinae, Phycitinae, and Pyraustinae (*Blepharomastix magualis* [Guenée]); this list includes four families not known as prey of the two other bembicine genera that specialize on Lepidoptera (Table 6.3). Prey appeared to be completely paralyzed.

Glenostictia

This genus contains 21 species, most of which occur in the western United States and northern Mexico (Pulawski 2006a); seven were described as recently as 1983 (Bohart 1983). Evans (1966a) contains reports on three species, *G. pulla* (Handlirsch), *G. gilva* Gillaspy, and *G. scitula* (Fox). A further brief report has appeared on *G. gilva,* along with observations on three other species.

Glenostictia argentata (C. Fox)—Nearctic (California, Mexico)

In Maricopa County, Arizona, Alcock (1975c) found a nest with a 24 cm long burrow and 15 cm deep cell that contained a small larva and several flies, mostly Bombyliidae. The nest was evidently being provisioned progressively.

Glenostictia gilva Gillaspy—Nearctic (Arizona, California)

In Maricopa County, Arizona, Alcock (1975c) found a nest with a 30 cm long burrow and 21 cm deep cell that contained a bombyliid fly with an egg attached vertically to the side of its thorax. A second nest, located in a mound of sand in the same county, had a burrow only 10 cm long and a cell 5 cm deep with a small larva, prey remains, and seven Bombyliidae.

Glenostictia pictifrons (F. Smith)—Nearctic (eastern and southern United States to California)

This short-winged species is widely distributed in the United States. In flight, the wasps produce a distinct whining sound that changes pitch as the insect alternates between hovering and darting forward (Alcock 1975c). Alcock found no nests, but did observe females hunting and capturing flower-visiting flies. His description of hunting behavior was as follows: the female "hovers in one area for many seconds, changing position slightly while it scans the location. Failure to detect a fly causes the wasp to dart to a new hovering station. Upon spotting a fly, the female attempts to maneuver to a point where it has a clear *upward* route to the fly. It then moves *backward* 5–15 cm while continuing to orient its body on line with the intended victim; the wasp then charges rapidly forward and tries to pluck the fly from its perch or from the air, if the prey endeavors to escape."

Alexander et al. (1993) published a detailed study of *G. pictifrons*, conducted near Lawrence, Kansas, where several nests occurred in bare, sandy areas in a hay field. Nests were unicellular (Figure 6.1H), with cells at a mean depth of 4.1 cm and a mound at the entrance that was not leveled until the final closure. When a female completed one nest, the next one she dug was generally within 0.5 m of the first, but sometimes as far as 2–3 m away. Provisioning was progressive, over a period ranging from 4 to 9 days per nest (mean = 5.5), the higher values perhaps being caused by periods of rain. Prey included 18 species of Diptera in seven families: unknown species (Anthomyiidae); *Bombylius* sp., *Chrysanthrax cypris* (Meigen), *Exoprosopa meigenii* (Wiedemann), *Hemipenthes seminigra* (Loew), *Rhynchanthrax parvicornis* (Loew), *Villa lateralis* (Say) (Bombyliidae); *Neorhynchocephalus volaticus* (Williston) (Nemestrinidae); *Eristalis transversus* Wiedemann, *Toxomerus geminatus* (Say) (Syrphidae); *Musca autumnalis* DeGeer (Muscidae); *Senotainia trilineata* (Wulp), *Senotainia* sp. (Sarcophagidae); three different *Phasia* sp., *Distichona kansensis* (Townsend), and *Prosenoides assimilis* Reinhard (Tachinidae). The egg was laid on the side of the first fly placed in the cell (Figure 6.5).

Glenosticta satan Gillaspy—Nearctic (New Mexico)

In southwestern New Mexico, Bohart (in Bohart and Gillaspy 1985) found a large, tight aggregation that included >2000 males and a large number of females producing a loud, high-frequency hum that could be heard 10 m away. At a similar (or possibly the same) New Mexico site, Longair et al. (1987) studied a dense aggregation estimated to contain at least 2000 individuals. Marked males returned each day to the same restricted area, where they hovered and darted at intruders, and chased one another and other insects. Upon detecting an emerging female, several males would converge on her, forming a cluster of up to 35 males, some of which were damaged in the melee. When one male seized a female, the pair walked away undisturbed, while the other males dispersed; this contrasts with interactions with *Bembecinus quinquespinosus* (Chapter 5), in which males continue to harass the pair until it leaves the emergence area in flight. Males often mis-

Figure 6.5. Egg of *Glenostictia pictifrons* attached to base of left wing of prey (*Toxomerus* sp.). From Alexander et al. (1993); used with permission.

directed copulation attempts at one another and sometimes attempted to mate with dead wasps, fly puparia, or even bits of plant tissue. Nests of this species remain to be found.

Overview of *Glenostictia*

Evans (1966a) and Alexander et al. (1993) have reviewed the published studies of *Glenostictia* behavior. Soil type within nesting areas has been reported as powdery sand in *G. gilva,* "sandy" in *G. pictifrons,* "powdery sand or loam" in *G. pulla,* and powdery clay/sand in *G. scitula.* Nests always consist of a single cell that is generally shallow (always less than 8 cm in both *G. pictifrons* and *G. scitula*), and at the end of a burrow that is never longer than 30 cm and typically shorter than 15 cm. Nests are always closed at the entrance when the female is away hunting, but no inner closure is maintained. Females of three species discussed by Evans (1966a), *G. gilva, G. pulla,* and *G. scitula,* level the mound while digging, using "distinctive rotational leveling movements." *Glenostictia pictifrons,* in contrast, levels the mound only at final closure.

In all five species for which information is available, females provision progressively with Diptera. The best-studied species, *G. scitula,* has a truly remarkable diet breadth for a bembicine, taking prey from at least 27 families of Homoptera, Hemiptera, Diptera, and Hymenoptera; all four prey orders may be taken by a single female. The flies taken by *G. scitula* include the typical range of families preyed upon by many Bembicini (e.g., Anthomyiidae, Bombyliidae, Sarcophagidae, Syrphidae, Tabanidae, and Tachinidae), but also include several tiny Diptera (Chironomidae, Chloropidae). Similarly, the non-Diptera used are also small (e.g., Cicadellidae, Psyllidae, Miridae, Torymidae, Formicidae, and small andrenid bees of the genus *Perdita*).

Xerostictia

Xerostictia includes a single known species confined to the extreme southwestern United States and Baja California, Mexico. Our only information on the biology of *Xerostictia* appeared in 1975.

Xerostictia longilabris Gillaspy—Nearctic (Arizona, California)

In Maricopa County, Arizona, Alcock (1975c) found two nests, each with a burrow about 42 cm long, one with two cells at 28 and 30 cm depth, the other with a single cell at 30 cm depth (Figure 6.1C). Prey-laden females dropped down to the nest entrance, in a series of steps, from a height of about 40 cm, the pitch of the loud "buzzing, whining" sound made by their wings changing as they descended. Nests were provisioned progressively. One of the cells in the first nest contained a large larva, along with the remains of five adult antlions, *Brachynemurus longipalpis,* stacked head-first into the cell, as well as four flatid bugs, *Ormenis saucia.* The other nest contained a medium-sized larva and the remains of two antlions and two flatid bugs. Other than several Australian species of *Bembix* (Evans and Matthews 1973; Evans et al. 1982), *X. longilabris* is the only bembicine reported to prey upon antlions; species in certain genera of Gorytini and Stizini also use Flatidae. However, the use of both prey types by *X. longilabris* is remarkable, because antlions and flatid bugs not only belong to quite unrelated groups but are grossly dissimilar in body form.

Steniolia

The 15 species of *Steniolia* collectively range from southern Canada to northern South America. Wasps of this genus are noted for their exceptionally long mouthparts, allowing them to take nectar from flowers with deeper corollas than those accessible to most other Bembicinae (Evans and Gillaspy 1964).

Steniolia Provancher—Nearctic (southwestern United States, Mexico)

Evans (1966a) reported on nine nests of *S. duplicata* from California and Texas whose burrows were 7–18 cm long, 7–10 cm deep, and progressively provisioned with Bombyliidae, Muscoidea, and Syrphidae. One further observation since then confirms the earlier studies. HEE (previously unpublished) found a nest in dunes in Imperial County, California, with a 15 cm long burrow and 8 cm deep cell. It was being provisioned progressively with Bombyliidae and Syrphidae.

Steniolia elegans J. Parker—Nearctic (western and southwestern United States, Mexico)

On the slag pile of an abandoned mine in Larimer County, Colorado, Evans (1973) discovered an aggregation of about 100 individuals. Females reopened nests in the morning, hovering over the entrance and then digging through the closure; within 20–60 s they emerged and made a fresh closure. These were evidently "inspection trips," serving to inform the female of the needs of the larva. Within a few minutes to an hour, females returned with their first prey, removing the closure and restoring it as they left less than a minute later. Prey consisted of 20 diverse Bombyliidae, 3 Syrphidae, and 1 Asilidae. Nests were shallow and unicellular, burrow lengths 7–17 cm, cell depths 4.5–9 cm. Males patrolled the nesting site and sometimes hovered over females working at their nests, and on several occasions attempted copulations.

Steniolia longirostra (Say)—Nearctic (Mexico)

In Guanacoste Province, Costa Rica, 200–300 females occupied a 55 m length of sandy road, several hundred meters from the ocean (Larsson 1990). Females provisioned their nests progressively with Bombyliidae, mostly in the morning, when temperatures were low and humidity relatively high. A second activity period occurred in late afternoon, when females were digging and closing nests for the night. When body temperatures of females were recorded, females with low wing load were active at higher temperatures and, as a result, could provision nests for longer periods during the day, and ultimately complete more nests than those with higher wing load.

Steniolia nigripes J. Parker—Nearctic (California, Baja California)

In the Granite Mountains of southern California (1300 m elevation), females and males aggregated in areas of creosote *(Larrea tridentata)* and burrobush *(Ambrosia dumosa)* (Thomas and Nonacs 2002). Females nested in "hard packed dirt and gravel . . . not noticeably different from the surrounding areas." Many males patrolled close to the ground (2 cm high) in wide overlapping paths, often chasing and contacting other insects. Copulations ensued when multiple males contacted females and one male succeeded in coupling and left attached to a female. Unlike most aculeate wasps, males are larger than females on average, both in linear dimensions and in body mass. However, it is not clear from descriptions in the paper whether males carry females in flight during mating, as occurs in other species either with larger males or without sexual size dimorphism (O'Neill 2001).

Steniolia obliqua (Cresson)—Nearctic (Colorado to Pacific states)

Evans and Gillaspy (1964) and Evans (1966a) provided detailed accounts of this species, which digs with short burrows (6–11 cm) and shallow cells (3–5 cm) (Figure 6.1I) provisioned with Bombyliidae, Muscoidea, and Syrphidae. That information came from observations in Jackson Hole, Wyoming. HEE (previously unpublished) has since obtained data on nine nests from two sites in Boulder and Larimer counties, Colorado, from 2100 to 2400 m elevation. The observations confirm that nests are unicellular and provisioned progressively with Diptera (Bombyliidae in all nests). In the Colorado nests, burrow length varied from 7 to 10 cm and cell depth from 3 to 6 cm.

Steniolia tibialis Handlirsch—Nearctic (western United States)

In Sierra County, California, Tyler (1986) observed 11 sleeping clusters of 2–5 individuals, including both males and females on plants. Some individuals returned to the same sleeping location for up to three nights. These clusters are much smaller than those of *Steniolia obliqua,* which can contain hundreds of wasps (Evans 1966a).

Table 6.4 Families (or superfamilies) of Diptera reported as prey of *Steniolia* (updated from Evans 1966a). See text and Evans (1966a) for references and details.

Steniolia species	Asilidae	Bombyliidae	Muscoidea	Stratiomyiidae	Syrphidae
S. duplicata		X	X		X
S. elegans	X	X			X
S. eremica	X				
S. longirostra		X		X	X
S. nigripes		X			
S. obliqua		X	X		X
S. tibialis		X			X

Overview of *Steniolia*

Steniolia females nest in a variety of soil types, including "powdery earth, often filled with small stones and roots" with low sand content *(S. obliqua)*, "hard packed dirt and gravel" *(S. nigripes)*, "very dry powdery earth and crushed rock" *(S. longirostra)*, and sand or "coarse sandy soil" *(S. duplicata)*. Aggregations tend to be small. Nests are unicellular and shallow with short oblique burrows, always <20 cm long and <10 cm deep. Without exception, prey for the seven species for which records are available (*S. duplicata, S. elegans, S. eremica, S. longirostra, S. nigripes, S. obliqua,* and *S. tibialis*) are Diptera (Asilidae, Bombyliidae, Muscoidea, Stratiomyidae, and Syrphidae; Table 6.4). As is the case with *Glenostictia*, prey are either deeply paralyzed or killed, so that they often quickly desiccate. The eggs are apparently always laid on the first prey in the cell. One notable feature of *Steniolia* behavior is the formation of dense sleeping clusters in which they spend the nights and periods of cool weather. Clusters, which are often formed away from nesting aggregations, may include less than a dozen individuals, but clusters of 55–73 wasps have been recorded for *S. duplicata* and may reach sizes of ~500 individuals in *S. obliqua*.

— 7 —

Bembicini:

The Cosmopolitan Genus Bembix

With nearly 350 species worldwide (Pulawski 2005), *Bembix* is the largest genus of the Bembicini and one of the largest of the Crabronidae, trailing only *Cerceris* (>850 species), *Trypoxylon* (>600 species), and *Tachysphex* (~400 species), and being about equal in size to *Liris*. About two dozen species are known from North America, though much of our biological information comes from this area. *Bembix* occur on all continents (except Antarctica), as well as on some islands of the East and West Indies. Literature on the group is extensive. Early major contributions include those on *Bembix rostrata* (L.) by Nielsen (1945) and on *Bembix niponica* (Smith) by Tsuneki (1956–58), as well as studies on orientation in *B. rostrata* by van Iersel (1952) and Chmurzynski (1964, 1967). Lengthy reviews that include information on the biology of this genus include Evans (1957b, 1966a) and Evans and Matthews (1973). Because of the size of this genus, the organization of this chapter differs from that of preceding chapters: here, species of *Bembix* are treated by zoogeographic region.

Nearctic *Bembix*

Bembix americana F.—North America

Bembix americana is a polytypic species, its seven subspecies occurring throughout temperate North America and adjacent islands to the southeast and west. Five subspecies are discussed in this section, whereas a sixth *Bembix americana antilleana* is discussed in the section on Neotropical *Bembix*.

Bembix americana spinolae Lepeletier, of the United States and Canada east of the west coastal states, is one of the best studied of sand wasps. Most available information was reviewed by Evans (1957b, 1966a), who listed Asilidae, Bombyliidae, Calliphoridae, Muscidae, Otitidae, Sarcophagidae, Stratiomyidae, Syrphidae, Tabanidae, Tachinidae, and Therevidae as prey from eight localities in New York and Kansas. Among the 120 prey records for the Kansas population (30 species in 10 families), no species made up more than 15% of the prey. More than half the prey from the New York samples were Calliphoridae. Evans (1970) documented the prey from another aggregation at Huckleberry Hot Springs in Jackson Hole, Teton County, Wyoming (just south of Yellowstone National Park). Here about one-third of the prey were from two species of *Chrysops* (Tabanidae), but the overall list included 52 species of 12 families and no other species made up more than 6% of the list. The 12 families of prey (and number of species each) were: Asilidae (2), Anthomyidae (3), Bombyliidae (6), Calliphoridae (3), Dolichopodidae (2), Muscidae (6), Sciomyzidae (2), Stratiomyidae (1), Syrphidae (15), Tabanidae (5), Tachinidae (6), and Therevidae (1). Although over 80% of the prey listed in 1964 were deer flies of two species of *Chrysops, B. americana* used a wide variety of other flies in 1967, when deer flies were uncommon.

Evans (HEE, previously unpublished) continued to gather data from sites in Massachusetts, Colorado, and Wyoming. In northwestern Wyoming, he observed an especially large "sun dance" of perhaps 300–400 males; the humming sound they produced could be heard several meters away. A mating pair was seen mating on the ground; when other males approached, the pair flew off and landed several meters away, still in mating posture. All of the 15 nests dug at diverse sites were unicellular, with burrow lengths 15–35 cm, cell depths 6–16 cm. There were no unusual prey records or behavioral traits not already described for this species.

Alcock (1972, 1973) published several papers on a population of *B. americana* at Lake Washington, Seattle. Females tended to build unicellular nests first, then switch to nests with 2–3 cells. A record of 140 prey consisted of flies of six families, most of which were within three species: 1 *Hylemya* sp., 1 *Spaziphora cincta* (Loew) (Anthomyiidae); 1 *Villa* sp. (Bombyliidae); 19 *Phaenicia sericata* (Meigen), 44 *Pollenia rudis* (F.) (Calliphoridae); 62 *Coenosia tigrina* (F.) (Muscidae); 1 *Sarcophaga* sp. (Sarcophagidae); 1 *Eristalis tenax*, 1 *Helophilus* sp., 3 *Melanostoma* sp., 1 *Metasyrphus subsimus* Fluke, 4 *Sphaerocephala sulphuripes* (Thomson), and 1 *Syrphus* sp. (Syrphidae). Some of the cells contained an unusually large number of prey, up to 44. Females brought in prey at an average rate of one fly per 8.2 min, sometimes stealing from conspecifics.

Evans (1966a) reviewed information on parasitism of *B. americana spinolae,* which is apparently most commonly due to miltogrammine flies. Recently Spofford and Kurczewski (1990, 1992) reported that 2 of 23 *B. americana* nest cells were parasitized by the miltogrammine flies, including one by *Senotainia trilineata.* When trailed by a *S. trilineata,* some females made diversionary flights or actually chased away the pursuing fly.

For the same population studied by Evans in Jackson Hole, O'Neill (1985) found that that the mean size (head width) of females exceeded that of males. Evidence for an advantage to large size in females came from the fact that there is a significant positive correlation between female size and (1) the length of the largest egg carried in the ovarioles and (2) the combined length of the six terminal oocytes. During this study, O'Neill (KMO, previously unpublished) collected what he first thought was a female as she was being pounced upon by a male. Close examination revealed that the "female" was a gynandromorph. Many of its external features were clearly female: relatively massive mandibles, 10 antennal flagellomeres, and rake spines on both foretarsi. However, it also had male genitalia and a pseudosting. The coloration was mixed. Its facial coloration was of the female type, but although the left side of the dorsum of its abdomen was whitish (as in females), the left side (excluding the terminal band) was yellowish (as in males). Finally, the three ovarioles on the left side of the abdomen were fully developed with nurse cells and one large oocyte (unfortunately, the existence of testes was not determined). There is at least one previous report of a *Bembix* gynandromorph, one of *Bembix rostrata* in Europe (Augener 1927).

Another subspecies, *B. americana comata* J. Parker, occurs in the west coastal states and appears to intergrade with *B. americana spinolae* in some areas. Lane et al. (1986) studied a population in a sand dune ecosystem at Point Reyes National Seashore, California, where wasps were active on foggy days and at temperatures as low as 12°C (though more commonly when it was above 18.5°). Nests were unicellular, as is typical, burrow length averaging about 18 cm, cell depth 11 cm; Parker (1925) had reported three-celled nests, but this was not confirmed by Evans (1957b). Lane et al. observed females stalking and capturing horse flies at sites where the flies were hovering and mating. Paralyzed prey lived up to 10 days following capture. The 113 prey taken included flies of 32 species in 12 families (Anthomyiidae, Asilidae, Bombyliidae, Calliphoridae, Dolichopodidae, Muscidae, Sarcophagidae, Sciomyzidae, Syrphidae, Tabanidae, Tachinidae, and Tephritidae), of which only several made up more than ~10% of the prey: 20 *Orthellia caesarion* (Meigen) (Muscidae); 17 *Brennania hera* (Osten Sacken) (Tabanidae); and 11 *Nicocles aemulator* (Loew) (Asilidae). A further 20 prey taken from females include four other species. Out of the above list, Anthomyidae, Asilidae, Bombyliidae, Sciomyzidae, and Tephritidae were not reported in the list compiled by Evans (1957b).

Menke (in Rust et al. 1985) reported on four *Bembix americana* subspecies of the Channel Islands, off the coast of southern California. He found *B. americana comata* on Santa Catalina Island, where he recorded diverse flies as prey, including Muscidae, Sarcophagidae, Stratiomyidae, Syrphidae, and Tachinidae. Evans (1957b) noted that this subspecies nests in aggregations on the Pacific coast of North America in "areas of bare, powdery soil, in smaller sandy areas, and in the periphery of large dunes." However, in a broad survey of a coastal dune ecosystem of Humboldt Bay near Arcata, California, Gordon (2000) also found *B. americana comata* nests in vegetated pastures, inland dunes, and lowlands. Menke also found *Bembix americana hamata* C. Fox as an inhabitant of San Miguel, Santa Rosa, and Santa Cruz islands (California Channel Islands) and reported its prey as the syrphid *Copestylum avidum* (Osten Sacken). The third subspecies found in the California Channel Islands was *Bembix americana nicolai* Cockerell, which occurs only on San Nicolas Island; it was found to prey on flies of the families Bombyliidae, Syrphidae, and Tachinidae. Finally, Menke found a large nesting population of *Bembix americana dugi* Menke in an area of sandy soil on San Clemente Island. Prey here consisted of flies of the families Bombyliidae, Calliphoridae, Sarcophagidae, Syrphidae, and Tachinidae. One nest was excavated and found to have a burrow 28 cm long, the cell at 13 cm depth.

Bembix amoena Handlirsch—western North America

One of the best-studied species of *Bembix* in North America, *B. amoena* is a predator of at least 11 families of flies, and usually makes two-celled nests "in flat or sloping areas of coarse sandy gravel or finely pulverized rock" (Evans 1966a). A list of 254 prey from two localities in Yellowstone National Park, Wyoming, contains at least 72 species in 11 families of

Diptera. The most common prey families were Bombyliidae, Tabanidae, and Tachinidae. One further study of a very different type has appeared. G. E. Bohart et al. (1970) studied this species and other sphecids to determine their effectiveness as pollinators of onion on farms in the vicinity of Logan, Utah. Like many wasps, *B. amoena* was less effective as a pollinator relative to bees and flower flies (Syrphidae). However, it was rated as the best pollinator among the 44 species of apoid wasp recorded on the plots.

Bembix boharti Griswold—Baja California

The type series of this species, described from diverse localities in Baja California, Mexico, included specimens studied 16 km west of La Paz (Evans 1976c). That publication referred to these wasps as *Bembix sayi,* but Griswold (1983) was able to point out several differences from the very similar *B. sayi.* Evans found about 20 nests scattered over the bottom of an arroyo, about 0.3 km from Bahia Pichilingue. No nests were closer together than 0.5 m and most were more widely spaced. One female brought in four flies over a 30 min period, in one instance requiring only 3 min to obtain prey. The nest was excavated in the afternoon, when it was closed from the inside. The 46 cm long burrow led to a cell 22 cm deep, which is surprisingly shallow considering the extreme heat and very dry soil at this locality. Flies in this cell and others belonged to the families Calliphoridae (*Callitroga* sp.), Otitidae (*Diacrita costalis* [Gerstacker]), Sarcophagidae (3 unidentified species), Syrphidae (*Copestylum isabellina* Williston and two unidentified species), and Tachinidae (*Peleteria neotexensis* Brooks). Nest entrances were left open during provisioning, perhaps leaving them susceptible to the *Exoprosopa sima* Osten Sacken (Bombyliidae) that were seen ovipositing in nest holes.

In two instances, females made large mounds of sand (17 × 25 × 2.5 cm deep) following the final closure. They landed on the top of the mound, then turned off to one side for a distance of 7–10 cm, kicking sand behind them, then made a brief flight and landed on top of the mound to repeat the performance over a slightly different track. One female made 40 such lines over a period of 50 min, then started a new nest 40 cm away. One mound measured 25 cm long by 17 cm wide and was 2.4 cm deep in the center. This "mound building" behavior was reminiscent of that of *Stictia maccus* and *Bembix littoralis.*

Bembix dentilabris Handlirsch (= *B. u-scripta* Fox)—Texas to southern California, northern Mexico

Bembix dentilabris is among a minority of *Bembix* that have their anterior ocelli preserved (i.e., not reduced to "mere arcuate grooves"; Evans 1966a). The preserved ocelli may be functional in this crepuscular species, whose females provision nests under relatively low light conditions relative to other *Bembix*. In May 1979, Evans (HEE—previously unpublished) revisited the nesting site of *dentilabris* near Port Isabel, Texas, first discovered in 1956 (Evans 1957b, 1966a). In 1979 there were approximately 60 females (Figure 7.1) nesting at the same site, their nest entrances 6–21 cm apart. Females brought prey starting at 0700–0830 h, and during this time the males flew about the nest area, 3–6 cm high, producing a distinct humming sound. One male seized a female and flew off with her into a clump of bushes. Two nests excavated were unicellular (though nests of this species are typically multicellular, with up to five cells [Evans 1957b]), bur-

Figure 7.1. Female *Bembix dentilabris* digging nest, Port Isabel, Texas. Photo by H. E. Evans.

row lengths 30–35 cm, cell depths 13–21 cm. Prey at this site consisted of diverse Diptera, mostly Bombyliidae. In Wyoming, Lavigne and Holland (1969) reported *B. dentilabris* as prey of the robber fly *Diogmites angustipennis* (Loew).

In southeastern Arizona, Cazier and Linsley (1974) found males of *B. dentilabris* this species patrolling flowers of Arizona poppy, *Kallstroemia grandiflora* (Torrey), in the early morning before the flowers were open. Males perched beneath blossoms and fed on nectar by inserting their tongues between the sepals to reach the nectary, presumably fueling themselves for a morning "sun dance" at a nearby nesting site. Cazier and Linsley concluded that *B. dentilabris* is a nectar robber and cannot act as a pollinator because it does not pick up pollen while feeding.

Bembix inyoensis Kimsey and Kimsey—southern California

Kimsey et al. (1981) found this recently described (Kimsey and Kimsey 1981) species nesting in Death Valley National Monument, California, during March and April, when *Larrea* and *Prosopis* were flowering. *Bembix inyoensis* is another crepuscular species, its adults becoming active at sunrise and completing their activities before noon. Thus it is noteworthy that the anterior ocellus of this species is not completely obliterated but is present as a "narrow translucent lunule." Nests, which were dug in damp sand, were unicellular and shallow, burrow lengths averaging 19 cm and cell depths 8 cm; the authors provided sketches of a typical nest and of the pattern of movements during nest closure. Prey extracted from five nests, which were provisioned progressively, included flies of four families: *Furcilla* sp., *Ablautus flavipes* Coquillett, *Laphysta martini* (Asilidae); *Aphoebantus tardus* Coquillett, *Aphoebantus leviculus* Coquillett, *Aphoebantus vulpecula* Coquillett, *Aphoebantus* spp., *Lordotus sororculus* Williston, *L. junceus* Coquillett, *Oligodranes trochilus* (Coquillett), *Heterostylum robustum* (Osten Sacken), *Villa aenea* (Coquillett), *Paravilla syrtis* (Coquillett) (Bombyliidae); *Phaeocera* spp., *Ammonaios* sp. (Therevidae); and an unidentified Tachinidae. The egg was laid on the wing base of one of the flies in each cell.

Bembix melanaspis Parker—southwestern United States, Mexico

Alcock and Gamboa (1975) found the species, which departs at least occasionally from a strict diet of flies, nesting near the Gila River in Arizona.

Nests were unicellular and quite deep, the burrow being 49–114 cm long, the cell being 25–60 cm deep. The egg was laid erect in the empty cell (Figure 7.2). Prey was provided after the egg hatched and consisted of flies of the families Calliphoridae, Conopidae, Syrphidae, Tabanidae, and Therevidae. However, one cell contained the wings of a damselfly, the first record of a North American species preying on Odonata (or, in fact, any nondipteran). According to Alcock and Gamboa, the nest entrance is "usually, but not always, kept closed between provisioning trips. There is always an inner closure. Final closure involved filling the entrance of the nest as the wasp kicks sand back down the burrow. The female then walks outward from the entrance repeatedly kicking sand backward toward the closed opening. In the course of this activity the wasp produces a cluster of

Figure 7.2. Egg in empty cell of *Bembix melanaspis.* From Alcock and Gamboa (1975); used with permission.

rays in the sand each about 15–20 cm long emanating from the concealed entrance."

Bembix occidentalis W. J. Fox—western United States, northern Mexico

Evans (1957b) provided many details on the biology of this species, which inhabits open dunes, constructs complex single-celled nests, and preys on flies of at least 11 families. In Baja California Sur, Evans (1976b) later observed females capturing flies from dead fish on a beach, but nests were not located.

Bembix pallidipicta F. Smith (= *B. pruinosa* W. Fox)—United States, Mexico

Rubink (1978, 1982) studied nest-site selection in this dune-inhabiting species, working at the LaJoya State Game Refuge in central New Mexico. Most of Rubink's observations address edaphic factors, such as soil texture, that are assumed to play a key role in nest-site selection in digger wasps. The following summary of his dissertation (Rubink 1978) is from O'Neill (2001): "In comparing 22 nest sites (separated by as much as 4 km) with similar adjacent habitats, [Rubink] found that nesting areas had (1) higher soil surface temperatures in the late morning (~6°C higher on average) and (2) flatter slopes (20% vs. 36% on average) that tended to face east to northeast (rather than north). Dropping to a lower spatial scale, he found that nest density within a large aggregation was correlated with (1) morning but not afternoon soil surface temperatures (areas of high density were ~2–3°C warmer in the morning than areas of low density) and (2) soil particle size (areas of high nest density had less-variable particle sizes). Although soil moisture was not an important characteristic in nest-site selection, *B. pallidipicta* females apparently adjust nest depth in response to subsurface conditions, digging deeper in drier parts of the dunes. Two other factors apparently influence nest-site selection on a smaller spatial scale. First, using nearest-neighbor analysis, Rubink showed that females are probably sensitive to overcrowding: each tries to maintain some minimum distance between her own nest and her neighbor's nests. Second, it appears that the sand surface at the point of nest initiation must have some minimum cohesiveness to provide structural stability to the upper reaches of the nest tunnel (i.e., females like their sand crusty on top)." The latter is particularly important because females of this species dig an extended preliminary tunnel that is 20–51 cm long and just 2–4 cm beneath the surface (Evans 1957b). Rubink also followed 50 marked females to determine how far they moved between consecutive nests within an isolated nest aggregation. Only four females made a second nest within 10 meters of the first. The other 46 females were not seen again within the aggregation, perhaps because they moved to other aggregations; Rubink admits "marking the wasps [possibly] traumatized them so that new nests were made elsewhere, although this seems unlikely."

Evans (1966a) mentioned that individuals in the population of *B. pallidipicta* at Great Sand Dunes National Monument, in southern Colorado, were strikingly different in color from those in other parts of the range, having yellow markings instead of the usual greenish-white. In July 1974 and August 1986, he returned and excavated 10 nests to determine if there were notable differences in behavior compared with that of other sites (HEE—previously unpublished). He found no such differences. All nests had a preliminary burrow that was later abandoned and a new exit made; and all had an elongate cell (10–20 cm in length) that formed a broad curve, with the egg laid at the extremity. Diverse flies were introduced progressively as the larva moved along the cell. Nests were quite deep, as might be expected in an area of extensive dunes. Burrow length varied from 44 to 80 cm, cell depth from 17 to 38 cm. Figure 39 in Evans's (1957b) account represents a nest of this species dug in an area where there was a good surface crust and the burrow was just beneath and parallel to the surface. At the Great Sand Dunes, however, preliminary burrows formed a 15–30° angle with the sand surface.

Bembix rugosa J. Parker—Arizona, Baja California, Utah

The first biological studies of this species (Evans 1976c) were conducted in Baja California Sur. Here females nested in dunes and made deep, unicellular nests in which the burrow had occasional vertical or lateral turns before reaching the cell. Burrow lengths were 47–127 cm, cell depths 22–78 cm. Cells were provisioned progressively with flies: *Ablautus flavipes* Coquillett (Asilidae), *Pseudonomoneura* sp. (Mydidae), and *Psilocephala* sp. (Therevidae); the use of Mydidae is extremely rare for *Bembix*. After digging nests, females made elaborate closures, forming numerous lines that radiated from the entrance (Figure 7.3).

Bembix sayi Cresson—United States, northern Mexico

A subject of a detailed report by Evans (1957b, 1966a), this widely distributed species has also been studied three times more recently. Alcock and Gamboa (1975) studied *B. sayi* along the Gila and Salt Rivers in Arizona, where, in contrast to the Florida nests described by Evans (1966a), some nests included a second or even a third cell from a common burrow. Nests were deeper than those in Florida, burrows 35–45 cm long, cells 19–27 cm deep. No important differences from the Florida population were noted in details of the unusual final closure. Other data comes from seven nests from two locations in Larimer County, Colorado (HEE, previously unpublished). All nests were unicellular and relatively shallow, burrow lengths 19–30 cm, cell depths 10–15 cm. Prey consisted of Tachinidae, with a few Syrphidae and Tabanidae. At both sites males performed a typical "sun dance," flying 0.5–1.1 m high above the soil in the morning hours, but occasionally landing on the soil with their legs extended. Recently R.

Figure 7.3. Pattern of radiating lines made by *Bembix rugosa* female during nest closure. Photo by H. E. Evans.

Matthews and J. Matthews (2005) found *B. sayi* nesting, sometimes along with *Cerceris fumipennis* Ashmead, in western Nebraska. They reported that their "limited observations agree with those of Evans . . . [from] Florida and Kansas."

Bembix stenebdoma J. Parker—western Texas to southern California

In sandy semidesert in Socorro County, New Mexico, Evans (1978) found the second bit of evidence that a North American species of *Bembix* preys on something other than Diptera. One nest excavated had a 30 cm long burrow and 20 cm deep cell that contained 10 lacewings (Neuroptera: Chrysopidae), all lying on their sides with their heads facing the entrance to the cell (Figure 7.4). An egg laid on the side of one of the lacewings deep in the cell was attached between the middle and hind coxae of the prey. The lacewings consisted of three species of the genera *Eremochrysa* and *Chrysoperla*. Evidently *B. stenebdoma* is a mass provisioner.

Figure 7.4. Egg and prey in cell of *Bembix stenebdoma*. From Evans (1978); used with permission.

Bembix texana Cresson—southern United States

Evans (1966a) reported extensively on the behavior of this species from his studies in 1961 in Florida, where he found that 94% of 125 prey were Tabanidae. Later Roberts and Wilson (1967), working in Louisiana, found that seven marked females provisioned with an average of 3.9 flies per day, a mixture of Sarcophagidae, Stratiomyidae, Syrphidae, and Tabanidae (see under *Stictia carolina* in Chapter 6 for comparison of the effectiveness of these to wasps in controlling horse flies around cattle).

Neotropical *Bembix*

Bembix americana F.—West Indies

Bembix americana americana F. This inhabitant of Puerto Rico and the Virgin Islands is one of two subspecies of *B. americana* that occur in the West Indies. Evans (in Evans and Matthews 1968) reported it nesting at sites on the coast of Puerto Rico. An aggregation of about 150 females and at least as many males occupied a sandy area just behind a beach, shaded by coconut palms. The "sun dance" of the males produced a humming sound audible some distance away. Nests, which were sometimes only 5–15 cm apart, were unicellular, burrow lengths 15–20 cm, cell depths 8–15 cm. Provisioning was progressive, and an outer closure was maintained between hunting trips. The egg was laid on the first prey placed in the cell. Thirty-seven flies recorded as prey were: *Villa* sp. (Bombyliidae); *Cochliomyia macellaria* (Calliphoridae); *Baccha clavata* F., *Eristalis albifrons* Wiedemann, *E. vinetorum*, and *Volucella* sp. (Syrphidae). *Eristalis* spp. made up nearly three-quarters of the prey.

Bembix americana antilleana Evans and Matthews occurs on Cuba, Hispaniola, Jamaica, Grand Cayman, and the Bahamas. Two recent studies from Cuba provide similar descriptions of its behavior. Near Havana, Alayo Soto (1989) found simple unicellular nests with an oblique burrow and cells at depths of 10–12 cm stocked with *Phaenicia* sp. (Calliphoridae); *Peckia praeceps* (Wiedemann) (Sarcophagidae); and *Palpada vinetorum* (F.) (Syrphidae). Genaro (1995) studied several Cuban populations nesting in diverse kinds of friable soil. Males engaged in a typical "sun dance," flying about 10 cm above the soil. When a male seized a female, he was often joined by others, forming a ball of competing males. Nests were unicellular, burrows 12–18.5 cm long, cell depths 9–11.5 cm. The egg was laid on the first prey in the cell, as in other subspecies of *B. americana.* Genaro presented a list of 227 prey flies from 10 families: Asilidae, Bombyliidae, Calliphoridae, Micropezidae, Muscidae, Sarcophagidae, Stratiomyidae, Syrphidae, Tabanidae, and Tachinidae. About two-thirds of these were from three species: the syrphids *P. vinetorum* (31% of prey) and *Ornidia obesa* (F.) (15% of prey), and the muscid *Musca domestica* (22% of prey).

Bembix citripes Taschenberg—South America

This widely distributed South American species is similar to *B. multipicta* and was, at one time, considered to be identical to that species. Evans and Matthews (1974) found 10 nests at two sites in or near Cafayate, Salta Province, and one site at Santa Maria, Catamarca, Argentina. No aggregations were found, merely scattered nests in sandy places. Nests were unicellular, burrow lengths 14–28 cm, cell depths 10–18 cm. The egg was laid erect in the center of the empty cell. Nests were provisioned progressively with flies of the families Bombyliidae (*Exoprosopa* sp.), Calliphoridae (*Cochliomyia macellaria*), Stratiomyidae (*Hedriodiscus pulcher*), and Syrphidae (*Myolepta apicalis* Fluke). After completing a new nest, females made elaborate outer closures, but when the larva was large, the closure was usually omitted while the female was foraging. An inner closure, just outside the cell, was maintained at all times.

Genise (1982d) found the species nesting in the province of Entre Rios, Argentina. The nests were in bare, sandy places also occupied by *Stictia arcuata* and *Bicyrtes variegatus.* Nests were unicellular and comparable in depth to those found by Evans and Matthews. Genise provided photos of females at nest entrances and of the plaster cast of a nest. Prey consisted of Tabanidae. Genise described leveling behavior at some length and found it slightly different from that described for *B. multipicta.*

Bembix multipicta F. Smith—Mexico to northern South America

Cane and Miyamoto (1979) published a report on a population of *B. multipicta* on the Osa Peninsula of Costa Rica. The nesting site, where *Stictia heros* also nested, was at the end of a grassy airstrip a short distance from the ocean. Nests were unicellular and quite shallow, burrow length averaging 25 cm, cell depth 14 cm. Females spent the night outside the nest, in contrast to most other species of *Bembix.* When they arrived in the morning, without prey, females quickly found and opened their nests even when overnight rains had disturbed the soil surface. The authors described three types of closures: thin, temporary closures constructed before foraging trips; thicker and better-concealed closures made after daily activities were over; and more elaborate final closures. The wasps observed hunting tabanids on horses may well have been *Stictia signata* rather than *Bembix multipicta,* as noted under the discussion of *S. signata* in Chapter 6. Some nests suffered from such severe marauding by ants (*Solenopsis* sp.) that they were completely emptied by the ants within an hour, and abandoned by the wasps. Females exhibited two forms of defense against ants. First, they attempted to rapidly plug the entrance with packed sand. Second, they attacked ants (and even a tiger beetle) by grasping them in their mandibles and carrying them in flight up to a meter away.

Palearctic *Bembix*

Bembix bidentata (Van der Linden)—Eurasia

Leclercq and Leclercq (1970) reported that at a site in Turkey, both males and females were attracted to a trap made in the form of a horse (a "Skufin trap") and designed to capture Tabanidae. Evidently the females were attracted to the silhouette of a horse, and the males followed them.

Bembix flavescens bolivari Handlirsch—southern Europe, northern Africa, Canary Islands

In Spain, this species nested in a sloping sandbank. Asís et al. (1992) provided a sketch of the nest, which was unicellular, with the burrow 48 cm long and the cell 25 cm deep. A nest closure was maintained and provisioning seemed to be truncated. The 15 prey collected consisted of *Pollenia leclercquiana* Lehrer (Calliphoridae); *Eristalinus aeneus* (Scolopi), *Eristalinus sepulchralis* (L.), *Metasyrphus latifasciatus* (Macquart), and *Sphaerophoria scripta* (L.) (Syrphidae).

Bembix merceti J. Parker—Spain

In Spain, Asís et al. (1992) found *B. merceti* nesting in flat, sandy soil and making unicellular nests, with burrows 21 cm long burrows and cells 7.5–11 cm deep. Nests had both outer and inner closures and females spent the night inside the nest. In a more recent and more extensive study near Soria, Spain, Asís et al. (2004) added further data, based on detailed observations of two females and prey collected from 15 females. All eight nests excavated were unicellular, with burrows 17.0–30.4 cm long and cells 8.2–14.8 cm deep that were 1.3–1.8 cm wide and 2.9–3.8 cm long. Each nest featured a sharp turn of 95–140° near the cell.

During provisioning, females maintained an inner closure separating the burrow from the cell and a temporary outer closure. The first prey brought into a cell was placed venter upward and the egg was laid in an erect orientation on the side of the prey. Unusually, one cell had two wasp eggs on the same prey item, but the origin of the second egg was unknown. Mutillids and chrysidids were observed entering nests, but the only concrete indication of the effect of a natural enemy was a partially provisioned nest abandoned by a female *B. merceti* after it was invaded by a worker ant.

Provisioning apparently did not commence until the day after the nest was completed. Two females provisioned progressively with 46 and 60 prey over a period of 5–6 days. The number of prey provided each day by these two females increased as the larvae got older. The progression in the number of prey provided per day was 1, 4, 14, and 27 for one female and 1, 2, 8, 22, and 27 for the other. Individual provisioning flights varied from 1.5–134.5 min, the average duration declining each day after the female initiated provisioning. Two specimens of prey collected in the first study (Asís et al. 1992) were both *Discachaeta maroccana* (Rondani) (Sarcophagidae). In the later study, prey taken varied between years. In 1993, 88% of 23 prey removed from nine nests were bombyliids along with two Tachinidae and one Muscidae. In 1994, however, six nests produced 33 flies from seven families, of which Muscidae constituted 49%, but Bombyliidae only (15%). Prey species identified by Asís et al. (2004) included *Anastoechus exalbidus* (Wiedemann), *Anthrax virgo* Egger, *Aphoebantus scutellatus* Loew, *Exoprosopa jacchus* (F.), *Hemipentes velutinus* (Meigen), *Heteralonia rivularis* (Meigen), *Thyridanthrax elegans* (Wiedemann), and *Usia aenea* (Rossi) (Bombyliidae); *Musca domestica, Neomyia cornicina* (F.) (Muscidae); *Sargus cuprarius* (L.) (Stratiomyidae); and *Haematopota csikii* Szilády(Tabanidae).

Bembix niponica F. Smith—Japan

Tsuneki (1969) followed up his earlier extensive studies (Tsuneki 1956–58) with a study of *B. niponica picticollis* Handlirsch in Inner Mongolia. He found no important differences from populations of *B. niponica niponica* in Japan. He excavated several nests, finding in some of them a short branch or "spur" not far from the cell, presumably a resting place for the female. The egg was laid on the first prey in the cell and nests were provisioned progressively with flies of at least three families: *Calliphora* sp., *Lucilia* sp. (Calliphoridae); *Eristalis cerealis* (F.), *E. tenax, Syrphus* sp. (Syrphidae); and unidentified Muscidae (= Stomoxyidae).

Bembix oculata Panzer—western Palearctic

This common species in southern Europe was studied earlier by Nielsen (1945) and others. In Spain, Asís et al. (1992) found females nesting in sandy soil either solitarily or in diffuse aggregations. Nests were unicellular, burrow lengths 18–37 cm, cell depths 6.5–17 cm, but burrow length and cell depth varied with the consistency of soil. Those in loose sand had burrows averaging 29.5 cm long and cells averaging 15.5 cm in depth, but those is hard-packed soil had burrows averaging 20.5 cm long and cells averaging 9.8 cm in depth. The egg was laid erect on the first prey in the cell. Asís et al. (1992) observed *B. oculata* in Spain from June to September, whereas El-Banna et al. (1999) observed it from April to August near Port Said, Egypt (peaking in abundance in June). One female in Egypt dug a 21–22 cm long oblique burrow in 15–20 min.

Radović (1985) noted that the sting apparatus of *B. oculata* has a structure typical of those species of apoid wasps that prey on highly mobile prey, with a "powerfully developed and curved stylet." The several reports of prey of this species do indicate that *B. oculata* (like many *Bembix*) prey on "highly mobile" species of flies. The 32 prey extracted from nests by Asís et al. (1992) consisted of flies of six families: *Exhyalanthrax afer* (F.), *Villa circumdata* (Meigen) (Bombyliidae); *Blaesoxipha rufipes* (Macquart), *Heteronychia pandellei* (Rondani) (Sarcophagidae); *Stratiomys longicornis* (Scopoli) (Stratiomyidae); *Estheria cristata* (Meigen), *Thelaria nigripes* (F.) (Tachinidae); *Thereva sp.* (Therevidae); and one unidentified Muscidae. Nazarova and Baratov (1981) reported three species of Tabanidae as prey in Tadzhikistan: *Hybomitra peculiaris* Szilady, *Tabanus accipiter* Szilady, and *Tabanus leneani* Aust. Larvae of brood parasitic flies *Protomiltogramma fasciatum* (Meigen) and *Craticulina tabaniformis* (F.) were found in five cells, four of which also contained wasp larvae. *Craticulina tabaniformis* were seen entering nests when female *B. oculata* opened temporary closures.

Bembix olivacea F.—southern Europe, northern Africa

Leclercq and Leclercq (1970) reported that in Morocco, females of *B. olivacea* flew to the entrance of a stable; when the bulls kept there were let out, the wasps followed them, no doubt as a means of being close to high concentrations of flies.

Bembix rostrata (L.)—Eurasia

Bembix rostrata is widely distributed in Eurasia and has been the subject of research for more than a century. Details of its life history and behavior are now well known. Females make unicellular nests, lay the egg erect on the first prey in the cell, and provision progressively with diverse flies. Asís et al. (1992) also studied the nesting behavior of five species of *Bembix* in Spain. Since the biology of *B. rostrata* is already well studied, they presented only nine prey records in eight species in four families: *Calliphora vicina* Robineau-Desvoidy, *Chrysomyia albiceps* (Wiedemann) (Calliphoridae); *Mesembrina meridiana* (L.) (Muscidae); *Chrysotoxum elegans* Loew, *Eristalis similis* (Fallén) (= *E. pratorum* [Meigen]), *Eristalis tenax* (Syrphidae); and *Atylotus quadrifarius* (Loew), and *Dasyrhamphis ater* (Rossi) (Tabanidae).

Larsson (1986) studied the effect of increased nest density of *B. rostrata* as a response to parasites (*Metopia* flies) and predators (*Formica* ants). The relative number of ants and flies per active nest declined as nest density increased, even though overall predator and parasite numbers increased as nest density increased. This was interpreted as support for Hamilton's (1971) selfish herd hypothesis, whereby individuals are predicted to experience reduced parasitism when the activities of close neighbors distract natural enemies.

Recently, most attention on *B. rostrata* has been devoted to studies of orientation, mate-finding, and thermoregulation. Chmurzynski, in Poland, has devoted a series of papers to studies of spatial orientation of females returning to the nest (see bibliography in his 1967 paper). He has also (1977) studied the approach of males to females, using various models.

Schöne and Tengö (1981), working on the island of Oland, Sweden, described male mate-searching in detail. Males aggregate at spots in the soil where females are about to emerge from the soil above their cocoons; the number of cocoons per m^2 of nesting area can be high because of the high density of nests during the previous year. At points of female emergence, males dig, sometimes in groups of up to 50, and when a female emerges she is grasped by one male. Others join in, often forming a "mating ball" that may roll over the ground. When matings occur after one male has carried away a female, couplings last 30–60 s. Sexual interactions continue after females have begun their nests, when males may land close to females or pursue them in flight. Females begin to nest four to eight days after their first emergence. The authors induced wasps to perform in a greenhouse. Here, males are apparently able to differentiate between males and females in the soil, and between virgin and nonvirgin females. The female's attractiveness appears to be chemically based and to arise from the body rather than from the head.

Larsson (1989b) and Larsson and Tengö (1989) studied the mating behavior of *B. rostrata* with respect to weather, population density, and male body size. They found that during times of high ambient temperatures, small males have an advantage, as larger males run a greater risk of overheating. But the advantages to smaller males are probably offset by the greater success of larger males in mating clusters. During periods of high temperature, some males become territorial and hover over restricted areas, as Larsson and Larsson (1989) also observed in *Stictia heros.*

Larsson and Tengö (1989) also studied female nesting success in relation to size, also on Oland Island, which is close to the northern extremity of the range of *B. rostrata.* They marked and measured many females and found that those with greater head width and wing length spent more days with a single nest than did smaller females. Larger and smaller females have about the same life span. Thus in an area of very short summer seasons, smaller females may complete more nests than larger females, at least in some years. This contrasts to many other Hymenoptera, in which larger females are able to fly at lower temperatures than smaller ones, and thus may have a reproductive advantage.

Ghazoul and Willmer (1994) studied activity of both sexes of *B. rostrata* and of *Bembix zonata* Klug, using thermocouples inserted into the thoraces. They found that the wasps were able to warm up endothermically at a rate comparable to that of bees of similar weight, the first demonstration of endothermy in apoid wasps. The wasps also warmed up behaviorally by basking on the sand surface. Nevertheless the wasps are not active on cloudy days or at temperatures below 22°C, even when adequate nectar is available to power warm-up. Absence of activity at lower temperatures may reflect lower availability of prey or increased risk of nest parasitism if females are away from the nest for long periods. Periods of elevated body temperature during normal activity might, however, lead to more efficient prey transport and enhanced ability of males to seize and fly off with females.

Bembix sinuata Panzer—southern Europe, northwestern Africa

In Spain, Asís et al. (1992) found 12 nests in flat sandy soil in an area 3 × 4 m. Nests were unicellular, burrow lengths 13–33 cm, cell depths 7–22 cm. Outer and inner closures were usually maintained. Provisioning was progressive, prey consisting of flies of four families: *Lucilia sericata* (Meigen) (Calliphoridae); *Neomyia cornicina* (F.) (Muscidae); *Eristalinus sepulchralis, Eristalis arbustorum* (L.), *Eristalis pratorum* (Meigen), *Sphaerophoria scripta* (Syrphidae); *Eriothrix apennius* (Rondani), *Germaria barbara* Mesnil (Tachinidae); and several unidentified Muscidae. One *B. sinuata* nest contained a *Parnopes grandior* Pallace (Chrysididae), and three of the eight nests were parasitized by sarcophagid flies: *Hilarella hilarella* (Zetterstedt), *Protomiltogramma fasciatum* (Meigen), and *Taxigramma multipunctatum* (Rondani).

Bembix weberi Handlirsch—China, Mongolia

In Inner Mongolia, Tsuneki (1969) discovered shallow nests (burrow lengths 12–15 cm), with both outer and inner closures. The burrow was unusual in that there was an expansion outside the cell where the female stayed when in the nest. The egg was laid on the initial prey and provisioning was progressive, prey consisting of flies: Anthomyidae, Muscidae, and Syrphidae. Tsuneki presented sketches of several nests.

Bembix zonata Klug—southern Europe

Bernard (1934, 1935) reported *Lathyrophthalmus aeneus* Scopoli (Syrphidae) as prey, to which Asís et al. (2004) added a male *Sphaerophoria scripta* (Syrphidae) collected from a prey-carrying female. Asís et al. also included a few details on nest structure and provisioning. Two nests excavated near Soria, Spain, were both unicellular, with burrows 14 and 17 cm in length and cells 8.0 and 11.5 cm in depth. Females provisioned progressively, one providing 13 prey to a completely provisioned cell over a period of seven days. The identity of the prey was not reported.

Oriental *Bembix*

Krombein and van der Vecht (1987) published reports on the ecology and behavior of several species of *Bembix* in Sri Lanka and South India. Krombein and his colleagues studied four species in the field in Sri Lanka, and he included data from the literature on three other species. All of the species made unicellular nests, laid the egg on the first prey in the cell, and provisioned progressively. Besides the prey records listed below for Sri Lankan species, notes taken by Krombein and van der Vecht from the literature included a record of *B. tranquebarica* (Gmelin) preying on Diptera in Pakistan, a record of *Bembix lunata* F. preying on biting flies (Muscidae) in India, and a record of *Bembix budha* Handlirsch preying on Diptera in India.

Bembix antoni Krombein and van der Vecht—Sri Lanka

Krombein and van der Vecht (1987) found females nesting at two sites in southwestern Sri Lanka, both areas of high annual rainfall. At one site in the Indurawa jungle, Gilimale, nests were located in "level, coarse riverine sand . . . [that was] . . . shaded by taller trees during much of the day." Because of the high rainfall, "soil below the surface was a damp sandy loam." At the second site, females nested 20 m from a river in the courtyard of a small Hindu shrine where there was a "flat surface of coarse, firmly compacted, riverine sand." Here also the soil was moist below the surface. The authors speculate that the *B. antoni* aggregation at the shrine may have originated when sand that contained wasp cocoons was brought up from the river to the courtyard. Males engaged in sun dances within emergence/nesting areas at both sites. At one site, 15–20 patrolled within an area of ~100 m^2 but did not attempt to mate with nesting females, so may have been seeking newly emerged virgin females; one copulation was observed.

One nest had a burrow length of 39 cm, while four others ranged from 16.9 to 24.8 cm, and then leveled off for 3.8–9.1 cm and terminated in a cell. Burrows ranged from 7 to 11 mm in diameter, whereas the cells of the this large species ranged from 2.4 to 3.5 cm long and from 2.0 to 2.5 cm wide. The egg was laid on the first prey in the cell, at the base of a wing (Figure 7.5). The 26 prey (or fragments of prey) found in cells consisted of flies of four families (Calliphoridae, Sarcophagidae, Syrphidae,

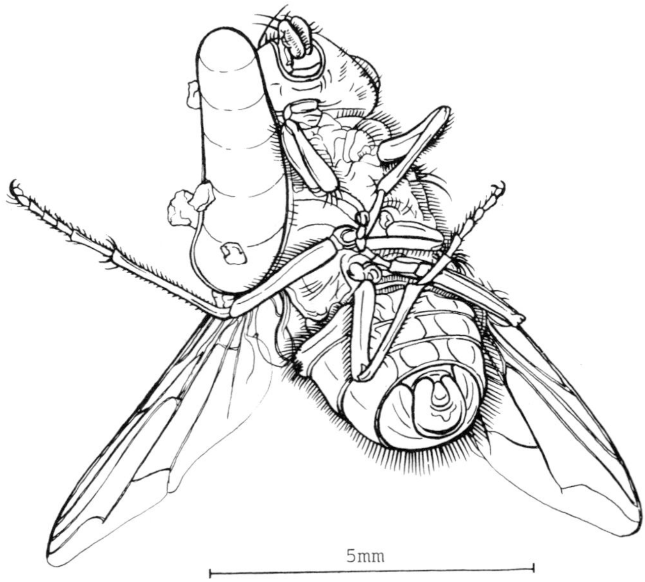

Figure 7.5. Position of egg on prey of *Bembix antoni*. From Krombein and van der Vecht (1987); used with permission.

and Tachinidae), most of which seemed to be associated with "decaying organic matter or filth." Identified prey included: *Calliphora* spp., *Chrysomyia megacephala, Hemipyrellia liguriens* (Wiedemann) (Calliphoridae); *Sarcophaga* sp. (Sarcophagidae); *Eristalinus arvorum* (F.) and *Eristalinus quinquestriatus* (F.) (Syrphidae). One cell contained four fresh calliphorid prey along with fragments of another and a small wasp larva. The lack of wings in another cell that contained a small larva and two prey upon excavation suggests that the provisioning female had removed wings from the cell sometime before the second prey was brought in. However, the presence of 46 wings and a cocoon in a third cell indicates that this does not always occur. The use of a temporary closure during prey flights was inconsistent.

Bembix borrei Handlirsch—India, Sri Lanka, Thailand, Indonesia

Bembix borrei, the most common Sri Lankan *Bembix*, is also quite widespread geographically, having been studied previously in India, Java, and Thailand (references in Krombein and van der Vecht 1987). In Sri Lanka, it was found to be common in fine sand either near the coast or on riverbanks, often in sand piles at construction sites. Burrows were 11.4–27 cm long. Females maintained outer closures while provisioning.

Prey consisted of flies of seven families. The list of prey identified was quite extensive and, as is typical of many *Bembix,* numbers were relatively evenly distributed among species: *Bombylisoma* sp., *Exhyalanthrax absalon* (Wiedemann), *Exoprosopa* sp., *Villa* sp. (Bombyliidae); *Chrysomyia megacephala, Chrysomyia rufifacies* (Macquart), *Lucilia illustris* (Meigen), *Phaenicia cuprina* (Wiedemann), *Thoracites abdominalis* (F.) (Calliphoridae); *Lispe* sp., *Musca domestica, Musca formosana* Malloch, *Musca lusoria* Wiedemann, *Musca pattoni* Austen, *Orthellia lauta* (Wiedemann), *Orthellia indica* (Robineau-Desvoidy), *Stomoxys calcitrans* (L.), *Xenosia* sp., (Muscidae); *Sarcophaga* sp. (Sarcophagidae); *Oplodontha rubithorax* (Macquart) (Stratiomyidae); *Eristalinus megacephalus* (Rossi), *Ischiodon scuttelaris* (F.) (Syrphidae); *Prosena sibirita* F., and *Sturmia convergens* (Wiedemann) (Tachinidae). Among the more than 110 prey collected, no species was represented by more than 20 specimens. In one cell containing a partly grown larva, the authors found 13 species of Bombyliidae, Calliphoridae, Muscidae, and Stratiomyidae. But another cell's 28 prey consisted only of three species of *Musca.*

Bembix glauca F.—India, Sri Lanka

Bembix glauca proves of unusual interest because of its long and tortuous burrows. At Pamunugama along the coast of Sri Lanka, females nested in several small aggregations in gently sloping pure beach sand that was moist just below the surface; at Palatupana, nests were on the lee side of small sand hills. Two nest profiles were illustrated, one with a burrow 80 cm long, reaching a cell at 60.5 cm, the other with a burrow 78.6 cm long, reaching a cell at 48 cm. Nests were unicellular, and an outer closure was maintained during provisioning. Females brought in flies at intervals of from 2 min to more than 1 h. Identified prey included: *Musca domestica* L., *Musca inferior* Stein (Muscidae); *Physiphora aenea* (F.), *Physiphora* sp. (Otitidae); Sarcophaga sp. (Sarcophagidae); and *Tabanus griseifacies* Schuurmans Stekhoven (Tabanidae). Some of the tabanids and muscoids may have been captured around water buffalo that were grazing nearby.

Bembix orientalis Handlirsch—Burma, India, Sri Lanka

Krombein found females nesting in a sandy loam field next to an airport in Sri Lanka. Males did not perform a typical "sun dance," but flew back and forth low to the ground and occasionally pounced on nesting females. Females made shallow nests, one of which had a burrow that "went downward at an angle of 45° for 11.5 cm, then turned at right angles and terminated in a cell 3.8 cm long and 1.3 cm wide. The burrow diameter varied from 5 to 10 mm in width depending on the looseness of the sand." Cells were provisioned with flies of three families: *Chrysomyia megacephala* (Calliphoridae), *Stomoxys calcitrans* (L.) (Muscidae), and *Plagiostenopterina* sp. (Platystomatidae).

Krombein and van der Vecht reported that 3 of 10 *B. orientalis* collected in Sri Lanka were stylopized (the first record of *Bembix* being parasitized by 1 to 3 stylopids outside of Australia). The stylopid was later determined to be a previously undescribed species by Kifune and Hirashima (1987), who named it *Paraxenos krombeini* Kifune and Hirashima (Strepsiptera).

Afrotropical *Bembix*

All of the observations reported below for African species of *Bembix* come from the work of F. W. and S. K. Gess working in South Africa. Besides the information provided in the individual species accounts, F. W. Gess (1986) presented prey records for four African species that were derived from labels on museum specimens: *Bembix capicola* Handlirsch (Diptera: Calliphoridae, Muscidae, Syrphidae, Tachinidae), *Bembix flavicincta* Turner (Diptera: Mydidae; *Nomoneuroides natalensis* Hesse), *Bembix fraudulenta* Arnold (Diptera: Bombyliidae, Sarcophagidae), *Bembix moebii* Handlirsch (Diptera: Tabanidae—taken at cattle). Gess also reviewed published accounts and found records of seven other African species that take flies: *Bembix bequaerti dira* Arnold, *Bembix braunsii* Handlirsch, *Bembix fascipennis* Lepeletier, *Bembix forcipata* Handlirsch, *Bembix massaica* Cameron, *Bembix olivata* Dahlbom, *and Bembix ugandensis* Turner. Gess found clear evidence that one African *Bembix* species, *Bembix regnata* Parker, is a predator on butterflies. R. H. R. Stevenson, working in Southern Rhodesia (now Zimbabwe), found females regularly taking butterflies, usually Pieridae and Hesperiidae, but sometimes Nymphalidae (reported by Benson 1934). Gess points out that about 90 species of *Bembix* have been reported from tropical Africa, so there may be other species that have departed from use of dipterous prey.

Bembix albofasciata. F. Smith—southern Africa

Of the 14 nests excavated in loose sand at several sites by F. W. Gess (1981), all were unicellular nests and very shallow, with burrows 7.5–13.5 cm long and cells 4–7 cm deep. Nests had a sharp change in the angle of the burrow prior to reaching the cell, and in six nests there was a spur at the angulation. A closure was maintained while the female foraged, and provisioning was progressive. Prey extracted from 11 nests and taken from females at nest entrances included flies of seven families: *Acnephalum andrenoides* (Weidemann), *Stenopogon dilutus* (Walker), *Synoclus* sp., *Xenomyza* (?) sp. (Asilidae); *Exoprosopa* sp., *Geron* sp., *Henica longirostris* (Weidemann), *Lomatia pictipennis* (Weidemann), *Lomatia pulchriceps* Loew, *Systoecus* sp., *Villa* sp. (Bombyliidae); *Chrysomyia* sp. (Calliphoridae); *Musca lusoria, Musca* sp. (Muscidae); *Sarcophaga* sp. (Sarcophagidae); *Nomoneura caffra* Hesse (Stratiomyidae); and *Chrysops obliquefasciatus* Macquart (Tabanidae). Further prey records from museum specimens also included species of Syrphidae and Tachinidae.

Bembix arnoldi Arnold—southern Africa

Bembix arnoldi was found nesting in dunes near the ocean, and one female was observed taking an *Adersia oestroides* Karsch (Tabanidae) at a pile of seaweed (F. W. Gess 1986). After capturing the fly, the female stung it ventrally while hovering.

Bembix bubalus Handlirsch (Gess and Gess)—southern Africa

In 1989, S. K. and F. W. Gess added this to the list of South African fly predators. On the investigators' first visit to the site, males were common and were seen flying rapidly over the ground. One male approached a female while producing a high-pitched buzzing sound, then grasped and flew off with her. Gess and Gess also found a thousand or more females nesting in level, friable sandy soil. Unfortunately, on a second visit the area had been trampled by livestock, and only a few wasps remained. Nest-building females threw up a tumulus up that was 65 mm in diameter and that remained in place throughout provisioning. Outer nest closures were maintained while the females were away from nests or inside the nest for a long period. Cell depth varied from 13 to 17 cm. Of the 12 nests excavated, nine were unicellular, whereas one had two cells, one three, and one four.

In the four-celled nest, two of the cells contained cocoons and fly remains, the third a large wasp larva, and the fourth a small larva and fly remains. Remarkably, this nest was found to be occupied by three females, although only two cells appeared to be being provisioned. Several instances were noted in which nests were being visited by more than one female, and in no case was one of the females evicted. Gess and Gess illustrated a one-celled and four-celled nest and provided excellent photographs of females with prey. They pointed out that nests of more than one cell have been reported for other *Bembix* species, but the presence of three females in one nest, with two cells being provisioned, is unique in the genus. Clearly this species deserves further study, and one wonders what other African species may exhibit nest-sharing.

Identified flies taken from females carrying prey (Figure 7.6) or from nests were of four families: *Bombylius discoideus* F., *Bombylius ornatus* Wiedemann, *Exoprosopa* spp., *Systoechus* sp., *Villa* spp. (Bombyliidae); *Musca* sp. (Muscidae); *Allographa calopus* Wiedemann, *Eristalinus taeniops* (Wiedemann), *Eristalis tenax* (Syrphidae); and *Chrysops obliquefasciatus* Macquart (Tabanidae). In addition to these, there were unidentified Bombyliidae, Sarcophagidae, Stratiomyidae, Syrphidae, and Tachinidae.

Figure 7.6. Female *Bembix bubalus* carrying prey. Photo by Harold Gess, from S. K. Gess and F. W. Gess (1989); used with permission.

Bembix cameronis Handlirsch—southern Africa

One female nested in a sandpit and dug a burrow 21 cm long, with a cell 9 cm deep. One female captured with prey carried a *Systoechus* sp. (Bombyliidae); a single museum record of prey was of the same genus.

Bembix capensis Lepeletier—southern Africa

Two prey records reported are *Sarcophaga* sp. (Sarcophagidae) and an undetermined species of Tachinidae (F. W. Gess 1986).

Bembix melanopa Handlirsch—southern Africa

B. melanopa was found nesting in the sloping banks of a sandpit, where two completed nests were excavated. They were unicellular, the burrow passing downward for 21–24 cm before changing direction and, after a further 2 cm, terminating in a cell. Prey from nests consisted of flies of five families: *Musca* sp. (Muscidae); *Sarcophaga* sp. (Sarcophagidae); *Eristalinus taeniops* (Syrphidae); and unidentified Calliphoridae and Tachinidae. A further record of a *Philoliche (Phara) flavipes* Macquart was obtained from a museum record (F. W. Gess 1986).

Bembix sibilans Handlirsch—southern Africa

Near Grahamstown, South Africa, females nested on the sides of a sandpit. Two nests were excavated and found to be unicellular, with burrows 8 and 13 cm in length. Prey in the two nests consisted of flies of four families: *Bombylius delicatus* Wiedemann, *Exoprosopa* spp., *Lomatia oreoica* Hess, *Villa vitripennis* (Loew) (Bombyliidae); *Atriadops vespertilio* (Loew) (Nemestrinidae); *Sarcophaga* sp. (Sarcophagidae); and *Amanella minor* Oldroyd (Tabanidae). Nests also include unidentified Bombyliidae, Calliphoridae, and Tachinidae.

Australasian *Bembix*

The Australian *Bembix* form a diverse lot, because the genus has undergone a major adaptive radiation in Australia, with more than 80 species, some widespread, some apparently of more local distribution. As far as *Bembix* is concerned, Australia was virgin territory in 1966. Except for a few intriguing hints (Wheeler and Dow 1933), the biology of Australian

Bembix was practically unknown. And about two-thirds of the species that we recognize today remained undescribed until the early 1970s (Evans and Matthews 1973). By comparison, only a handful of new North America species have been described during the same interval (Evans and Matthews 1968; Kimsey and Kimsey 1981; Griswold 1983).

Evans and Matthews (1973, 1975) discussed the ethology of 21 species, with fragmentary notes on five others. Evans et al. (1982) added five species and supplemented knowledge of five others that had been studied previously. Evans and Matthews (1973) recognized 12 species-groups. The eleven for which there are ethological data are treated here in the same order in which those authors discussed them. Although these groups were defined on morphological grounds, to some extent they also reflect behavioral differences. Roughly speaking, species of the first six groups may be considered "large," with body lengths generally 16–22 mm, so that they are actually comparable in size to most *Bembix* species from other parts of the world. The remaining five species-groups contain "small" wasps, with body lengths usually of 10–16 mm. Most examples of unusual behavioral traits occur among these smaller species. It goes without saying that only a few salient features of each species can be reviewed here. Unless otherwise noted, the information presented is derived from Evans and Matthews (1973, 1975) or Evans et al. (1982).

Pectinipes SPECIES-GROUP. The five species of this group have the fore basitarsi expanded in both sexes, the females' forelegs bearing an unusually high number of rake spines. For example, the front basitarsus of *Bembix mareeba* bears 14 long spines (Evans and Matthews 1973), whereas that of *Bembix americana* bears just six, more widely spaced spines (Evans 1966a). Compared with their North American counterparts, the forelegs of males of this group are even more widely divergent, being colorful and shield-like, and bearing 22 short spines. Their shape and color is reminiscent of those of male *Crabro*, though *Crabro* shields bear no spines (Bohart 1976). Presumably, the forelegs of the females are used for digging nests, whereas those of the male play a role in courtship (as is thought to be the case for *Crabro*), and perhaps also in digging.

Bembix mareeba Evans and Matthews—coastal Queensland

On the coast of Queensland, there were over 30 nests in a sandy field (in soil containing many roots), and in the firm sand of a road. The unicellular nests had large tumuli, 10–19 cm long, 8–15 cm wide, and 1–3 cm deep in the center. There was no evidence that these tumuli were ever leveled, and no accessory burrows were found. Burrow lengths in the firm sand of a vehicle track were 32–47 cm, cell depths 15–17 cm, while nests in the much looser sand elsewhere were deeper, burrow lengths 80–95 cm, cell depths 32–40 cm. Females carrying prey plunged directly into open nest entrances and the egg was laid on the first fly in the cell. Thirteen prey records included flies of six families: *Musca terraereginae* Johnston and Bancroft (Muscidae); *Euprosopia tenuicornis* Macquart (Platystomatidae); *Dasybasis* sp., *Scaptia aureohirta* (Richards) (Tabanidae); and *Tritaxys* sp. (Tachinidae); as well as one unidentified species each of Asilidae and Syrphidae.

Bembix pectinipes Handlirsch—Northern Territory, Queensland

Nothing is known of this species, except that a female in the collections of the Australian museum is labeled as being a "Wasp which kills March flies" (Tabanidae) (Evans and Matthews 1973).

Palmata species-group. The two species making up this group have fore basitarsi somewhat expanded in both sexes, but less strongly expanded than in the preceding group.

Bembix palmata F. Smith—eastern Australia from northeast to southeast

Along the shores of Lake Burley Griffin, Canberra, females nested in "bare, sloping gravel or coarse sand . . . on hillsides," on a "hard, gravel road," and on a "hard-packed sandy road." At one location, about 50 females were dispersed along 50 m of road, nesting alongside females of *Bembix trepida.* Females initiated burrows with their mandibles, rotating while digging to create circular holes in the compact soil and making "gnawing sound audible half a meter away." Loosened soil was thrust beneath the wasp as far as 10–12 cm behind (Figure 7.7). The 32 nests examined entered the soil at 35–60°, had burrows 14–28 cm long and cells 8–14 cm deep, and took about four hours for the female to dig. Most nests were unicellular, but there is some evidence of two- and three-celled nests. Females provisioned nests progressively, gluing the egg in an erect position on the first (usually large) prey in the cell. Successive nests of individual females were often as close as several centimeters apart.

After completion of the nest burrow, the female levels the mound of soil that has accumulated at the nest entrance. "When leveling, the females

Figure 7.7. Female *Bembix palmata* clearing soil from entrance of nest. Photo by H. E. Evans.

starts near the middle of the mound and works forward scraping sand while turning slightly from one side to the other; when she reaches the closed entrance, she takes flight briefly and lands near the middle of the mound and works forward. Leveling requires only 3–5 minutes, some of that time being spent in the closing and camouflaging of the nest entrance. After the mound is dispersed, 4–6 radiating lines are made beside the nest entrance. Although the mound is thoroughly dispersed, the soil can often be detected for a day or two because the subsoil is darker than the sun-baked surface crust."

Evans and Matthews considered an earlier report that this was a predator of grasshoppers to be in error. Eighty-one prey from three locations in the Australian Capitol Territory, two in Queensland, and one in New South Wales were from seven families of Diptera: *Comptosia* sp. (1 individual) (Bombyliidae); *Calliphora accepta* Mall., *Calliphora augur* (F.), *Calliphora fuscofemorata* Mall., *Calliphora stygia* (F.), *Calliphora tibialis* Macquart, *Chrysomyia rufifacies, Chrysomyia incisuralis* Macquart, *Stomorhina subapicalis* Macquart (Calliphoridae); *Taylorimyia iota* J. & T., plus one undetermined species (Sarcophagidae); *Odontomyia* spp. (Stratiomyidae); *Eristalis punctulatus* Macquart, *Melangyna* sp. (Syrphidae); *Ectenopsis australis* Ric., *Scaptia berylensis* Ric., *Scaptia neoconcolor* Mack. (Tabanidae); and *Anagonia* sp., *Froggattimyia lasiophthalma* Mall., *Prodiaphania* sp., *Rutilia* sp., plus four undetermined species (Tachinidae); over one-half of the prey were Calliphoridae and one-third were *C. tibialis.* Females apparently hunted Calliphoridae on nearby trees.

Males, which were observed at only two of six localities studied, patrolled in flights that were "unusually slow, with much hovering a few centimeters above the ground and with prolonged pauses on the ground." They were active quite early in the morning (as early as 0800 h in one location), retiring to their oblique, 10–12 cm long sleeping burrows by noon each day.

Bembix vespiformis F. Smith—widespread, except northwest Australia

This species has nesting behavior similar to that of *B. palmata.* Both species are rather noisy nesters, producing a low buzzing sound as they approach the nest with prey, and both species make elaborate leveling movements at the final closure. However, four of the five *B. vespiformis* females observed nested in "powdery soil in a sloping bank." Nest burrows of *B. vespiformis* were also somewhat longer (29–38 cm) and cells were deeper (15–18). Eggs were glued in an erect position to the side of the first prey placed in the cell (Figure 7.8). Although just 17 prey were documented, they were from seven species in three families of Diptera: *Exoprosopa* sp. (Bombyliidae); *Calliphora nociva* Hardy, *Chrysomyia rufifacies, Metallea* sp. (Calliphoridae); *Prodiaphania* sp., *Rutilia* sp., and one undetermined species (Tachinidae). One cell containing a fairly large larva also housed 13 still uneaten prey, but the total number of prey provided to a larva is unknown. Provisioning was fully progressive and females maintained an outer (but not inner) closure during hunting.

Promontorii species-group. Four species of this group have been studied in the field. All have somewhat expanded fore basitarsi.

Bembix gunamarra Evans and Matthews—coastal Northern Territory and Queensland

At Newell Beach, Queensland, a group of ~40 nests occupied a slope with "coarse, gravelly sand"; all were at the base of bushes. Nests tended to be

Figure 7.8. Egg on prey of *Bembix vespiformis* (photographed after being removed from nest). Photo by H. E. Evans.

at least 50 cm apart, and no active nests were closer than 15 cm. On a beach near Yeppoon, Queensland, an aggregation of ~30 nests occupied "pure, fine-grained sand" on the slopes of an old dune. The area containing nests was "now well anchored with grass, herbs, and a stand of small Casuarina trees," so that most nests were in at least partial shade. At Yeppoon, females dug deep nests on sandy hillocks in "rather pure, fine-grained sand close to the base of grass and herbs, and some were actually in the middle of clumps of low composites." One nest had a burrow 125 cm long, reaching a depth of 85 cm, and five cells at depths of 50–65 cm. A second nest had three cells 55–62 cm deep, and a burrow 112 cm long. Provisioning females left the entrance open while hunting, so apparently plunged directly into the nests with their prey. The 55 prey consisted of flies of six families: *Orthellia lauta* Wiedmann (Muscidae; 45% of the prey); *Cydistomyia avida* (Bigot) (Tabanidae); and *Rutilia* sp. (Tachinidae); as well as one unidentified species each of Sarcophagidae, Stratiomyidae, Syrphidae, Tabanidae, and Tachinidae. Eggs are evidently laid on the first fly in the cell.

Males patrolled the nesting area, flying in irregular patterns, 10–30 cm high. Their usual flight was rather slow, but they would pick up speed abruptly when pursuing another wasp. When two males approached each other they would circle and butt one another briefly. Sometimes a male would land on the ground or on a leaf, perching with the antennae extended rigidly forward.

Bembix kamulla Evans and Matthews—southeastern and southwestern Australia

Thirty-one nests found at one locality in the interior of New South Wales were "well-scattered along little used sandy tracks through *Eucalyptus-Acacia-Callitris* woodland, most of them separated by 1–3 m" (Evans et al. 1982); some nests were in clusters, some well separated from their neighbors. Nests were relatively shallow, burrows 18–30 cm long, cells 8.0–15.5 cm deep. The majority of nests had a "sharp lateral bend part way down the burrow." Most nests appeared unicellular; in some nests, there were 3–4 cells containing cocoons close to a cell being provisioned, but they may not have been part of one nest. Each nest had two accessory burrows, 2.0–2.5 cm apart and 0.2–7.0 cm long, on each side of the closed true burrow, rendering nests easy to locate.

The 21 prey collected consisted of both flies (Diptera) and antlions (Neuroptera). The four flies were from two genera of Asilidae (*Neoitamus armatus* (Macquart) and *Ommatius* sp.), whereas the antlions (Myrmeleontidae) included *Campsoleon* sp., *Glenoleon pulchella* (Rambur), *Heteroleon* sp., and *Myrmeleon* sp. The egg was laid erect on the first prey. Fragments found in cells all appeared to be either antlions or asilid flies, chiefly the former, and cocoons were always enclosed in the wings of antlions.

Bembix octosetosa Lohrmann—southeastern Queensland, New South Wales, Victoria

At three sites in Queensland and New South Wales, females nested in bare soil of varying degrees of friability, where they left tumuli at nest entrances and often made accessory burrows. Burrows were apparently ended blindly, but just before the end there was a branch that passed upward or horizontally and then downward to a cell. Nests varied greatly in depth at the different sites; at one site cells were 10 cm deep, at another 14–16 cm, at another 31.5 cm deep. The egg was laid on the first fly in the cell, and provisioning was progressive. The five prey collected were flies of different species: *Apiocera* sp. (Apioceridae); *Exoprosopa* sp., *Thyridanthrax* sp., *Villa* sp. (Bombyliidae); and *Chrysomyia rufifacies* (Calliphoridae).

Bembix promontorii Lohrmann—coastal Queensland and New South Wales

In dunes along the eastern subtropical coast of Australia, females dug deep burrows, often within several centimeters from other nests "in vertical sandbanks resulting from man-made excavations and in steep slopes of protected dunes" (Evans et al. 1982). One nest was traced to 65 cm, where the female was still digging. Mounds at nest entrances were not leveled. Males patrolled the nesting area at heights of 0.5–1.0 m, occasionally interacting with conspecific males, and both sexes visited *Eucalyptus* flowers for nectar. No further data on nests or prey are available.

Furcata species-group. This "group" contains only *Bembix furcata*, a common species in southern Australia and the only *Bembix* to occur on Tasmania.

Bembix furcata Erichson—southeastern to southwestern Australia, Tasmania

Most studies of *B. furcata* were conducted along the Cotter River in Australian Capital Territory (A.C.T.), where ~50 females dug nests in fine- to coarse-grained sand along the river. Nests were remarkably shallow for so large a wasp (body lengths 15–19 mm). The species generally nested in montane habitats that were cooler and moister than those of other *Bembix* in Australia. In 22 nests excavated, burrow length varied from 9 to 17 cm, cell depth from 5 to 9 cm. Nests were unicellular and were closed at the entrance while the female was away. The egg was laid on the first fly placed in the cell, and nests were provisioned progressively with diverse Diptera of 17 species, 12 genera, and seven families: *Bombylius* sp., *Geron* sp., *Villa* sp. (Bombyliidae); *Calliphora hilli* Patton, *Calliphora tibialis, Stomorhina subapicalis* (Calliphoridae); *Trichophthalma biritta* Walker, *Trichophthalma harrisoni* Mackerras, *Trichophthalma nicholsoni* Mackerras (Nemestrinidae); *Chrysogaster* sp., *Melangyna* sp. (Syrphidae); *Scaptia auriflua* (Donovan), *Scaptia anomala* Mackerras, *Scaptia maculiventris* (Westwood) (Tabanidae); *Myothiria armata* Malloch, three undetermined species Tachinidae; and *Anabarrhynchus* sp. (Therevidae); over 40% of the 75 prey were *Bombylius* sp. One of 22 cells contained maggots of an unidentified species.

Evans and Matthews (1973) observed males patrolling a nesting area, "flying 10–20 cm above the sand, back and forth or in irregular patterns over some portion of the nesting area." But Dodson and Yeates (1989), who studied male behavior in *B. furcata* on a hilltop in southeastern Queensland, found something quite different. They marked and measured numerous males and found them to be territorial, each male occupying a space and remaining in nearly constant flight, while making frequent darting movements toward other nearby insects. Their flight was 10–20 cm above the ground, moving for several meters in one direction and then reversing their course. Most males returned to the same territories each day. Dodson and Yeates removed males from territories on 16 occasions and in most cases replacement males arrived (in a mean time of 7.8 min). In eight of 11 replacements in which the size of the males was known, the replacing male was smaller than the original male. The researchers found no females nesting at this site and interpreted this as a landmark-based mating system ("hilltopping"). It is unclear whether the difference in male behavior recorded in the two studies represents geographic variation or alternative mating tactics that may both turn out to be present in all populations.

LOBIMANA SPECIES-GROUP. Of the four species in this group, only one has been studied in the field.

Bembix trepida Handlirsch—southeastern Queensland to Victoria, southeastern South Australia

Females of this common species nested in hard-packed sand, especially in dirt roads, but also in urban picnic areas, along with digger wasps of the genera *Cerceris* and *Sericophorus.* Here *B. trepida* sometimes formed dense aggregations, where nests were often separated by only a few cm. Tumuli are left more or less intact, and nest entrances are left open during periods of provisioning. Burrow length varied from 14 to 25 cm, and cell depth from 7 to 13 cm. Provisioning was fully progressive.

The 141 prey collected from five sites consisted of flies of six families: *Calliphora accepta, Calliphora augur, Calliphora hilli, Calliphora nociva, Calliphora sternalis* Malloch, *Calliphora stygia, Calliphora tibialis, Calliphora* spp., *Chrysomyia rufifacies,* Stomorhina subapicalis (Calliphoridae); *Helina* sp. (Muscidae); *Trichophthalma punctata* Macquart (Nemestrinidae); *Eristalis punctulatus* (Syrphidae); *Dasybasis circumdata* (Walker), *Dasybasis neobasalis* (Taylor), *Dasybasis oculata* (Ric.) (Tabanidae); *Chaetophthalmus* sp., *Rutilia* sp., and eight undetermined species of Tachinidae. *Calliphora* made up 75% of the prey, *Calliphora tibialis* being most common (45% of prey). Having completely stocked and closed a nest, females often started a new nest only a few cm away. There was evidence that some new burrows were started from the walls of a previously active nest, resulting in a nest of two to three cells.

Males patrolled the nesting area 10–20 cm above the ground in irregular patterns, where they would frequently "land on the sand, their legs spread widely and antennae extended rigidly forward, alert to passing *Bembix* and other insects." They spent nights in sleeping burrows, either digging a new burrow up to 13 cm long in mid- to late afternoon or using existing holes (perhaps old sleeping burrows).

CURSITANS SPECIES-GROUP. Of the two species in this group, one has been studied in the field.

Bembix cursitans Handirsch—coastal Western Australia

Nests were usually found on the higher portions of partially vegetated dunes 0.5–2.0 km from the ocean. *Bembix cursitans* is a large species, and both males and females made a loud humming noise as they went about their activities. Females left large tumuli at nest entrances that were not leveled, and nest entrances were left open during provisioning. The straight or gently curved burrows entered the soil at angles of 40–70° and were 32–50 cm long. Two to three cells were found in some nests, the new cells being constructed at the ends of side burrows that were 12–17 cm long. All cells occurred at depths of 22.5–42.5 cm. During final closure, females filled the burrows with sand directly in front of the nest entrance, so that a trough formed; neither the trough nor the remaining mound was leveled before the female departed. In one nest excavated following the female's final closure, the burrow was so solidly packed that its course could not be followed.

The 203 prey collected included 36 species, of which 15 were unidentified Tachinidae. Identified prey came from a broad array of Diptera, and included *Apiocera latifrons* Paramonov, *Apiocera* sp. (Apioceridae); *Bombylius* sp., *Compostia* sp. *Phthiria* sp., *Villa* spp. (Bombyliidae); *Calliphora nociva, Chrysomyia rufifacies, Lucilia cuprina* (Wiedemann), *Metallea gracilipalpis* (Macquart), *Stomorhina subapicalis* (Calliphoridae); *Blaesoxipha pachytyli* (Skuse), *Miltogramma* sp., *Protomiltogramma plebeia* Malloch (Sarcophagidae); *Odontomyia* sp. (Stratiomyidae); *Simosyrphus grandicornis* (Macquart) (Syrphidae); *Anagonia* spp., *Prodiaphania cygnus* (Malloch) (Tachinidae); *Acraspisa* sp., and *Psilocephala* sp. (Therevidae). Flies were brought in rapidly once the egg hatched, the females leaving the nests open during hunting (though they maintained inner closures separating cell from burrow). One nest with a recently hatched larva contained 17 flies, while another with a larva only 10 mm long contained 39 flies. Two females provisioned a second cell while the larva in the first cell was still consuming its store of flies (truncated progressive provisioning). No parasites were found in nests, but several females were stylopized.

Males patrolled the nesting sites, flying in circuitous patterns 20–40 cm above the sand. Each appeared to patrol a limited area in bare spaces among bushes, and males sometimes flew at and butted each other. Males made unusually deep sleeping burrows later in the day, the burrow length averaging 16 cm, the terminus 7.5 cm deep.

Atrifrons species-group. This is the first of five species-groups of generally smaller wasps having relatively slender fore tarsi in both sexes. The *B. atrifrons* group contains 18 species, of which five have been studied in the field to varying extents. Several members of this group have departed from the strict use of Diptera as prey, a fact first discovered by Wheeler and Dow (1933), but not confirmed until much later.

Bembix allunga Evans and Matthews—northern and northeastern Australia

This orange-banded species occurs near the east and north coasts of Australia, but also at some inland localities, in some cases along with *Bembecinus* sp. and *Bembix littoralis*. In one location, the soil in the nesting areas was a thin layer of sand over loam, whereas in the other it was "flat sand among bushes behind a row of dunes paralleling a beach" (Evans et al. 1982). Nests were unicellular, burrow lengths 39–60 cm, cell depths 21–35 cm. Mounds at nest entrances are leveled by the female, who twists from side to side and sprays the sand widely, and an outer closure is maintained while the female is hunting.

Prey consisted of Diptera, Neuroptera, and Odonata, making this the only *Bembix* known to use prey from three insect orders. The 28 Diptera identified included *Neoitamus* sp., *Leptogaster* sp. (Asilidae); *Geron* sp. (Bombyliidae); and *Tabanus nigrimanus* Walker (Tabanidae). The 14 Neuroptera included *Suphalasca* sp. (Ascalaphidae); *Italochrysa fascialis* (Banks) (Chrysopidae); and *Leptoleon* sp. and *Myrmeleon* sp. (Myrmeleontidae). Both Odonata were *Diplacodes bipunctata* (Brauer) (Libellulidae). Intriguingly, one cell contained a cicada wing, perhaps the remnants of the wasp's prey, perhaps the remnants of the prey of the asilid. Some of the prey exceeded the wasps in length and extended well behind them as they flew to their nests. The egg was laid on the first prey in the cell. Cells containing a partly grown larva were sometimes crammed with flies, indicating that provisioning was not fully progressive. One cell contained small unidentified maggots along with a wasp larva, and several contained mites.

Bembix atrifrons F. Smith—widespread in southern half of Australia

Females nested chiefly in interior areas of extensive, fine-grained sand. While digging, females leveled accumulated soil in a very characteristic manner, forming a series of radiating lines from the entrance, in a pattern that is quite different from those of the sympatric *Bembix littoralis* (Figure 7.9, top, *atrifons;* middle, *littoralis*). Thus it is fairly easy to identify fresh nests. The one- to five-celled nests were quite deep, burrow lengths 30–60 cm, cell depths 14–35 cm. The egg is laid on the side of the first prey and provisioning is progressive, but the number of flies brought in before the egg hatches is variable.

The 134 prey collected at eight sites consisted of 43 species of Diptera of seven families, with 27 species of Bombyliidae contributing 64% of the prey: *Bathypogon* spp. (two species) (Asilidae); *Anastoechus* sp., *Anthrax* sp., *Bombylius* spp. (five species), *Compostia* spp. (three species), *Dischistus* spp. (two species), *Exoprosopa* spp. (three species), *Phthiria* spp. (four species), *Systoechus* sp., *Villa* spp. (eight species) (Bombyliidae); *Metallea gracilipalpis, Metallea* sp. (Calliphoridae); *Lispe* sp., *Musca vetustissima* Walker (Muscidae); *Aenigmetopia fergusoni* Malloch, *Blaesoxipha pachytyli* (Sarcophagidae); *Anagonia* (two species), *Hyalomyia* sp., *Palpostoma* sp., one undetermined species (Tachinidae); *Acraspisa* sp., and one undetermined species (Therevidae). The only reported natural enemies are ants (*Solenopsis* sp.) that were observed raiding a nest cell (Evans and Matthews 1973) and a single record of *Paraxenos occidentalis* Kifune and Hirashima (Strepsiptera) parasitizing a female (Kifune and Hirashima 1987).

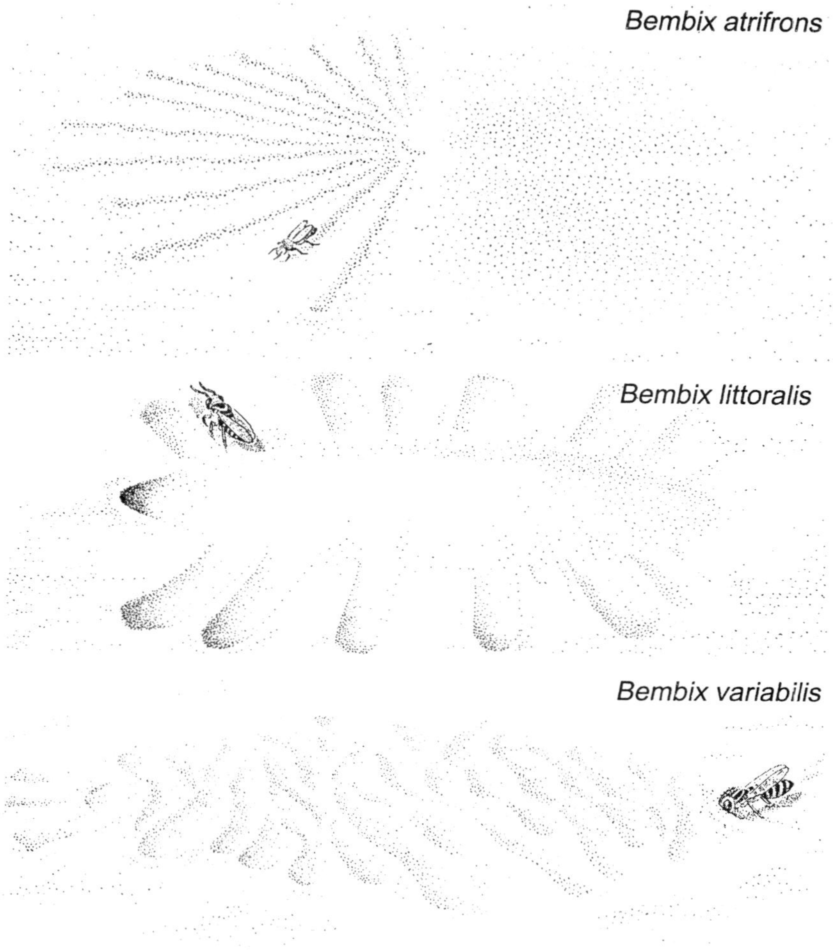

Figure 7.9. Patterns created by females of *Bembix atrifrons* (top); *Bembix littoralis* (middle); and *Bembix variabilis* (bottom) during nest construction. Original drawings by Sarah Landry in Evans and Matthews (1973); used with permission of the American Entomological Institute.

Bembix cooba Evans and Matthews—interior of New South Wales and Victoria

This "small, robust, and almost wholly black species" nests on sandy ridges, often in the company of *B. variabilis* and *B. littoralis.* Before initiating a burrow, females "pepper a considerable area of sand with shallow holes, only a fraction of a cm deep," likely as a means of testing soil texture. *Bembix cooba* females digging nests threw sand into a diffuse mound that was leveled by zigzag movements across it. An outer closure was maintained while the female was away from the nest. In the three nests studied, all of which were unicellular, burrow lengths were 18–35 cm long, cells 14–17 cm deep. All prey in the small sample consisted of Asilidae, six *Bathypogon* sp. and 13 *Ommatius* sp. Cells were filled rapidly once the egg hatched.

Males were plentiful at one site, seemingly concentrating their activities atop the dune ridges. Their flights were exceedingly swift, 10–15 cm above the soil in the late morning, producing a low hum.

Bembix coonundura Evans and Matthews—Western Australia

Bembix coonundura is evidently the species that Wheeler and Dow (1933) reported preying on damselflies, but the females collected by those authors and deposited in the collection of the Museum of Comparative Zoology at Harvard were earlier misidentified as being *B. atrifrons* (Evans 1966a). Evans and Matthews did not actually find it nesting, but they dug several cocoons from the soil at depths of 17–25 cm that were surrounded by the wings of damselflies. One of these cocoons gave rise to a male wasp clearly different from *B. atrifrons,* and this led to a search in museums for similar specimens that were subsequently described as *B. coonundura.* The two prey species thus far found in cells were *Austrolestes annulosus* (Selys) (Lestidae) and *Xanthagrion erythroneurum* (Selys) (Coenagrionidae).

Bembix minya Evans and Matthews—southcentral Queensland to Victoria and southeastern South Australia

In the Adelaide Hills of South Australia, five unicellular nests occupied a pile of "builder's sand" that had been present for seven days (Austin 1999). Two nests examined had 20–25 cm oblique burrows, with entrances about 50 cm above the base of the sand. At the end of the burrows were horizontal cells, 10–12 cm long and containing 15–19 prey each. The 79 prey collected were from two families of damselflies (Odonata): 1 *Ischnura aurora* (Brauer), 17 *Xanthagrion erythroneurum* (Coenagrionidae); 30 *Austrolestes analis* (Rambur), and 31 *Austrolestes annulosus* (Lestidae). The closest source of prey was 200 m from the sand pile.

Bembix wilcannia Evans and Matthews—inland Queensland and New South Wales

The conclusion that *B. wilcannia* preys on antlions (Myrmeleontidae) is based on a single record, provided by T. F. Houston, who captured a female carrying prey tentatively identified as *Myrmeleon diminutus* Esben-Peterson). Whether this species also uses other kinds of prey besides antlions remains to be seen. This was the first reported record of a *Bembix* provisioning with Neuroptera (Evans and Matthews 1973).

Moma species-group. Three of the nine species in this group have been studied in the field. Several have departed from a strict diet of flies.

Bembix moma Evans and Matthews—widespread in drier parts of Australia

This small, pale species, whose name derives from the aboriginal word for "ghost," blends in well with the sand on which it nests. It is widely distributed in Australia, where it forms circumscribed, populous aggregations, sometimes numbering in the thousands of individuals. Most aggregations occupied areas of fine-grained sand, but one was found in firm, dark loam along an irrigation ditch. Tumuli at nest entrances were well dispersed and an outer closure was maintained. Most of the nearly 50 nests examined had a single cell, but one appeared bicellular. Burrow length at five sites varied from 15 to 33 cm, cell depth from 11 to 32 cm. However, at one site where the soil was very dry and powdery, burrows were 50–72 cm long, cells 23–45 cm deep. Nests at some sites were very close together. One 20 cm square excavation unearthed 16 cells at depths of 15–19 cm. The egg was attached to the first prey in the cell, always a bee ($N = 17$).

Provisioning was progressive, but cells were filled rapidly once the egg had hatched. Prey consisted of a diverse mix of Diptera ($N = 28$) and Hymenoptera ($N = 363$), including both wasps and bees. Flies among prey were *Docidomyia* sp. (Bombyliidae); *Chrysomyia rufifacies, Metallea* sp., *Rhinia pallida* Malloch (Calliphoridae); *Musca domestica* (Muscidae); *Australosepsis niveipennis* Becker (Sepsidae); *Odontomyia* spp. (two species) (Stratiomyidae); *Hyalomyia* spp. (two species), *Toxocnemis* sp., and an undetermined species (Tachinidae). The families of wasps among the prey were quite diverse: *Labium* sp. (Ichneumonidae); *Hyptiogaster* sp.

(Gasteruptiidae); *Epactiothynnus* sp, *Gymnothynnus* sp., *Lestricothynnus* sp., *Thynnoturneria* sp. (Tiphiidae: Thynninae); *Telostegus* sp. (Pompilidae); *Larisson* sp., *Liris* sp., *Lyroda* sp., *Pison* sp., *Tachytes* sp., and an undetermined Miscophini (Crabronidae). One cell contained a female *Bembix littoralis,* a larger wasp that nested nearby; presumably it had been taken as prey. Thynninae were the most common wasps, constituting over 13% of the prey. Only three families of bees were found, but they included 24 species: *Trigona essingtoni* Cockerell (Apidae); *Euryglossa* spp. (two species), *Hylaeus albonitens* Cockerell, *Hylaeus huselus* Cockerell, *Hylaeus lateralis* Smith, *Hylaeus* spp. (two undetermined species), *Meroglossa* (two species), *Xanthesma* spp. (two species) (Colletidae); *Homalictus dampieri* Cockerell, *Homalictus dotatus* Cockerell, *Homalictus* (three undetermined species), *Lasioglossum* (five undetermined species), and *Nomia (Austronomia)* spp. (two species) (Halictidae). Bees made up 75% of the prey, with *H. albonitens* (16% of prey), *H. damperi* (20%), and *H. dotatus* (22%) being most common. Individual females were clearly generalists. One cell contained 34 halictids (3 species), 1 ichneumonid, and 1 sepsid. Another cell contained four colletids, one *Trigona,* one tiphiid, two apoid wasps, and five stratiomyids. The diversity of Hymenoptera families among prey is reminiscent of prey of some species of beewolves, such as *Philanthus basilaris* Cresson and *Philanthus zebratus* Cresson (Evans and O'Neill 1988).

In the nesting area, males flew rapid, circuitous patterns close above (3–10 cm) the soil surface, sometimes pouncing on females and investigating open holes. One observation suggested the possibility of a form of mate-seeking behavior among males perhaps different from that typically observed in *Bembix.* Evans and Matthews (1973) state that "at Katherine Gorge, N.T., we found a swarm of 50 males flying about a sandy area covered with tall, dried grass, producing a low humming sound. There were no females present here, and it seemed liked an unusual place for a nesting site."

Bembix mundurra Evans and Matthews—central and southern Western Australia (and one record from New South Wales)

In sand dunes in central Western Australia, brief observations suggested that nests ("dug in moderately compact, rather dry sand") were unicellular and quite shallow, burrows being 15–18 cm long and cells 8–10 cm deep. An outer closure was maintained whether the female was hunting or was inside the nest; the egg was laid on the first fly placed in the cell. The presence of many flies in a cell containing a small larva suggests that provisioning is not fully progressive. During final closure, females filled the burrow completely, but rather loosely. The 47 prey recorded were from 13 species in seven families of Diptera : *Geron* sp. (Bombyliidae); *Calliphora nociva, Chrysomyia rufifacies, Metallea gracilipalpis, Stomorhina subapicalis* (Calliphoridae); *Lamprolonchaea brouniana* Bezzi (Lonchaeidae); *Protomiltogramma plebeia* (Sarcophagidae); *Odontomyia* sp. (Stratiomyidae); *Acupalpa* sp., *Psilocephala* sp. (Therevidae); *Chaetophthalmus* sp., *Hyalomyia* spp. (two species) (Tachinidae); and undetermined Tachinidae. A majority of prey were Tachinidae. Male *B. mundurra* were common at two localities, flying circuitously close to the ground and also visiting flowers. One mating was observed in progress.

Bembix thooma Evans and Matthews—interior of Victoria and New South Wales to central Western Australia

Nests of *B. thooma* occupied bare soil amid vegetation on sandy tracts. Two nests were found with burrows 35 cm long and cells 16–17 cm deep. Eight prey records, from two widely separated sites in Western Australia and New South Wales, consisted of Tiphiidae (Thynninae) of five species tentatively identified as members of the genera *Aspidothynnus* (including *Aspidothynnus rostratus* Turner) and *Thynnoturneria.* Eggs were apparently laid on the first prey in the cell. Females have an unusual modification of the metasomal sternites, including a "high arching crest" on sternite 6, possibly an adaptation for handling this unusual prey.

Flaviventris species-group. This is a group of 11 species of small wasps with short wings that produce a high-pitched whine as the wasps dart about within foliage or going to a nest; two species have been studied in the field.

Bembix flaviventris F. Smith—widespread across southern Australia

This species nested in firm sand near dunes or sandy ridges, where nests were unicellular, burrow lengths 16–30 cm, cell depths 8–15 cm. The 26 prey in the progressively provisioned cells consisted of flies of seven families: *Ommatius* sp. (Asilidae); *Phthiria* spp. (two species), *Systoechus* spp. (two species), *Villa* spp. (two species) (Bombyliidae); *Australophrya rostrata* Robineau-Desvoidy (Muscidae); *Blaesoxipha pachytyli,* an undetermined species (Sarcophagidae); *Odontomyia* spp. (two species) (Stratiomyidae); *Hyalomyia* sp. (Tachinidae); and *Camaromyia* sp. (Tephritidae).

Males were observed both in the nesting area and swarming around flowering trees. Some were seen "in areas of hard soil at some distance from any known areas of sand" where the females were likely to be nesting.

Bembix mianga Evans and Matthews—widespread across southern Australia

Bembix mianga is similar to *B. flaviventris,* but more often nests in very friable soil. Although no large aggregations were found, pairs of nests were sometimes just 3.5 cm apart. Nests were shallow and unicellular, entering the ground at 30°, with burrows 13–28 cm long and cells 7.5–17.0 cm deep. While digging, females undertook extensive leveling of the mound of soil at the nest entrance, so that a "freshly completed nest of *B. mianga* shows little or no evidence of a mound." However, leveling results in a fairly conspicuous pattern of radiating lines that remains visible until dispersed by the wind. One nest, which the female was in the initial stages of provisioning, had a pattern of nine radiating lines, up to 25 cm long.

Prey consisted of flies of five families: *Anastoechus* sp., *Anthrax* sp., *Bombylius* spp. (three species), *Dischistus* sp., *Docidomyia* sp., *Geron* sp., *Phthiria* spp. (three species), *Villa* spp. (four species) (Bombyliidae); *Melangyna* sp. (Syrphidae); *Hyalomyia* sp. (Tachinidae); *Spathulina* sp. (Tephritidae); and an undetermined species of Therevidae, with the bombyliids making up 73% of 44 prey identified. As in *B. flaviventris,* the egg is laid on the first fly placed in the cell and provisioning is fully progressive. Females maintained outer closures except briefly when within nests after returning with prey. Males patrolled flats and blowouts among the dunes, "often hovering over certain spots and occasionally alighting on the sand."

Tuberculiventris species-group. The three species of this group all prey exclusively on bees, thus showing behavioral convergence with apoid wasps of the genus *Philanthus* (Evans and O'Neill 1988), which do not occur in Australia. It appears that all three of these species come very close to mass provisioning their nests.

Bembix flavipes F. Smith—northern, northeastern, eastern Australia

Females nested in sloping sand as close as 2–3 m from water. At one site, the sand had a crust on the surface, at another it was firm, fine-grained with many small roots. All three aggregations were within wooded areas, so that nests were often partly shaded, a relatively rare situation for a *Bembix,* although *Bembix tuberculiventris* nested at the same sites. Nests were also comparable in structure and depth to those of *tuberculiventris,* being unicellular with burrows 15–40 cm long and cells 10–24 cm deep. Females bring in prey rapidly, taking only a minute or two to capture prey and return to the nest; however, a closure is maintained between trips. At all three locations, females preyed only on male *Trigona essingtoni* (Evans and Matthews 1973). It seems probable that prey are taken in swarms outside hives (in contrast to *Bembix tuberculiventris,* which probably hunts on flowers). Cells containing eggs also contained from 1 to 26 bees, the last figure close to the full complement. Two further nests, with 27 total prey, excavated in Western Australia also contained only *T. essingtoni* (Evans et al. 1982). Fully provisioned cells contained 30–35 prey and many were probably completely provisioned within three days.

Bembix musca Handlirsch—eastern Queensland and New South Wales, Northern Territory

Bembix musca occurs primarily along the east coast of Australia. Nests tend to be deeper in dry, friable sand than in moist firm sand. Nests of this species are shallow, densely packed, and comparable in depth to those of *B. flavipes* and *B. tuberculiventris.* Nests are unicellular, burrow lengths 10–45 cm, cell depths 5–26 cm. As a result of leveling movements, during which the females frequently rotates to change the direction of soil dispersal, a semicircular pattern of darker-colored soil can be seen near each nest entrance for several days. However, there was some minor variation in leveling behavior among sites.

Provisioning is often very rapid, although females returning to nests must burrow through the temporary closures that they always erect when away from nests (or when inside it for a prolonged period). One nest dug out after final closure had a larva only 6 mm long and 48 bees. Of 114 prey collected in total, 110 were male *Trigona carbonaria* Smith, whereas four were workers and one was a young queen. The speed of provisioning and the sex ratio of prey suggest that females hunt in nearby mating swarms of the bee. Kifune and Hirashima (1987) found a single *Paraxenos australiensis* Kifune and Hirashima (Strepsiptera) parasitizing a female *B. musca.* At three of the five sites at which this species was observed, males flew "close above the ground somewhat apart from the main nesting area." At night and during cloudy periods, they retreated to short sleeping burrows dug in more friable sand than that used by females.

Bembix tuberculiventris R. Turner—widespread in Australia

Females nested in firm sand near dunes or along sandy tracks. Nests of *B. atrifrons, B. lamellata,* and *B. flaviventris* occur in similar places, their nests sometimes intermingled with those of *B. tuberculiventris.* At Amby, Queensland, digging females scraped sand into a mound that was periodically leveled as the female moved to the top of the mound and rotated her body, kicking sand to the sides. Nests were unicellular, burrow lengths 24–37 cm, cell depths 11–23 cm. Most entrances were left open during provisioning, the female bringing prey rapidly. As a result of the final leveling of the mound, the 5 × 7 cm soil mound near the entrance was completely dispersed, leaving "a weak, irregular pattern of lines over the dispersed soil as well as a few short radiating lines on the opposite side of the entrance."

In Northern Territory and Western Australia, prey consisted of stingless bees (*Trigona essingtoni* and *T. carbonaria*). In the south of the continent (outside the graphic range of stingless bees), prey consisted of 11 species of bees of two families (Halictidae and euryglossine Colletidae), 59% of which were *Lasioglossum purnongense* (Cockerell) (Halictidae) (Evans and Matthews 1973). All 57 prey at Amby, including 46 in one cell, were Colletidae (Euryglossinae): 3 *Brachysema* sp., 53 *Euryglossa* spp., and 1 *Xenohesma* sp. (Evans et al. 1982). The egg was laid erect on the side of the first bee placed in the cell.

Early in the day, males flew about close to the ground in the nesting areas, occasionally "landing with legs and antennae rigidly extended." Attempts to mate with digging/provisioning females were rejected, suggesting perhaps that only newly emerged females are sexually receptive.

Littoralis species-group. This group includes some of the most common and widely distributed of Australian species. Three of the seven species have been studied in some detail, and there are brief notes on two others.

Bembix kununurra Evans and Matthews—northeastern Western Australia

Females were found making shallow nests along a sandy road in Western Australia. The one nest studied was closed at the entrance and had a small tumulus beside it. The burrow was 25 cm long, reaching a cell at a depth of 15 cm. The cell contained an egg attached to the side of a fly, and there were nine additional flies in the cell, all Calliphoridae.

Bembix lamellata Handlirsch—widespread, interior and southern Australia

An inhabitant of areas of coarse-grained sand or sandy gravel, the *B. lamellata* studied nested in the soil that had eroded from a slag heap left from an abandoned mine. Females left a small tumulus outside the nest entrance, which was normally closed while the female was away. Nests were simple and shallow, burrow lengths 15–39 cm, cell depths 9–18 cm. The egg was laid on the first prey placed in the cell, after which provisioning was fully progressive; nests that had received a final closure contained larvae that were nearly fully grown.

The 81 prey records from six localities consisted of 28 species in 10 families of Diptera: *Compostia* sp., *Geron* sp., *Phthiria* spp. (two species) (Bombyliidae); *Aphyssura* sp., *Calliphora tibialis, Chrysomyia rufifacies,* Metallea spp. (Calliphoridae); *Helina* sp., *Passeromyia* sp. (Muscidae); *Austrometopia burnsi* Malloch, *Blaesoxipha pachytyli, Miltogramma regina* Malloch (Sarcophagidae); *Simosyrphus grandicornis* (Syrphidae), *Odontomyia* spp. (two species) (Stratiomyidae); *Dasybasis hebes* Walker (Tabanidae); *Anagonia* spp. (three species), *Chaetophthalmus* sp., *Hyalomyia* spp. (three species), *Hydromyia* sp., undetermined species (Tachinidae); *Tephritis pelia* Schiner (Tephritidae); and *Anabarrhynchus* sp. (Therevidae). No single species made up more than 20% of the prey.

As a result of a stereotyped sequence of movements during the initial stages of final closure, a female created a circular, shallow pit outside the entrance as she packed the burrow with soil. Following this, she repeatedly dispersed soil while backing away from the nest entrance, this time forming an irregular radial pattern of 5–15 cm long lines. Finally, she dug a 1–2 cm deep accessory burrow, opposite of the true burrow.

Within nesting areas, males spent most of their time flying "in circuitous patterns 2–6 cm high and only rarely landed on the ground." Later they dug sleeping burrows up to 17.5 cm in length in areas where soil was more friable than in the nesting area.

Bembix littoralis (R. Turner)—widespread across Australia

Behaviorally, this species of *Bembix* is one of the best-studied in Australia. It is ubiquitous throughout the continent, both on the coasts and at sites in the interior where there is bare sand: beaches, streamsides, dunes, sandy ridges, or artificial sand pits. Nests were often interspersed with those of other species, especially *Bembix variabilis.* Females digging nests leveled mounds periodically, but when the burrow was at full length, their behavior changed and the soil was piled in an elongate mound, 11–17 cm long, 4–7 cm wide, and 1–2 cm deep. Each mound had several grooves emanating from it, produced by the female as she kicked sand behind her (Figure 7.9, middle). The mounds, which made fresh nests easy to spot, are unique among Australian *Bembix* and suggest the mounds at nests of *Stictia maccus* (South America) and *Bembix boharti* (Baja California). Nests were unicellular and quite variable in depth. In 50 nests from coastal sites, burrow length was 14–44 cm, cell depths 7–27 cm; but at sites in the interior of New South Wales, burrow lengths were 30–65 cm, cell depths 18–54 cm. The egg is laid on the first fly placed in the cell, and cells are filled rapidly once the egg has hatched.

One single stingless bee (*Trigona* sp.) was found among flies in one nest. However, this may be an aberration (and we have not included Hymenoptera among prey of this species in our summary tables—see below). The 232 prey from 15 localities consisted of >60 species of flies of 14 families (listed in order of decreasing frequency): Calliphoridae (≥18 species, 42.7%), Muscidae (9 species, 12.5%), Asilidae (1 species, 12.1%), Tachinidae (≥8 species, 7.8%), Tabanidae (6 species, 5.2%), Stratiomyidae (5 species, 4.3%), Platystomatidae (3 species, 3.9%), Bombyliidae (2 species, 2.2%), Sarcophagidae (4 species, 2.2%), Syrphidae (4 species, 2.2%), Tephritidae (1 species, 2.2%), Therevidae (one species, 1.3%), Dolichopodidae (2 species, 0.9%), and Lauxaniidae (2 species, 0.9%), an average of less than four records per species (range: 1–30). The most common prey genus was *Calliphora* (47 records), the most common species *Calliphora fuscofemorata* Malloch (30 records, all from one site). Other geographic variation in prey records was evident in all 28 asilids being from N.S.W. and all 12 tabanids from Queensland.

Males patrolled the nesting area, "flying back and forth or in irregular patterns close to the ground near the nesting sites." When they landed, which was only occasionally, they appeared very alert and were difficult to capture.

Bembix variabilis F. Smith—widespread throughout Australia

Bembix variabilis, another abundant, widespread, and well-studied species, has a number of unique behavioral traits. Nest sites include inland sand dunes, sand behind sea beaches, alluvial sand along streams, and man-made sand pits or mounds. Evans and Matthews (1973) excavated 89 nests in 21 localities. Females made deep nests that sometimes took more than one day to complete. Burrows, many of which were quite crooked, entered the sand at an angle that was at first small (30–40°) and then increased to 45–60°. The burrow leveled off to an elongate brood chamber 9–12 mm in diameter that was horizontal or nearly so. Brood chambers, which were generally 6–12 cm long (rarely up to 25 cm), were sometimes straight but were more often curved gently or, rarely, S-shaped. The 89 nests excavated showed remarkable variation in depth. Nests not far from water were relatively shallow (cell depths 10–30 cm), while those in drier soil of the interior were deeper (cell depths 23–50 cm), and at one site in Western Australia very deep (cell depths 43–67 cm). When the burrow was complete, the female leveled the soil with characteristic movements back and forth across the tumulus, creating a zigzag pattern in the sand (Figure 7.9, bottom). She then made 1–3 accessory burrows well back on the tumulus. Both outer and inner closures were maintained.

The egg was laid in the empty brood chamber, near its terminus, glued to a few sand grains so that it was directed toward the center of the bore. As the larva fed, it moved along the chamber, consuming fresh prey and leaving debris behind. In numerous sites in various parts of the continent, prey (N = 370) consisted of flies in about 90 species and 14 families (listed in order of decreasing frequency): Calliphoridae (≥19 species, 18.6%), Ephydridae (7 species, 18.6%), Muscidae (15 species, 18.6%), Tachinidae (≥5 species, 10.3%), Stratiomyidae (≥2 species, 7.6%), Asilidae (7 species, 6.8%), Dolichopodidae (6 species, 4.9%), Sarcophagidae (8 species, 4.6%), Bombyliidae (7 species, 3.8%), Tabanidae (4 species, 2.2%), Syrphidae (4 species, 1.9%), Therevidae (≥4 species, 1.6%), Anthomyiidae (1 species, 0.3%), and Chloropidae (1 species, 0.3%). No single species made up more than 10% of the dipteran prey, so that the average number of specimens per prey species was about four; the most abundant species in the records was *Ochtera pilimana* Becker (Ephydridae).

Bembix variabilis, like *B. littoralis,* is clearly a generalist with regards to its dipteran prey, and at certain sites in northern Australia, damselflies were also used in provisions (Figure 7.10). We acquired 44 records of five species of Coenagrionidae: *Austroagrion exclamationis* (Campion) (12 specimens), *Ischnura aurora* (Brauer) (6), *Pseudagrion aureofrons* Tillyard (2), *Pseudagrion cingillum* (Brauer) (4), and *Pseudagrion microcephalum* (Rambur); 36 of the records were males, including 25 of 26 *Pseudagrion.* At two sites near Darwin, nests were stocked with both flies and damselflies, but at a site on the Ord River, Western Australia, cells contained nothing but damselflies. One nest with a small larva contained 23 damselflies. Nests containing damselflies were near water courses, where these insects were abundant and perhaps more readily accessible than flies. One nest cell contained a cicadellid leafhopper among 10 flies, but this was undoubtedly placed there by mistake by a *Bembecinus* female.

Males made short "hops" 1–5 cm above the sand, moving in a "low, swift, irregular flight" from one spot to another, 5–10 cm apart, before taking flight again. These "hopping flights" were very similar to those of the North American *Bembix pallidipicta,* and are possibly a response to high temperatures on the soil surface. Several other behavioral features are shared with *B. pallidipicta:* (1) nesting on dunes and other broad expanses of sand; (2) the contours of the burrow, the upper part somewhat analogous to the "preliminary burrow" of *B. pallidipicta;* (3) the elongate brood

Figure 7.10. Female *Bembix variabilis* carrying damselfly prey. From Evans and Matthews (1975); used with permission.

chamber, (4) the egg laid in the empty chamber at its extremity, and (5) prey placed in series in the chamber, the larva eating its way toward the entrance.

Bembix wangoola Evans and Matthews—interior of South Australia and New South Wales

The two completed *B. wangoola* nests examined had burrows 55 cm long and single cells at depths 38 and 45 cm. When the nest was examined, the cell contained 13 bush flies *(Musca vetustissima)* and a tachinid. There were also remains of several other flies, including two Asilidae. Females were seen capturing bush flies from a human. The wasp would fly at him with a buzzing sound and seize a fly in the air or pluck it off clothing with a quick rap. If the wasp failed to take a fly, she would land on the ground nearby, facing the person, then resume hunting.

Overview of *Bembix*

Habitat. As a group, *Bembix* are "sand wasps" in the true sense of the term. Although more widespread studies of individual species need to be undertaken, the specific types of sandy substrate used by females seem to vary among species. Evans (1957b) noted that *B. occidentalis* and *B. pallidipicta* occupied "large tracts of open, loose sand, and more or less blowing sand," whereas *B. amoena* and *B. sayi* were found in "loose sand, generally in small tracts where there is less blowing," *B. americana* in "firm sand, soft earth, or fine gravel, *B. belfragei* and *B. hinei* in "firm coarse sandy soil," and *B. cinerea* in "hard-packed soil" of coastal salt flats. More recent descriptions of nesting substrate include *B. variabilis* nesting in "open expanses of sand," *B. comata* in "in flat or sloping areas of coarse sandy gravel or finely pulverized rock," *B. texana* in the hard-packed vegetation-free soil of a parking lot (Evans 1966a), and *B. antoni* in "level, coarse riverine sand" (Krombein and van der Vecht 1987). The common feature of the nesting substrate of most *Bembix* seems to be its high sand content and (with some notable exceptions) its high friability.

Nest structure. Table 7.1 updates a table presented in Evans (1966a) that listed species by the number of cells within nests. Then, as now, unicellular nests are most common, though further data for some of the species listed might reveal multicellular nests on occasion. For some species, notably *B.*

Table 7.1 Number of cells per nest in different *Bembix* species. For references, see individual species accounts in text as well as Evans (1957b, 1966a).

Location	One cell	Often or always two cells	Often or always with ≥3 cells	Insufficient evidence
N. America	*B. americana* *B. inyoensis* *B. melanaspis* *B. occidentalis* *B. pallidipicta* *B. rugosa* *B. sayi* *B. texana* *B. troglodytes*	*B. amoena* *B. belfragei* *B. cinerea*	*B. dentilabris* *B. hinei* *B. nubilipennis*	
C. and S. America	*B. citripes* *B. multipicta*		*B. brullei*	*B. cameronis* *B. truncata*
Palearctic	*B. merceti* *B. oculata* *B. rostrata* *B. sinuata* *B. zonata*		*B. niponica*	*B. flavescens*
Africa	*B. albofasciata*		*B. bubalus*	*B. melanopa* *B. sibilans*
Orient	*B. antoni* *B. borrei* *B. glauca* *B. orientalis*			
Australia	*B. allunga* *B. cooba* *B. flaviventris* *B. furcata* *B. lamellata* *B. littoralis* *B. mareeba* *B. mianga* *B. minya* *B. mundurra* *B. musca* *B. tuberculiventris* *B. variabilis*	*B. moma* *B. trepida* *B. vespiformis*	*B. atrifrons* *B. flavipes* *B. palmata*	*B. cursitans*[a] *B. gunamarra*[a] *B. kamulla* *B. octosetosa*

[a]In *B. cursitans* and *B. gunamarra,* no unicellular nests were documented among the three completed nests found.

littoralis and *B. variabilis,* extensive studies in many locations have yet to reveal anything but unicellular nests. On the other hand, although Evans (1966a) felt confident in concluding that *B. americana* females consistently construct unicellular nests (just a single report among many indicated otherwise), Alcock (1972) later found a population in which two- and three-celled nests were common. In many species, both unicellular and multicellular nests are constructed, with as many as five cells having been found in nests of *B. atrifrons, B. dentilabris,* and *B. gunamarra,* and six being reported in *B. niponica.*

Nests of *Bembix* are generally relatively simple, an oblique burrow leading to one or more cells. However, there are variations on this theme, as illustrated by the variety of nest types dug by Australian *Bembix* (Figure 7.11). Perhaps the most common variant is the construction of a short, blind burrow or "spur" along the shaft of the true burrow, presumably providing a place for the female to sit when she is in the burrow (e.g., in *B. albofasciata, B. bubalus, B. multipicta, B. niponica,* and *B. octosetosa*). It is noteworthy that *Bembix* females (in contrast to members of most other genera of Bembicini) spend the night in the nest; thus it is not unexpected that several species have been found to have apparent resting areas within the nest.

More distinctive variations are found in *B. belfragei, B. occidentalis,* and *B. pallidipicta,* all of which were described by Evans (1957b) (Figure 7.12). In *B. belfragei,* the cell is located on a lateral branch that rises upward from the burrow before leveling off to the cell. In *B. occidentalis,* the "main burrow, which usually starts at a moderate angle, 25–40°, and then goes down at a sharper angle, around 50°, terminates blindly and without a cell at from 28 to 64 cm. At some distance before the end of the burrow (3–10 cm) a second burrow passes upward 2–3 cm. and then sharply downward, deeper into the earth." The secondary burrow, which is separated from the primary burrow by a sand plug, may also terminate blindly, the cell being located at the end of a tertiary branch. Finally, the nest of *B. pallidipicta* is "one of the most remarkable structures of any digger wasp," having a long (21–51 cm) preliminary burrow that lies parallel to and just beneath the soil surface. Surprisingly, this tunnel is eventually abandoned, and a new entrance is made near its end after the oblique burrow is constructed. In addition, just below the point in the oblique burrow where the cell branches off, there is a short spur. Although most nests are simple burrows with a terminal cell, or branched with a cell at the end of each branch,

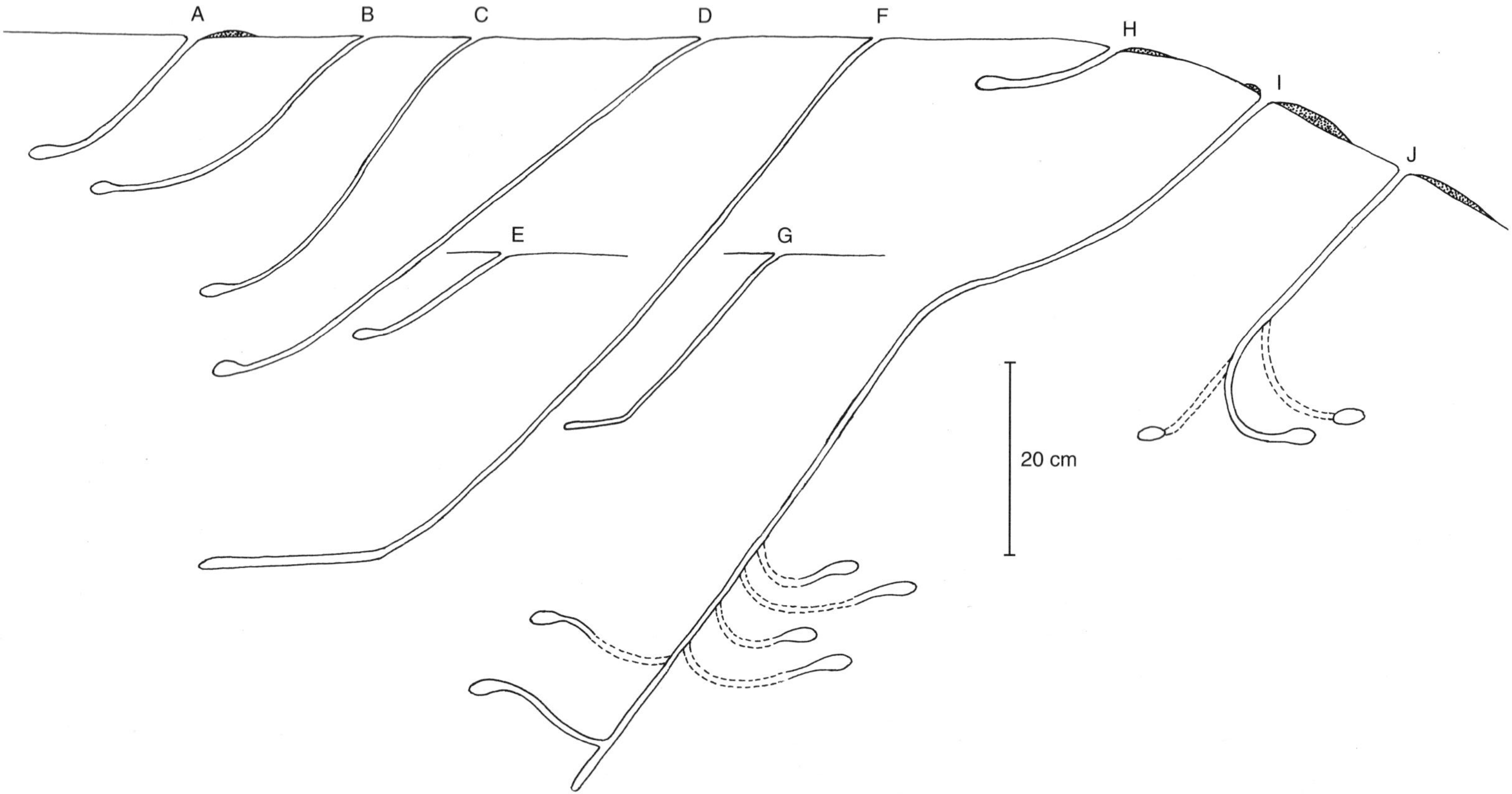

Figure 7.11. Nests of Australian *Bembix:* (A) *Bembix trepida,* (B) *Bembix thooma,* (C) *Bembix atrifrons,* (D, E) *Bembix littoralis,* (F, G) *Bembix variabilis,* (H) *Bembix furcata,* (I) *Bembix gunamarra,* and (J) *Bembix cursitans.* All redrawn to same scale from Evans and Matthews (1973).

nests of unusual structure have been reported. In *B. pallidipicta,* the brood cell is exceeding long (11–33 cm) and usually gently curved or even S-shaped when viewed from above. Elongate brood cells are also present in a few other species, including *B. minya* (10–12 cm long) and *B. variabilis* (4–25 cm).

Cell depth can be quite variable both between and within species, depth perhaps being correlated with soil type and soil moisture (see Chapter 8). In North America, species such as *B. melanaspis, B. pallidipicta,* and *B. rugosa* tend to have relatively deep nests. The depths of 91 nests of *B. pallidipicta* reported in Evans (1957b) and earlier in this chapter ranged from 16 to 56 cm. On the other hand, species such as *B. americana, B. cinerea,* and *B. sayi* cells are relatively close to the soil surface. The depths of 71 *B. sayi* nests (Evans 1966a; unpublished notes in this chapter; Alcock and Gamboa 1975) ranged from 10 to 27 cm. The data just mentioned for *B. pallidipicta* give a picture of the intraspecific variation in nest depth, much of which can be found within populations (e.g., a range of 21.5–53.5 cm among 38 nests excavated in Reno, Kansas, by Evans 1957b). Similarly, in extensive studies of *B. variabilis* in Australia, Evans and Matthews (1973) documented quite wide geographic variation in cell depth. At one extreme, cell depths in five nests at Ivanhoe Crossing near Kununurra in Western Australia ranged from 43 to 67 cm. At the other extreme were 34 nests from various sites in Australia where nest depths never exceeded 23 cm and were as shallow as 9 cm.

Leveling. The behavior used to level soil around the nest entrance following completion of the burrow and cell is one of the most variable and stereotyped in the genus *Bembix.* A large number of species from several geographic areas apparently perform no leveling whatsoever (or else it is rare and irregular): *B. amoena, B. belfragei, B. cinerea, B. occidentalis, B. sayi,* and *B. texana* in North America; *B. albofasciata, B. bubalus,* and *B. melanopa* in Africa; *B. cursitans , B. flaviventris, B. gunamarra, B. lamellata, B. mareeba, B. mundurra,* and *B. promontorii* in Australia. Considering the species that do level the mound to a greater or lesser extent, Evans (1957b) and Evans and Matthews (1973) have made attempts to classify the types of mound-leveling behaviors. However, they have also made it clear that species in the same category may differ from one another somewhat (and other species may overlap several categories). Several Australian species make only rudimentary efforts to level the tumulus; *B. furcata* females made only "a few hasty leveling movements," whereas *B. trepida* females

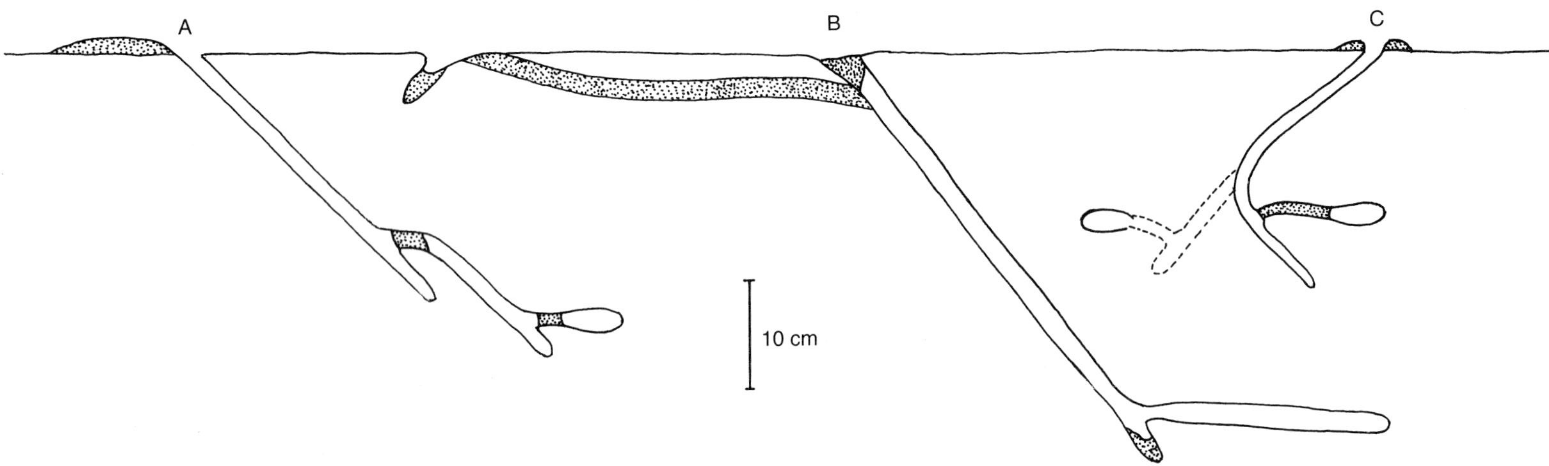

Figure 7.12. Nests of three North American *Bembix*: (A) *Bembix occidentalis*, (B) *Bembix pallidipicta*, and (C) *Bembix belfragei*. All redrawn to same scale from Evans (1957b).

"leveled only in a desultory manner" (Evans and Matthews 1973). A number of other species put more effort into the process using either irregular zigzagging movements across the mound (e.g., *B. americana, B. palmata,* and *B. variabilis*), patterns of radiating lines outward from the nest entrance (e.g., *B. multipicta, B. citripes,* and *B. mianga*), or rotational movements that disperse the soil in a wide arc away from the entrance (e.g., *B. atrifrons*). Finally, and most curiously, there is "reversed" leveling or "mound-building behavior," first described in *B. littoralis* (Evans and Matthews 1973) and, more recently, in *Bembix boharti* (Evans 1976c).

Accessory burrows. Among the Bembicinae, accessory (false) burrows are found only among certain species of *Stizus, Bembecinus,* and *Bembix;* among other apoid wasps, they are found in some *Sphex* and *Philanthus* (Evans 1966b; Evans and O'Neill 1988). Among North American species of *Bembix,* they have been observed in *B. amoena, B. pallidipicta, B. sayi, B. texana,* and *B. troglodytes.* Similarly, in Australia, accessory burrows have been observed in *B. kamulla, B. lamellata, B. octosetosa,* and *B. variabilis.*

Oviposition. In the vast majority of *Bembix,* females lay the egg on one of the prey items in the cell, usually the first prey placed there; that prey item usually remains uneaten, serving only as a pedestal for oviposition. The placement of the egg, which is often laid in an erect position on the fly, results in the dislocation of the wing and middle leg on one side of the prey. Ten *Bembix* species are known, however, to lay the egg in the empty cell prior to provisioning. In eight of these species, the egg is glued to substrate attached to a few sand grains in an erect position, generally near the center of cell. These eight include three Nearctic species *(B. melanaspis, B. texana, B. troglodytes),* three Neotropical species *(B. brullei, B. citripes, B. multipicta),* one Palearctic species *(B. olivacea),* and one Australian species *(B. variabilis).* Two other variants in egg position are represented by one species each, *B. pallidipicta,* in which the egg is placed in semierect position against the apical end of cell, and *B. occidentalis,* in which the egg is placed flat on the bottom of the cell.

Prey and provisioning. Most *Bembix* are progressive provisioners. In North America, *B. hinei* and, perhaps, *B. stenebdoma* are mass provisioners. All Australian species are progressive provisioners, but *B. atrifrons, B. flavipes,* and *B. musca* bring in prey so rapidly that they come close to being mass provisioners. As is usual for progressive provisioners, the number of prey provided to each larva is difficult to estimate, and is usually done either by using prolonged observation of an individual or by estimating prey num-

Table 7.2 Species of *Bembix* that prey on non-Diptera.

Bembix species	Location	Number of prey families from each order					Total number of prey records	Proportion non-Diptera in prey records
		Odonata	Neuroptera	Lepidoptera	Hymenoptera	Diptera		
B. allunga	Australia	1	3			2	44	0.36
B. citripes	South America				1	5	?	Very small
B. coonundura	Australia	2					?	1.00
B. flavipes	Australia				1		191	1.00
B. kamulla	Australia		1			1	21	0.81
B. littoralis	Australia				1	14	232	<0.01
B. melanaspis	North America	1				5	?	?
B. minya	Australia	2					79	1.00
B. moma	Australia				8	6	410	0.93
B. musca	Australia				1		317	1.00
B. regnata	Africa			3			?	1.00
B. stenebdoma	North America		1				10	1.00
B. thooma	Australia				1		?	1.00
B. tuberculiventris	Australia				3		57	1.00
B. variabilis	Australia	1				14	414	0.11

bers from prey fragments in cells; but the latter is not possible in species that clean prey remains from cells. Among North American species, Evans (1957b, 1966a) has estimated the maximum number of prey per cell to be, for example, 13 for *B. hinei,* 17 for *B. cinerea,* 26 for *B. troglodytes,* 30 for *B. pallidipicta,* and 41 for *B. nubilipennis.* For Australian species, Evans and Matthews (1973) gave estimates ranging from 23 to >50. Whether the range of values observed represents consistent interspecific differences, or whether the variation is due to other factors (e.g., site-specific differences in the sizes of prey available), needs to be determined. For example, Evans and Matthews (1973) estimated that perhaps twice as many prey were needed per cell when *B. variabilis* provisioned with small Diptera, compared with when they used damselflies; much of the variation in prey use was geographic.

Among the most interesting findings to appear since 1966 is the extent to which many *Bembix* species have diversified prey use beyond Diptera. One species, *B. allunga,* takes prey from three (and perhaps more) orders, and several species have apparently abandoned the use of flies altogether. Altogether, 15 species from a variety of geographic areas are now known to prey on non-Diptera (Table 7.2). With the exception of two brief reports, that of *B. regnata* taking butterflies (Benson 1934) and of *B. coonundura* taking damselflies (Wheeler and Dow 1933), all reports on non-Diptera prey of *Bembix* appeared after 1970.

The fact that a number of *Bembix* throughout the world have expanded into the use of non-Dipteran prey suggests that there is something about this genus that allows such evolutionary shifts. Clearly this is the case to some extent, or *Bembix* would not have the flexibility to take both small robust flies and large, elongate Odonata and Neuroptera. But such a flexibility to take morphologically and behaviorally diverse prey is also evident in the records of species that take Diptera only. Are *B. americana* and *B. littoralis,* which take both tiny, slow-flying Dolichopodidae and large, fast-flying Tabanidae, really much less behaviorally flexible than *B. variabilis,* which also use damselflies? Unfortunately, we know very little about the intra- and interspecific diversity of the hunting techniques associated with different prey types (and nothing about the pharmacological effects of venoms on different prey used by the same species). We return to this subject in Chapter 8.

Perhaps there is also something about the environments that *Bembix* inhabit that promotes evolutionary diversification of diet. Do these environ-

Table 7.3 Families of Diptera prey known to be taken by Australian *Bembix*.

Bembix species (smaple size)	Anthomyiidae	Apioceridae	Asilidae	Bombyliidae	Calliphoridae	Chloropidae	Dolichopodididae	Ephydridae	Lauxaniidae	Lonchaeidae	Muscidae	Nemestrinidae	Platystomatidae	Sarcophagidae	Sepsidae	Stratiomyiidae	Syrphidae	Tabanidae	Tachinidae	Tephritidae	Therevidae
B. allunga (28)[a]			X	X																	
B. atrifrons (134)			X	X	X						X			X					X		X
B. cooba (19)			X																		
B. cursitans (203)		X		X	X									X		X	X		X		X
B. flaviventris (26)			X	X							X			X		X			X	X	
B. furcata (75)				X	X							X					X	X	X		X
B. gunamarra (55)											X			X		X	X	X	X		
B. kamulla (21)[b]			X																		
B. kununurra (10)					X																
B. lamellata (81)				X	X						X			X		X	X	X	X	X	X

B. littoralis (232)[c]			X	X	X		X		X		X		X	X		X	X	X	X	X	X
B. mareeba (13)			X								X		X				X	X	X		
B. mianga (44)				X													X		X	X	X
B. moma (27)[c]				X	X						X				X	X			X		
B. mundurra (47)				X	X					X				X		X			X		X
B. octosetosa (5)		X		X	X																
B. palmata (81)				X	X									X		X	X	X	X		
B. pectinipes (?)																		X			
B. trepida (141)					X						X	X					X	X	X		
B. variabilis (370)[d]	X	X		X	X	X	X	X			X			X		X	X	X	X		X
B. vespiformis (17)				X	X														X		
B. wangoola (16)			X								X								X		
Number of species	1	3	8	14	13	1	2	1	1	1	10	2	2	9	1	9	10	8	16	4	8

[a]Also takes Neuroptera and Odonata; [b]also takes Neuroptera; [c]also takes Hymenoptera; [d]also takes Odonata.

Table 7.4 Families of Diptera prey known to be taken by North American *Bembix*.

Bembix species	Anthomyiidae	Apioceridae	Asilidae	Bombyliidae	Calliphoridae	Conopidae	Dolichopodididae	Ephydridae	Muscidae	Mydidae	Otitidae	Nemestrinidae	Sarcophagidae	Sciomyzidae	Stratiomyiidae	Syrphidae	Tabanidae	Tachinidae	Tephritidae	Therevidae	Trypetidae
B. americana comata	X		X	X	X		X		X		X		X	X		X	X	X	X	X	
B. americana spinolae	X		X	X	X		X		X		X		X	X	X	X	X	X		X	
B. amoena			X	X	X				X				X	X	X	X	X	X		X	
B. belfragei				X	X				X				X			X	X	X			
B. boharti					X						X		X			X		X			
B. cameroni				X					X							X	X	X			
B. cinerea				X	X			X	X				X		X	X	X	X			
B. dentilabris		X	X	X									X		X	X	X				
B. hinei															X	X	X				
B. inyoensis			X	X														X		X	

B. melanaspis[a]					X	X										X	X			X	
B. multipicta			X		X				X						X	X					
B. nubilipennis			X	X	X				X		X	X	X		X	X	X	X			X
B. occidentalis		X		X	X				X		X		X		X	X	X	X		X	
B. pallidipicta	X			X	X		X		X		X		X		X	X	X	X		X	
B. rugosa			X							X										X	
B. sayi			X	X	X				X		X		X			X	X	X		X	
B. texana				X	X								X		X	X	X	X			
B. troglodytes				X	X				X				X		X	X	X	X			
B. truncata			X																		
Number of species	3	2	10	14	14	1	3	1	12	1	7	1	13	3	11	17	15	14	1	9	1

[a]Also takes Odonata.

ments exhibit particularly wide temporal variation in prey availability, flies being abundant some years, but less so in others? Evans and Matthews (1973) suggested that the exclusive use of damselflies in some areas by *B. variabilis* may be due to a combination of two factors. One, perhaps, is competition with *B. littoralis*, which takes only flies. The other is the fact that *B. variabilis* nests near permanent water, where damselflies supply a reliable source of alternative prey. Temporal and spatial variation in prey use has also been documented in other solitary wasps (reviewed in O'Neill 2001). Or can prey diversification in *Bembix* be attributed to the fact that the genus has expanded into geographic areas where the lack of important competitors opens up new opportunities to also expand prey use? For example, the shift to Hymenoptera by at least four species of Australian species of *Bembix* could perhaps be attributed to the absence from Australia of the wasp genera *Philanthus and Palarus*, which are bee and wasp predators in many areas of the world (Evans and Matthews 1973). Certain species of the genus *Cerceris*, a group of beetle predators in most places, have also shifted to use of Hymenoptera in Australia. However, this is all evolutionary guesswork at this point, and competition as a cause of differences among species should only be invoked when direct evidence is available.

But flies are the mainstay of the diets of most *Bembix*. Of the 29 Australian species that have been studied, 18 take only flies (so far as is known), while four take both flies and insects of other orders (Neuroptera, Hymenoptera, Odonata, and Homoptera) (Tables 7.2 and 7.3). The remaining 7 species have not been found to take flies but take members of the orders Hymenoptera, Neuroptera, and Odonata. Similarly, of the 20 species of North American *Bembix* for which prey records are now available (Table 7.4), 18 take only flies. One species *(B. stenebdoma)*, for which only a few records are available, may take only lacewings, while another *(B. melanaspis)* takes both flies and damselflies.

Male behavior. In most *Bembix*, males find females by patrolling areas in which females first emerge as adults from their mothers' nests and later begin their own nests. The reason that these areas typically overlap, of course, is that populations often nest in the same places each year unless the habitat is disturbed or unless erosion or vegetational changes modify the habitat. Patrolling males are generally nonaggressive with one another, their mutual chases being best interpreted as investigatory flights (even when contact is made). The term "sun dance" to describe the flight patterns of males was originally used to describe the behavior of male *Bembix nubilipennis* (Rau and Rau 1918). Subsequent studies have revealed minor variations on the general pattern of the dances. Much variation in the descriptions of sun dances is in terms of the density of males. The dances also differ in the movement patterns of individual males. Whereas in most species males undertake smooth, sinuous flights above the ground, the height of the flight path may vary among species, being for example 2–5 cm in *B. cinerea* and 10–25 cm in *B. belfragei* (Evans 1957b). The sun dances are qualitatively different in *B. variabilis, B. occidentalis,* and *B. pallidipicta. B. pallidipicta*'s flights involve "hopping dances," or series of short hopping flights that make groups of males look like "aggregations of very small toads" (Evans 1957b). Two studies appearing since 1966 reveal two very different forms of male behavior in *Bembix* that somewhat expand the amount of behavioral variation in the genus. Male *B. rostrata* also patrol the emergence areas, but rather than waiting for females to appear above ground, they dig down to the female in attempts to contact her just at the point of emergence. Male *B. furcata* diverge even further from the standard behavioral mode for *Bembix,* exhibiting hilltopping behavior away from emergence areas.

Not only are students of *Bembix* confronted with relatively little interspecific variation in male activities, the range of intraspecific variation (when fully documented) may include much of the reported variation among species. Much of that variation could be attributable to conditions that happen to hold at the time and place of observation. At other times, species that at first may seem to have slightly different patterns of male behavior may in fact be indistinguishable in their male behavior. Factors that could induce intraspecific variation in male behavior may include local topography and vegetation (which could influence patterns of movement), temperature (which affects flight speeds and the duration of sustained flight), the size of the nesting area (which could affect patterns of movement, and which could vary within a season), male and female density (which influences the frequency of male responses to potential competitors and mates), and the operational sex ratio (the ratio of mate-seeking males to sexually receptive females, which affects the availability of receptive females and thus the number of likely targets of male attention).

— 8 —

Comparative Ethology of Sand Wasps

The foregoing tribe-by-tribe review of research done over the last 40 years makes it clear that the Bembicinae form a behaviorally diverse group. Here we will bring together recent information on behavioral diversity in the Bembicinae with that summarized previously by Evans (1957b, 1966a). Unless otherwise given, the material cited in this chapter can be found in these publications or in the preceding chapters of this book. Readers interested in pursuing sand wasp behavioral studies may want to consult the Appendix, which presents what we consider to be important questions that need to be answered.

Habitat

Sand wasps seem especially diverse in such places as the deserts of the southwestern United States, southern Africa, and Australia. Many Bembicinae occupy open, unshaded habitats, where conditions can be extremely harsh on occasion. Rubink (1978) recorded soil surface temperatures as high as 60–70°C within nesting areas of *Bembix pallidipicta* on sand dunes in New Mexico. Of course wasps are not necessarily active on soil surfaces at times of day when such temperatures occur. They may spend the hottest times of day underground or on flowers, or modify their activities to avoid attaining stressful body temperatures. Evans (1957b) noted that when temperatures on the soil surface in nesting areas of *Bembix pallidipicta* in Kansas exceeded 50°C, temperature at the depth of nest cells remained at about 30°C. Both males and females retreated underground near midday, males to sleeping burrows, females to their own nests, which they closed from inside. Willmer (1985) and Weaving (1989a) present quantitative analyses of the capacity of nests of ground-nesting wasps to buffer temperature and humidity.

Evans also noted that female *B. pallidipicta* digging on hot sand surfaces alternate short bouts of digging with short flights into the air. Similarly, on the grasslands in northern Colorado, we found that *Bembecinus quinquespinosus* and *B. strenuus* males were active on and near the ground when temperatures exceeded 50°C. But as temperatures increased during the day, males perched on the soil surface for shorter periods (as also happens in *Stizus continuus*); the ability of larger males of *B. quinquespinosus* to remain active on hot surfaces may be enhanced by their bright yellow body coloration, which reflects more solar radiation than the darker integument of smaller conspecific males. Temperatures exceeding 50°C are rapidly fatal to another digger wasp, *Philanthus psyche,* in the same habitat as *B. strenuus,* and they would assuredly be stressful to sand wasps as well. Chapman et al. (1926) noted that, in the lab, *Bembix* and *Microbembex* showed signs of stress at temperatures as low as 45°C. It is perhaps to avoid overheating that male *Stictia heros* abandon their sun dances, which occur within 15–30 cm of the soil surface, as temperature increases during the morning. At that time, some males switch to territorial behavior and remain active for another hour or so; but as ambient air temperature increases beyond 36°C (and soil surface temperatures surpass 55°C), all male activity ceases.

Evans and Matthews (1973) hypothesized yet a different type of adaptation of some Australian *Bembix* to weather conditions. They suspected that many *Bembix* can go through extended diapause in their cocoons, a diapause that ends only when cocoons are wetted by soaking rains that also increase prey and nectar availability. One example may be *Bembix coonundura,* the damselfly predator that lives around Lake Violet in Western Australia. Here damselflies emerge only when the lake has water, an event that does not occur every year. Although Evans and Matthews were not present during a wet spell, they were able to induce emergence of *B. coonundura* by moistening cocoons. Admittedly, the hypothesis remains conjecture, especially because it is possible that *B. coonundura* preys on flies when damselflies are not present; recall that *Bembix variabilis* also takes both Diptera and Odonata.

Parsivoltinism, in which some adults of a generation emerge only after two years have passed, may prove to be widespread in Bembicini, though it has been little studied. Martins et al. (1998) found that *Bicyrtes angulatus* that were dormant in the prepupal or pupal stage within the cocoon develop fully and emerge any time from 44 to 375 days from the date the egg was laid. Parsivoltinism may well be a form of "bet hedging," that is, a mechanism for survival during periods of unfavorable conditions, such as prolonged drought or submersion. Wasps of some species may need to delay emergence in flooded areas, though some sand wasps may die in saturated soil (see under *Bembecinus quinquespinosus*). In his book *The Naturalist on the River Amazons,* Bates (1863) reported that *Stictia signata* "sometimes excavates its mine solitarily on sandbanks laid bare in the middle of the river." Similarly Richards (1937) spoke of *S. signata* as "nesting in sands that are submerged in the wet season." Alcock and Gamboa (1975) found *Stictiella callista* nesting in a "sandy floodplain" of the Gila River in Arizona, and *Bicyrtes viduatus* on "a small island in the middle of a seasonally dry wash."

Not all Bembicinae are denizens of hot, dry habitats. A few species prefer cool climates or microclimates, including species of *Alysson* and *Didineis* of the Alyssontini, and *Trichostictia* and *Zyzzyx chilensis* of the Bembicini. Although most *Bembix* prefer relatively xeric habitats and high temperatures, there are a few exceptions. *Bembix rostrata* is probably a typical *Bembix.* Both males and females are inactive on cloudy days or when air temperatures are below 22°C, after which they must use endothermic warm-up to achieve the minimum thoracic temperatures required for flight (i.e., 36°C). At Point Reyes National Seashore in California, however, *B. americana comata* were active on foggy days and at temperatures as low as 12°C. Similarly, *B. furcata* generally nests in cooler and moister montane sites than other *Bembix* in Australia. Also in Australia, *B. flavipes* and *B. tuberculiventris* tend to differ from other species, because they nest in areas that are partially shaded.

As we noted in Chapter 1, the desert southwest of the United States harbors a great diversity of Bembicinae, including several genera that occur nowhere else. And sand wasp diversity and abundance decrease with latitude and elevation in North America. Nonetheless, Evans (1970) recorded 15 species at 44° north latitude in Jackson Hole, Wyoming, and adjacent Yellowstone National Park at sites exceeding 2000 m elevation. Although winters in both Jackson Hole and Yellowstone can be extremely harsh, prepupae of these wasps manage to survive in their shallow nests. Further north, an extensive four-year survey in the Little Belt Mountains of central

Montana (nearly 47° N and over 2100 m elevation) picked up only one Bembicinae, *Alysson triangulifer* (KMO and J. Fultz—unpublished data). But this is by no means the northern limit of Bembicinae on the continent. Steiner (1973) collected five species of Bembicinae, *A. triangulifer, Gorytes albosignatus* Fox, *Lestiphorus cockerelli* (Rohwer), *Nysson lateralis* Packard, and *Nysson subtilis* Fox, and in the Northwest Territories at 61–62°N latitude.

Nesting Substrate

It is undoubtedly true that the local distribution of sand wasp species is determined partly by the availability of suitable nesting substrate. The question largely remains, however, what precisely is "suitable nesting substrate" for each species. Ethologists are generally not deeply trained in soil science, so most of the characterizations of nesting substrate in the literature are qualitative descriptions by untrained observers. For that reason, we have quoted soil-type descriptions in many of the species accounts to avoid adding further imprecision by paraphrasing. Nevertheless, even entomologists can distinguish sand from clay, and silt from pebbles, so the information reported certainly has much value. At a minimum, the descriptions tell us something about the general type of habitat occupied by a species, and give us a baseline for comparing different populations, as long as different observers have the same general idea as to what is meant by such terms as "heavy clay loam" or "coarse riverine sand." However, anecdotal habitat descriptions can be idiosyncratic: what is "slightly sloping coarse sand" to one researcher might be "nearly flat fine gravel" to another.

Many Bembicinae certainly live up to the name "sand wasps," especially if we take a broad view of what "sand" means and include friable soils of different grain size. Entomologists cataloguing the insect inhabitants of sand dunes or other habitats in with sparsely- or nonvegetated sand could find, depending upon their location, species of such genera as *Bembecinus, Bembix, Bicyrtes, Glenostictia, Gorytes, Hoplisoides, Microbembex, Rubrica, Steniolia, Stictia, Stictiella, Trichostictia,* and *Zyzzyx.* Even so, the same sites would likely also harbor apoid wasps of subfamilies not called sand wasps. Furthermore, certain species in the above genera may nest in nonsand substrates. Within the spectrum of grain sizes found in dry, friable soils, some species of *Glenostíctia* and *Steniolia,* for example, prefer fine-grained powdery soil. And *Bembecinus quinquespinosus, Bembix amoena, B. belfragei, B. gunamarra, B. hinei, B. lamellata, B. palmata,* and *Clitemnestra plomleyi* are found in gravelly soil.

Wasps that nest in bare, friable, sandy substrates find the soil relatively easy to penetrate, and they do not have to deal with many plant roots or subterranean rocks; loose soils probably also inflict less wear and tear on mandibles and legs. But many Bembicinae construct nests in hard-packed soils. *Sphecius hogardii* nests in hard stony soil, *Sphecius nigripes* in hard-packed dirt and gravel, *Bembecinus posterus* in hard clay, and *Bembix truncata* in extremely hard, heavy clay loam. Other sand wasps, such as *Bembix cinerea, B. dentilabris, Hemidula singularis,* and *Stizus continuus* nest in saline soil where few other species live.

Even this range of soil conditions does not the exhaust the possibilities within the subfamily. We began our survey in Chapter 2 by noting that Alyssontini nest in firm soil that is often quite moist, as is also the case for *Pseudoplisus ranosahae* females nesting in damp clay banks. In addition, F. X. Williams (1928) found *Sagenista brasiliensis* nesting "banks of rich soil along the margin of a jungle," although it has been reported in sandy soil elsewhere. In Sri Lanka, *Bembix antoni* nested in "coarse, riverine sand along the bank of a rocky stream . . . [where] the soil below the surface was a damp sandy loam" (Krombein and van der Vecht 1987). With regard to the orientation of nesting substrate, most Bembicinae inhabit level to gently sloping substrates. However, an assortment of species dig nests in vertical banks, including *Bembix promontorii, Clitemnestra bipunctata, Editha magnifica, Gorytes simillimus, Hoplisoides jaumei,* and *Tanyoprymnus moneduloides.*

Information available in 1966 (Evans 1966a) indicated that no species of Bembicinae nest in preexisting cavities of any kind or burrow in soft wood. This led to the conjecture that the "lack of diversification in body form and gross behavior" in the Bembicinae, relative to the sphecid subfamilies Larrinae and Crabroninae, was due to the fact that all nest-provisioning Bembicinae are fossorial and dig their own nests. There may be a few exceptions, however, one of which was first noted by Ferton (1908, referenced in Krombein 1984), who found *Bembecinus fertoni* and *B. gazagnairei* nesting in preexisting cavities in fine-grained limestone. More recently it has been suggested that *Clitemnestra plomleyi* and *Tanyoprimnus moneduloides* nest in preexisting holes made by other insects in soil. In Chapter 3, we mentioned the oddest record of nesting substrate yet recorded for a bembicine, that of *Argogorytes nipponis* nesting in wood pith; but we also noted that the record should be considered tentative.

How specific are the requirements of sand wasps with regard to the soil type and vegetation cover of potential nest sites? Certainly no wasp species is going to be oblivious to the range of soil hardness, particle size, and moisture conditions present in its environment, and vegetation cover does seem to be important. No broad geographic studies have addressed this quantitatively, however, though many species seem to choose similar habitats in different locations. Examples include *Bembix pallidipicta* and *B. variabilis,* which are always reported to nest in open expanses of sand. Some of this consistency is perhaps due to substrate requirements associated with the nest type built by a wasp, the wasps' digging abilities, and morphological constraints (e.g., number, structure, and rigidity of rake spines, mandibular structure). But we cannot yet rule out other habitat qualities associated with soil type and vegetation cover (e.g., prey type, thermal environment). Furthermore, there are also examples of species that seem to have less specific preferences, or at least an ability to occupy varying types of habitat when their first choices are not available (e.g., *Bembecinus tridens, Bembix moma, B. octosetosa, Sagenista brasiliensis*).

Even relatively homogeneous habitats with generally similar soil conditions will not be completely homogeneous with respect to soil structure, temperature, or moisture. Thus one would expect sand wasps to be able to make fine-scale choices of nest sites or to modify nest structure to match microsite conditions. Examples of both types of adaptations have been observed among sand wasps; the latter will be discussed later. As far as we know, only three studies of digger wasps have applied quantitative techniques to analyze soil characteristics that influence nest-site selection. One of those was on a nonbembicine, *Sphex ichneumoneus* (Brockmann 1979), so it does not concern us here. The most extensive study on a sand wasp was conducted by Rubink (1978). Rubink demonstrated that *Bembix pallidipicta* nests are nonrandomly distributed on several spatial scales. Within those areas occupied by females, nest density increased with higher late-morning soil surface temperatures and more even soil-particle size distribution. Rubink found no influence of soil hardness on nest-site selection, a result confirmed by Pimenta and Martins (1999) in their study of *Rubrica nasuta.* Of course, their results apply only to the spatial scale and range of soil conditions over which their measurements were made; no wasp species is likely to ignore hardness conditions if they range from

Figure 8.1. Trial holes dug by a *Microbembex monodonta* female at Great Sand Dunes National Monument, Colorado. Photo by H. E. Evans.

loose sand to hardpan clay. There also exist various pieces of anecdotal evidence on nest-site selection within nesting areas. For example, Tsuneki (1956–58) noted that, across a season, *Bembix niponica* nests were constructed progressively higher on a cliff face in order to stay within more insolated sites as temperatures declined later in the season. Evans (1966a) and Rubink (1978) suggested that female *Microbembex* make decisions concerning soil conditions as they repeatedly attempt to initiate nests, leaving a series of small pockmarks on the sand surface (Figure 8.1); such "trial holes" or "false starts" have also been observed in *Bembix cooba, B. occidentalis,* and *Stictiella emarginata.*

Nest Dispersion: Causes and Consequences

One of the salient features of many sand wasps, and one that makes them attractive for ethological studies, is their tendency to nest in aggregations. Seasonal activities often begin when males, which start to emerge as adults earlier than females, gather in the previous year's nesting area to search for mates. Phil and Nellie Rau (1918) wrote of their amazement at first finding an aggregation of "one or two hundred" *Bembix nubilipennis* on a baseball diamond near St. Louis, Missouri:

> The very air above the surface of the bare ground seemed vibrant with the low-flying wasps, which formed a wavering, yellowish-green haze over the smooth, dusty earth . . . The whole was not a helter-skelter commotion, but a merry whirl to the music of a faint, eerie hum of many wings, with every few moments a rather musical crescendo.

Though nonquantitative, this description of a sun dance gives a clear idea of the high densities of recently emerged wasps that result from high densities of nests during the preceding year.

Less poetic attempts to describe aggregations have included estimates of nest densities, which often illustrate how closely packed aggregations can be. At one site in Western Australia, for example, 2000–3000 *Bembix moma* females nested in a 2 × 20 m area, where most entrances were only a few cm apart (50–75/m^2). In a smaller but even denser aggregation, ~300 nests of *Alysson melleus* were seen in a 0.75 x 1.5 m patch of soil (250/m^2). Estimates of the nest densities of *Bembecinus quinquespinosus* ranged from 84–288/m^2 at various locations in nesting areas. Whether one observes such high densities may be a matter of the year that a site is visited. In Kansas in 1953, for example, Evans (1957b) found a small, diffuse aggregation of *Bembix americana,* but one year later ~200–300 individuals were present, with some nests only a few centimeters from their neighbors.

Other species typically form discrete aggregations that have lower densities than those cited above: *B. americana* (50–100 nests, no closer than 0.5 m from one another), *B. sayi* (small groups of ~30 females with nests >1 m apart), *B. belfragei* (a group of 150 females with nests 10–50 cm apart), and *B. cinerea* (in Florida in 1954, two colonies separated by 30 m and estimated to contain 200 and 300 nesting females, 10–20 cm apart). Similarly, a mixed aggregation in a 0.65 × 35 m strip of soil studied by Martins and his colleagues included 147 *Bicyrtes angulatus* nests (~6.5/m^2) and 193 *Rubrica nasuta* nests (~8.5/m^2).

When there are no important disturbances to a site, a species may nest there year after year. Evans returned to the site of an aggregation of *Bembix dentilabris* after 23 years and found the wasps still thriving. One aggregation of *Bembix americana comata* had been present in a driveway in California for at least 15 years, and *Bembecinus tridens* continued nesting in the same location year after year, even though the vegetation cover changed.

On the other hand, *Bembecinus quinquespinosus* females may move their nesting sites every year, apparently to avoid harassment by conspecific males, and *Bembix cinerea* females may move to find soil of the required subsurface dampness.

The friable soil required by many sand wasps tends to occur in localized patches, with the result that wasps often nest in aggregations that include both conspecifics and other species. The tendency for females to remain close to conspecifics can also be observed when the locations of nests are followed through a season. Females may remain close to one another as an aggregation grows from a central core, as in *Bembix americana* and *Stictia carolina,* or as small islands of female nests change their location within areas of apparently suitable nesting substrate, as in *B. quinquespinosus.* Whether they nest in dense or diffuse aggregations, spacing among females may be influenced by social interactions in which females attempt to maintain some minimum distance from one another (Rubink 1982). There are likely to be costs to nesting in the vicinity of conspecifics. First, when nests are close together, there may increased possibility of parasitism by flies and predation by ants. Larsson (1986) found that the frequency of fly parasitism on *Bembix rostrata* increased with nest density, although the relative incidence of parasitism declined on a per-nest basis. He suggested that this supports the "selfish herd" hypothesis of Hamilton (1971). One potential advantage of having close neighbors is that they may distract the attention of parasites that follow moving wasps (Evans and O'Neill 1988; Spofford and Kurczewski 1992).

A second cost of nesting in aggregations is that neighbors may interfere with one another while digging, or after one accidentally enters the nest of another. Lüps (1973) noted that female *Bembecinus tridens* nesting within several centimeters of each other interfered with one another considerably while digging and provisioning. In a few cases, this led to abandonment of a nest. N. Lin (1963b) argued that fighting among female *Sphecius speciosus* was density-dependent, with the result that females kept away from one another's nests. However, Dambach and Good (1943) and Pfennig and Reeve (1989) discussed cases in which female *S. speciosus* entered the burrows of conspecifics, resulting in either a fight or temporary cohabitation. (Dambach and Good even found individual cells containing two *Sphecius* eggs.) Neighbors, which are more closely related on average than non-neighbors, generally engage in relatively nonviolent fights; attempted usurpations of nests were more likely between non-neighbors (Pfennig and Reeve 1989, 1993). Thus in this species, nesting in the close proximity of conspecifics, whether or not they are immediate neighbors, may impose costs such as nest usurpation and, possibly, brood parasitism. Entering the wrong nest by mistake may also result in the loss of prey, if a female deposits a prey in the cell of another female. However, this would be difficult to document without lengthy observations of individually marked females. The presence of a leafhopper among the normal prey of *Bembix variabilis* led Evans and Matthews (1973) to suggest that it had been deposited there accidentally by a *Bembecinus* female. Still, the ability of most female sand wasps to home to their own nests, even in dense aggregations, may lower this potential cost.

One well-documented disadvantage of colonial living and high nest density is the increased chance of having prey stolen by neighbors (cleptoparasitism), though perhaps we could just as well turn the argument around and claim that the opportunity to steal is one the *advantages* of colonial living. Prey theft was observed in Wisconsin over a century ago by George and Elizabeth Peckham (1905) during their work on *Bembix americana*. They noted that, after being gone on hunting trips, most "of the wasps bring nothing home with them, and these fall to robbing their unfortunate companions. Those that are carrying flies must pause a moment, burdened as they are, to scratch away the earth at the entrance to the nest. When unmolested they go in very quickly, but it is just at this point that the marauders fall upon them, displaying an amount of persistence and energy in their attacks that, were it properly directed, might easily enable them to secure flies for themselves." The Peckhams appeared to miss the point that, in attempting to steal prey, females were perhaps making up for being unsuccessful at obtaining their own prey, whether out of some lack of skill on their part or just bad luck. When all else fails, crime does pay if you are a wasp.

Theft from other species has been noted for *Stictia signata* (victimizing *Stictia heros*), *Bembix texana* (victimizing *Stictia carolina*), and *Microbembex monodonta* (stealing from ants and other digger wasps, including *Tachysphex terminatus*). In addition to *Bembix americana*, theft by conspecifics has been noted in *B. cinerea, B. nubilipennis, B. pallidipicta, B. texana, Microbembex monodonta,* and *Stictia heros;* theft is apparently common in the three *Bembix* species listed. Stealing of prey generally occurs when one female pounces on a prey-carrying female at a nest entrance, and tries to grasp the prey and fly off with it. *Stictia heros* frequently take prey from each other at nest entrances; the probability of a female *S. heros* being attacked is related to the size of her prey, larger prey eliciting more attacks perhaps because they are more visible when being carried (Larsson and Larsson 1989). Interactions may begin with an audible collision between would-be thief and victim, and can last for up to a minute, while drawing in other participants in the nesting area. Certain female *S. heros* are also "marauders" that enter other females' nests, remove fresh flies, and take them to their own nests. Some individuals engage in persistent cleptoparasitism for extended periods, and marauding females tend to concentrate on raiding nearby nests that have previously yielded prey.

Of course the closest association that female wasps can have with one another, without being truly social, is to share nests. Nest sharing and communal nesting have been reported in several taxa of otherwise solitary wasps (Brockmann 1997; O'Neill 2001), but until recently such behavior had not been found in Bembicinae, other than reports of *Sphecius speciosus* females temporarily cohabiting. Recently S. K. Gess and F. W. Gess (1989) found a four-celled nest of *Bembix bubalis* that had three females associated with it. Two cells were being provisioned at the same time. This nest sharing stands in contrast to the examples of prey stealing noted above, but the account also needs to be followed up to see how common such behavior is in *B. bubalus.*

Interspecific Nesting Associations

Because different species of sand wasps share general habitat preferences, including nesting substrate, they often nest in close proximity to one another. Such associations may include non-Bembicinae as well as prey and parasites. We began this book with a description of part of the apoid wasp community found at a site in central New Mexico. A few further examples should suffice to illustrate the types of sympatric associations that have been documented:

- In Jackson Hole, Wyoming, Evans (1970) mapped the nest distribution of a wasp assemblage at one site that included (among other species) the sand wasps *Bembix americana, Hoplisoides hamatus,* and *Steniolia obliqua,* as well as *Ammophila azteca* Cameron, *Diodontus argentinae* Rohwer, *Miscophus evansi* (Krombein), *Oxybelus uniglumis* Say, and *Stenodynerus papagorum* (Viereck). One other species nesting at the same site was *Philanthus pulcher* Dalla Torre, a predator of bees and wasps; judging from its prey records, which included nearly 20 species of apoid wasp, the local wasp community was quite diverse (Evans and O'Neill 1988). Evans (1970) also documents the overall assemblage of solitary wasps and their natural enemies in the broader Jackson Hole area. Figure 8.2, for example, illustrates some of the interactions known (or very likely) to occur among apoid wasps and other organisms in Jackson Hole, Wyoming. The qualitative food web portrayed suggests a number of hypotheses for how predation, parasitism, prey availability, and competition could influence the size of sand wasp populations and the structure of sand wasp communities.

- Along a little-used path through a pine forest in Pfinwald, Germany, Lüps (1973) recorded *Bembecinus tridens, Bembix integra, B. rostrata, Harpactus elegans* (Lepeletier), *Harpactus exiguus* (Handlirsch), and two *Nysson* spp., along with the nonbembicines *Ammophila sabulosa* (L.), *Ammophila pubescens* Curtis, *Cerceris rybyensis* (L.), *Cerceris* sp., *Oxybelus argentatus* Curtis, *Oxybelus* sp., *Philanthus triangulum* (F.), *Podalonia affinis* (W. Kirby), and *Prionyx kirbyi* (Vander Linden).
- As noted above, Martins et al. (1998) found, within an area 0.65 × 35 m along a sandy road in Brazil, nests of *Bicyrtes angulatus, Bicyrtes discisus,* and *Rubrica nasuta,* along with *Ammophila gracilis* Lepeletier, *Prionyx fervens* (L.), and *Trachypus* sp.
- In South Africa, the solitary wasp fauna in the vicinity of Grahamstown includes a diverse array of sand wasps: *Bembecinus argentifrons, B. cinguliger, B. dentiventris* (Handlirsch), *B. haemorhoidalis, B. oxydorcus, Bembix albofasciata, B. cameronis* Handlirsch, *B. capensis, B. fuscipennis* Lepeletier, *B. melanopa, B. sibilans, Hoplisoides aglaia, Hoplisoides thalia* (Handlirsch), and *Nysson braunsi.* Dozens of other apoid wasps, vespoid wasps, and spider wasps share the same habitat, and often the same nesting substrates (F. W. Gess 1981).

Even when different species nest within sight of one another, each may have its own specific requirements that keep it somewhat separated within the broader shared habitat. In discussing the nesting sites of Australian *Bembix,* for example, Evans and Matthews (1973) noted that "*mianga* nested primarily in slopes beneath bushes in the sandy ridges around Lake Violet, while *atrifrons* and *thooma* occupied flat sand between the bushes. Near Wilcannia, N.S.W., *mianga* nesting mainly in blowouts and lower

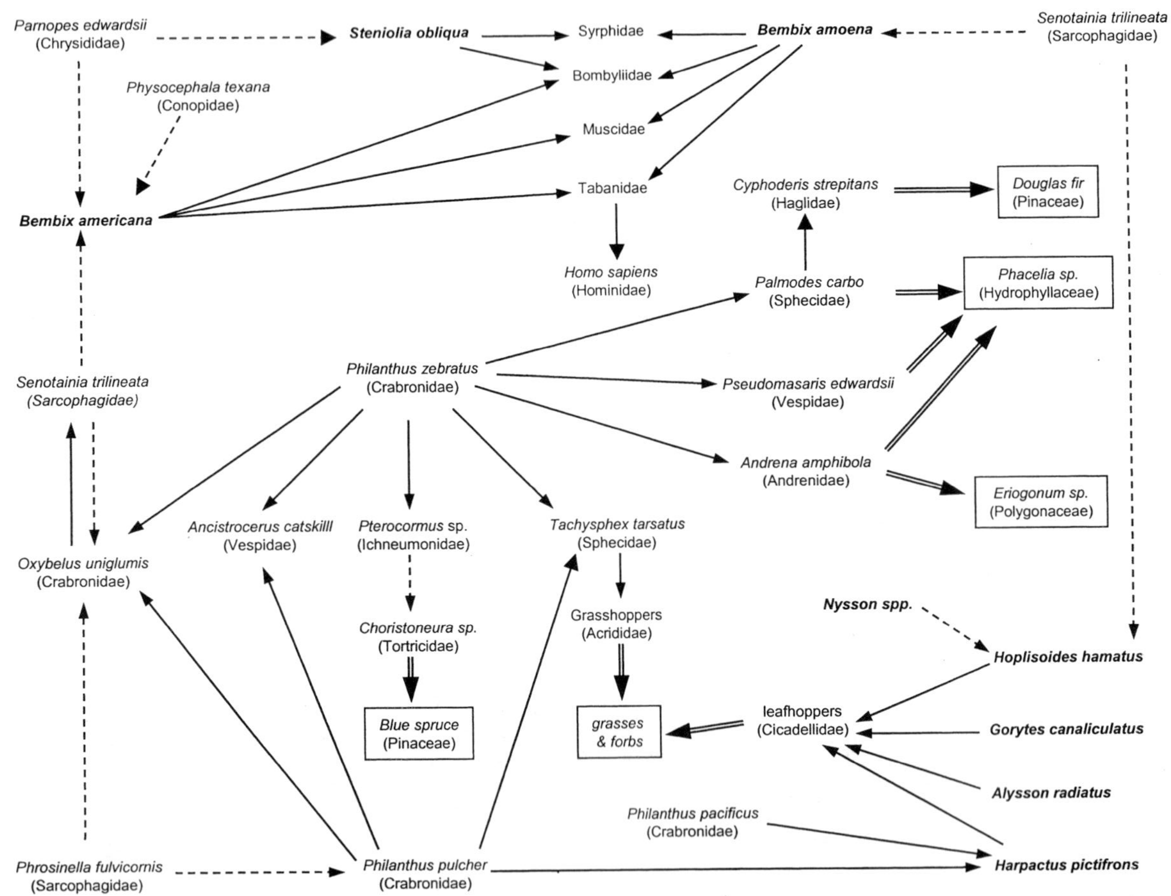
Parnopes edwardsii
(Chrysididae)
Steniolia obliqua
Syrphidae
Bembix amoena
Senotainia trilineata
(Sarcophagidae)
Bombyliidae
Physocephala texana
(Conopidae)
Muscidae
Tabanidae
Cyphoderis strepitans
(Haglidae)
Douglas fir
(Pinaceae)
Bembix americana
Homo sapiens
(Hominidae)
Palmodes carbo
(Sphecidae)
Phacelia sp.
(Hydrophyllaceae)
Philanthus zebratus
(Crabronidae)
Pseudomasaris edwardsii
(Vespidae)
Senotainia trilineata
(Sarcophagidae)
Andrena amphibola
(Andrenidae)
Eriogonum sp.
(Polygonaceae)
Ancistrocerus catskilll
(Vespidae)
Pterocormus sp.
(Ichneumonidae)
Tachysphex tarsatus
(Sphecidae)
Oxybelus uniglumis
(Crabronidae)
Grasshoppers
(Acrididae)
Nysson spp.
Choristoneura sp.
(Tortricidae)
Hoplisoides hamatus
Blue spruce
(Pinaceae)
grasses
& forbs
leafhoppers
(Cicadellidae)
Gorytes canaliculatus
Alysson radiatus
Philanthus pacificus
(Crabronidae)
Phrosinella fulvicornis
(Sarcophagidae)
Philanthus pulcher
(Crabronidae)
Harpactus pictifrons

slopes, while such species as *variabilis* and *cooba* nested toward the tops of the ridges." Similarly, F. W. Gess (1981) gives an extensive breakdown of the site preferences of a large wasp community in South Africa. Among the *Bembecinus* he studied, for example, *Bembecinus cinguliger* and *B. oxydorcus* nest in firm, sparsely vegetated clay soils, while *B. argentifrons* and *B. haemorrhoidalis* prefer the friable soil of a sand pit, and *Bembix melanopa* occupied a site with "steeply sloping firm and compacted sand."

Nest Construction

Digging behavior and its morphological correlates. All Bembicinae, at least those for which we have good data, are soil nesters. But there is still considerable variation in how they dig. Evans (1966a) noted interspecific (and intertribal) variation in (1) the relative role of the mandibles and legs, (2) the degree of synchrony in use of the legs, (3) the coordination of leg and abdomen movement, (4) the use of "bulldozing (particularly well-developed in *Sphecius*), and (5) the speed of digging.

Most Bembicinae dig with their mandibles and front legs, the mandibles being tools for initial loosening of the soil. (By observing females digging in clear-sided observation cages, Asís et al. (1988) found that *Stizus continuus* also use their mandibles to cut small roots.) The soil may then be removed in discrete clumps that are pushed backward beneath the body (e.g., in *Alysson*). In most species, the legs are deployed (usually synchronously) to rake loose soil backward, the abdomen often being raised with each stroke to allow soil to pass beneath the body. Evans (1966a) noted that the speed of digging is generally faster in the Bembicini than in the Gorytini, though there is also considerable intraspecific variation due to temperature-mediated effects on activity. Raking may be facilitated by the rake spines (forming a "pecten" or "tarsal comb") on the foretarsi. The rake spines are elongated and flattened in females of species that dig in fri-

Figure 8.2 *(opposite page).* Hypothetical (and incomplete) food web, including species of sand wasps (in bold) and other solitary wasps (Crabronidae, Sphecidae, and Vespidae), in Jackson Hole, Wyoming (based primarily on information in Evans 1970 and Evans and O'Neill 1988). Solid single lines indicate predator–prey relationships, dashed lines indicate parasitoid (or brood parasite)–host relationships, double solid lines indicate plant–herbivore relationships (including pollen feeders); plant species are outlined in boxes.

able soil, but short in species of *Bembecinus* of that nest in compact soil (F. W. Gess and S. K. Gess 1975). In the Alyssontini, which dig mainly with their mandibles and do not rake soil away from the entrance, the pecten is "very weakly developed," and there is a tendency for the excavated soil to form a circular tumulus initially centered on the nest entrance. Recently a different type of morphological aid to digging has come to light. The large spurs on the hind legs of *Sphecius speciosus* are used to "bulldoze" soil as the wasp backs out the burrow. Finally, "the fact that the [preapical tooth on the mandible of females] is missing in the two brood parasitic apoid wasp taxa Nyssonini and *Stizoides* [see Chapter 5], . . . strongly suggests that the preapical tooth has significance in the various nesting activities" of female ground-nesting and pith-nesting wasps (Prentice 1998).

Nest depth. In the best-studied genus of Alyssontini, *Alysson*, nest cells are as shallow as 5 cm in *Alysson melleus* and as deep as 24 cm in *A. cameroni.* Smaller species of Gorytini have been found with nest cells as shallow 0.8 cm *(Hoplisoides ater),* 1.8 cm *(Ammatomus icarioides),* and 2.5 cm *(Gorytes tricinctus),* while larger wasps in the tribe situate their cells as deep as 50 cm *(Sphecius pectoralis)* and 60 cm *(Exeirus lateritius).* The range of cell depths is much narrower among Stizini. *Stizus* and *Bembecinus* have not been found with nests exceeding 20 cm in depth, and those of *Bembecinus* are often as shallow as 3 cm. The deepest recorded cells of *Bembix,* on the other hand, are those of *Bembix pallidipicta* at 56 cm, *B. melanaspis* at 60 cm, *B. variabilis* at 67 cm, and *B. rugosa* at 78 cm. Because burrows are oblique, such cell depths require long burrows, as long as 84 cm in *B. pallidipicta,* 112 cm in *B. variabilis,* 114 in *B. melanaspis,* and 127 in *B. rugosa.*

Variation in nest depth within and between species appears to be related to the type of soil used. Compacted soils are difficult to dig, costing females both time and wear and tear on body parts. And in dry soils, a wasp must dig deep to find soil of the proper moisture level. Considerable observational evidence exists that sand wasp females respond to local soil conditions when determining the depths of their nests. Much of this evidence comes from various studies of *Bembix.* Females of *Bembix atrifrons, B. mareeba, B. moma,* and *B. oculata* have been reported digging deeper nests in looser soil, although the opposite was observed for *Bembix octosetosa;* in that species, nests in more "friable, loamy sand" at one site were shallower than those at a location where soil was "firm to moderately friable" (Evans and Matthews 1973).

Anecdotal evidence suggests that nest depth also correlates with soil moisture levels. For example, *Bembix variabilis* nests in the arid interior of Australia were deeper (43–67 cm) than those found near sources of water (10–30 cm), and Rubink (1978) noted that *Bembix pallidipicta* females apparently adjust nest depth in response to subsurface conditions, digging deeper where the sand is drier. A similar form of spatial variation in cell depth was found for *Microbembex argyropleura;* and temporal variation in cell depth correlated with soil moisture was reported for both *M. argyropleura* and *Microbembex nigrifrons.* In the latter, when the average depth of soil moisture was 5 cm, cell depth ranged from 19 to 29 cm, whereas cells were 31–49 cm deep during drier periods, when moisture could not be found without digging to depths of 11–18 cm (Alcock and Gamboa 1975). Evans (1966a) also suspected that *Hoplisoides nebulosus* dug deeper nests in drier soils. Depending upon the habitat, it may be the case that females sometimes dig deeper in dry soils to reach moisture levels high enough to prevent desiccation of egg, larvae, and prey. In places with moister soils, however, they may dig shallower nests to avoid placing cells in saturated zones.

Cell number and order of construction. The left-hand side of Table 8.1 presents an updated version of Table 44 in Evans (1966a), giving the known maxima for the number of cells found in nests of different bembicine genera. Most of the genera listed have at least one species with multicellular nests, but this is a somewhat misleading statistic. First, in some species known to have multicellular nests, most nests are actually unicellular. Second, in some genera that have species with multicellular nests, many other species make unicellular nests only. A notable example is *Bembix,* in which only about one-third of the species has been reported with multicellular nests. Third, multiple studies of a few species have revealed variation among colonies of one species. *Bembix nubilipennis,* for example, has only unicellular nests in some colonies, but multicellular nests in others; and in a study at one location in Texas, the spring generation made multicellular nests, whereas the autumn generation made only unicellular nests. Several authors have suggested that in cases where multicellular nests have been found in taxa that normally make unicellular nests, multicellular nests are adaptations to nesting in particularly hard soil (e.g., Krombein [1984] referring to *Bembecinus comberi* and Evans et al. [1974] referring to *Rubrica nasuta*).

In the Alyssontini, nests can have as few as one cell in some nests of *A.*

Table 8.1 Maximum reported values for the number of cells per nest and prey per cell in genera of Bembicinae.

Genus	Maximum number of cells per nest	Species	Maximum number of prey per cell	Species
Alysson	9	*A. cameroni*	23	*A. melleus*
Clitemnestra[a]	10	*C. bipunctata*	20	*C. bipunctata*
Exeirus	"Several"	*E. lateritius*	1[k]	*E. lateritius*
Sphecius	15.8 (mean)	*S. speciosus*	12	*S. pectoralis*
Tanyoprimnus	3	*T. moneduloides*	13	*T. moneduloides*
Ammatomus	7	*A. icaroides*	6	*A. icarioides*
Argogorytes	9	*A. campestris*	27	*A. mystaceus*
Harpactus[b]	15	*H. concinnus*	4	*H. gyponae*
Trichogorytes	2	*T. cockerelli*	10	*T. cockerelli*
Austrogorytes	6	*A. bellicosus*	27	*A. bellicosus*
Gorytes	4	*G. canaliculatus*	20	*G. canaliculatus*
Pseudoplisus[c]	2	*P. natalensis*	4	*P. ranosahae*
Sagenista[c]	3	*S. brasiliensis*	6	*S. brasiliensis*
Hoplisoides	5	*H. jaumei*	60	*H. punctuosus*
Hapalomellinus	2	*H. albitomentosus*	15	*H. albitomentosus*
Stizus	9	*S. pulcherrimus*[d]	13	*S. fuscipennis*
Bembecinus	3	*B. comberi*	74	*B. quinquespinosus*[j]
Bicyrtes	5	*B. fodiens, B. variegata*	24	*B. angulatus*[j]
Microbembex	8	*M. ciliata*[e]		
Hemidula	1	*H. singularis*		
Rubrica	3	*R. nasuta*	~50	*R. nasuta*[j]
Selman	1	*S. notatus*		
Stictia	2	*S. signata*[f]	63	*S. carolina*[j]
Editha	2	*E. integra*		
Trichostictia	1	*T. guttata*		
Zyzzyx	6	*Z. chilensis*[g]		
Stictiella	17	*S. formosa*[h]	21	*S. serrata*
Microstictia	7	*M. minutula*	11	*M. minutula*
Glenostictia	1	*G. scitula*	40[l]	*G. scitula*[j]
Xerostictia	2	*X. longilabris*		
Steniolia	1	Four species		
Bembix[i]	6	*B. brullei*	>50	*B. merceti,*[j] *B. musca,*[m] *B. variabilis*[j]

[a]*Clitemnestra* was considered distinct from *Ochleroptera* in 1966, but the two have since been combined; [b]listed as *Dienoplus* in 1966; [c]considered a subgenus of *Gorytes* in 1966; [d]a different study of *S. pulcherrimus* has since reported only unicellular nest in a Japanese population (1966 report was for a Korean population); [e]a more recent study found only unicellular nest; [f]even this species usually makes unicellular nest; [g]value needs confirmation as recent studies suggest unicellular nest only; [h]no other species is reported to construct more than 2 cells per nest; [i]majority of species make unicellular nest only (see Table 7.1); [j]progressive provisioner; [k]"most cases" (Evans 1966a); [l]rough estimate based on prey fragments; [m]truncated progressive provisioner.

melleus and as many as nine in *A. cameroni.* All 14 genera of Gorytini listed in Table 8.1 have at least one species with multicellular nests; females of *Sphecius speciosus* commonly construct nests with over a dozen cells. Unicellular nests are more common in the Stizini. The maximum number of cells in nests of *Stizus* species ranges from 1 to 9. In *Bembecinus,* strictly unicellular nests are most common (>10 species); a maximum of two has been recorded for five species, and three-celled nests have been documented for one. In the Bembicini, unicellular nests are characteristic of species of *Glenostictia* and *Steniolia,* as well as most species of *Microbembex* and *Bembix.* The reigning champion among Bembicini is *Stictiella formosa.*

The available data on cell numbers per nest may be somewhat complicated by the fact that researchers often excavate a nest before a female has completed it and made a final closure. So published values may sometimes be underestimates for species with multicellular nests. This problem would be particularly evident in populations of species such as *Sphecius speciosus,* if it is true that females construct just a single nest in their lifetimes, adding to it as the season progresses. However, it is much easier for an observer to excavate and accurately diagram a nest *before* the female has completed the nest and filled the burrow with soil. Another problem with determining nest structure occurs in dense nesting aggregations, where it may be difficult to associate adjacent cells with specific nests; this problem was frequently acknowledged by authors of studies that we have cited in this book.

Although unicellular nests are likely the ancestral state for apoid wasps as a whole, the building of unicellular nests is likely an evolutionarily derived state within the Bembicinae (Evans 1966a). It seems that most of the time and effort of digging is invested in construction of the main burrow of a nest, as a look at diagrams of the nests of most species would suggest. If so, why should females of so many species of Bembicinae construct unicellular nests and, therefore, have to repeat the construction of the main burrow multiple times? Perhaps the most obvious possibility is that species nesting in loose soils need to repeat the process because the main burrow eventually collapses. A second possibility is that high levels of parasitism favor construction of multiple burrows, if parasites that enter a nest can attack more than one cell. Social factors may also come into play. If there is some optimal nearest-neighbor distance between nests, the result of a balance between the advantages and disadvantages of having close neighbors, females may move to new burrows as the nest densities change through a season. Season-long studies of nest density will be needed to address this.

Among species with multicellular nests, there is variation in the order of cell construction relative to their distance from the nest entrance. Species that add new cells progressively closer to the nest entrance after first constructing an initial cell apically are said to have a *regressive* cell pattern; those that place them further and further from the entrance as new cells are added have a *progressive* pattern (Iwata 1976). Whatever the pattern, it can only be determined from careful dissection of a nest, while recording cell contents that give a hint as to the recency of provisioning (i.e., empty cells, cells with eggs, larvae of different ages, or cocoons). From such evidence, it seems that *Alysson melleus* females follow a regressive pattern (Evans 1966a), but this does not seem to be the case for *Alysson conicus* (on the basis of one nest drawn in O'Brien and Kurczewski 1982). The nest diagrams provided in Tsuneki's (1969) paper on *Alysson cameroni* also indicate no clear relationship between cell depth and the sequence in which cells were provisioned within a nest. In the Gorytini (except *Clitemnestra* and *Sphecius*), the first cell provisioned is generally the farthest down the burrow, and new cells are added regressively, as is also true of *Rubrica nasuta* (Pimenta and Martins 1999) in the Bembicini. Both *Sphecius speciosus* (Evans 1966a) and *S. pectoralis* have a progressive pattern, as is apparently the case in *Microbembex ciliata* and *Zyzzyx chilensis* (Janvier 1928). For the Korean population of *Stizus pulcherrimus,* the only population of this species known to have multicellular nests, there seems to be no consistent order of cell construction (Tsuneki 1965b). Tsuneki states that "although we can perceive a more or less general trend of cell construction from interior to exterior, there are too many exceptions to take this as a general rule."

Leveling. A great deal has been published on the leveling (or nonleveling) of soil that accumulates at the nest entrance. Leveling of the mound may occur initially when the nest burrow is first completed, periodically during periods of provisioning, or at final closure; or leveling may be unnecessary if soil is dispersed widely during digging, as is the case with *Bicyrtes* and *Microbembex.* Leveling is thought to remove a potential cue for parasites seeking nest entrances, but also has several disadvantages: it takes time and energy, it disperses loose soil that could later be used for fill during closure, and removes a landmark that could be used for homing. With regard to the latter, however, many sand wasps have a remarkable ability to find their own nests despite extensive leveling (e.g., van Iersal and van den Assem 1963; Evans 1966a).

Evans (1966a) noted that in "the majority of [Bembicinae] the mound of earth that accumulates at the nest entrance is left more or less intact." And some sand wasps, such as *Gorytes* and *Hoplisoides,* have weak and irregular leveling movements. Among those species that do level mounds extensively, the pattern left by leveling movements is often so species-specific that freshly prepared nests can be associated with a particular species simply by examining the pattern of lines. We described this earlier for many species of *Bembix.* However, there are also *Bembix* that do not level mounds. Curiously, females of several unrelated species have what might be called reversed-leveling: they actually build a mound at the nest entrance by scraping soil onto the tumulus. Mound-building has been described for *Stictia maccus, Bembix boharti,* and *B. littoralis.* Perhaps such mounds assist the female in finding her nest quickly, but why would these species need mounds when more subtle cues suffice for their relatives? Matthews (1991) suggests that erecting a mound over the entrance may obscure "any tell-tale wasp odor(s)" that linger at the nest entrance and attract marauding ants. More broadly, he also suggests that mound-leveling serves the same function. In at least some populations of *Rubrica nasuta,* females make a small, circular pile of soil at the nest entrance; these mounds are often used as perches by females (Evans et al. 1974), but it would be premature to conclude that this is their evolved function.

Temporary and final nest closures. Many sand wasps block off some portion of their nest burrow with soil before they leave to forage (temporary closure) and/or after the nest is completed (final closure). Closures may occur near the cell (inner closures) or at the nest entrance (outer closures). Certain groups, however, never have temporary outer closures, including apparently all Alyssontini, some Gorytini (*Ammatomus, Argogorytes, Clitemnestra, Exeirus, Sphecius, Tanyoprymnus*), some *Stizus (S. distinguendus, S. fasciatus), Rubrica denticornis* (Bodkin 1917), *Microbembex californica,* and several *Bembix* (*B. belfragei, B. boharti,* and *B. olivacea*).

In the Gorytini, temporary outer closures are made by *Gorytes, Hapalomellinus, Harpactus,* and most *Hoplisoides* (except *H. semipunctatus*). In the Stizini, certain *Stizus (S. perrisi ibericus, S. pulcherimus)* and nearly all *Bembecinus* make temporary closures. Krombein (1984) expressed doubt that *Bembecinus comberi* constructs temporary closures, and *B. cinguliger and B. oxydorcus* make them only at the end of each day. Most Bembicini construct temporary outer closures at least sometime during their tenure at a nest. In *Bembix*, the propensity to close during foraging may decline as larvae get older, perhaps because they are less susceptible to natural enemies that attack only cells with eggs or newly hatched larvae (e.g., miltogrammine flies) or because the existence of a closure slows provisioning at a time when prey are brought in rapidly.

Temporary inner closures are rarer, being apparently absent in all Alyssontini and most Gorytini and Stizini. It is only recently that exceptions have been found in the latter two tribes. Nest entrances of *Argogorytes carbonarius* are left open during foraging, but inner cell closures of compacted clay are present prior to the final closure of the nest. *Bembecinus cinguliger* and *B. oxydorcus* construct outer and inner closures, thin sheets of mud, only at the end of each day. Among the Bembicini, inner closures are made only by *Stictia* and most *Bembix* (except *B. nubilipennis* and *B. texana*). This much was known about *Bembix* as of 1966, and more recent studies have found inner closures in the North American *Bembix melanaspis* and European *B. citripes.* Evans (1966a) notes that many *Stictia* and *Bembix* have nests in which the burrow levels off near the cell, and it is here that the inner closure occurs.

Final closures are the most elaborate of the plugs placed in nests. They take more time and effort to construct, often extend through much if not all of the nest burrow, and are more compacted (and thus probably more difficult for natural enemies to penetrate). At final closure, many species use the deflected tip of the abdomen to pound the soil in place (e.g., *Bembix, Bicyrtes, Editha magnifica, Hoplisoides, Rubrica nasuta, Steniolia, Stictia signata*). Closures made at different times vary in the extent to which they fill the burrow. Female *Bembix multipicta,* for example, make temporary closures prior to foraging that are about "one body length in thickness, packed little (if at all), and . . . not concealed by general scattering of soil around the entrance." However, closures made each night are about twice as long, well packed, and more concealed (Cane and Miyamoto 1979). During final closure, the most extensive of all, the female *B. multipicta* fills the burrow while collapsing its walls, compacting the sand, and thoroughly concealing the entrance by leveling.

Perhaps equally important to the thickness of the plug is the degree to which the entrance of the nest is disguised during the final closure. In *Pseudoplisus ranosahae,* for example, final closure of the nest is apparently a quick affair, so that the nest entrance remains visible. In many other species, of course, leveling after the final closure is so extensive that the former entrance is essentially invisible. Some species also place bits of debris over the site of the burrow after final closure (e.g., *Rubrica nasuta, Stictia maculata*).

In apoid wasps in general, mound-leveling and nest closures are thought to have evolved as measures to deter brood parasites and parasitoids that lay their eggs at the nest entrance, somewhere deep inside the burrow, or in the cell itself (Evans 1966a; O'Neill 2001). The presence of both outer and inner closures in *Stictia* and *Bembix,* for example, may act as a double deterrent to nest-invading brood parasites, parasitoids, and prey thieves, but it is yet to be determined whether these two genera suffer lower parasitism relative to other sand wasps. In fact, it is generally the case that hypotheses concerning antiparasite adaptations in sand wasps—leveling, nest closures, and accessory burrows—enjoy only indirect evidence in their support. However, the importance of nest closure was demonstrated by Alcock and Gamboa (1975), who removed female *Microbembex monodonta* as they emerged from nests, leaving the entrance open before they could close it. The open burrows proved attractive to bee flies and cuckoo wasps, and especially to female *Microbembex.* A further relevant observation was made by Cane and Miyamoto (1979), who found that, at night, *Solenopsis* ants did not raid nests of *Bembix multipicta,* whose females made thick closures at the end of each day. However, at the same location ants frequently plundered nests of *Bembecinus bolivari,* which "apparently does not make such secure overnight closures." Further observations of this type on sympatric species would be useful. Both temporary and final closures take time to construct, but temporary closures also come with a further cost. A wasp returning with prey must take time to open the nest, during which time its prey is vulnerable to attack by flies that lay their eggs on the prey or by conspecifics that steal prey. *Gorytes tricinctus* females, for example, construct thick temporary closures that take from 30 to 60 s to clear when they return with prey.

Accessory burrows. Accessory burrows, the shallow, blind burrows adjacent to the true nest burrow, have been reported to occur at least some times in certain *Stizus* and *Bembecinus, Rubrica gravida, Selman notatus, Stictia heros,* at least two species of *Bicyrtes,* and many species of *Bembix.* A particularly striking pair of accessory burrows is prepared by *Bembix kamulla,* each female consistently placing one such burrow on each side of the true burrow (Evans et al. 1982). Accessory burrows, which also occur in some *Sphex* (Sphecidae) and *Philanthus* (Crabronidae), are hypothesized to have evolved to distract the attention of parasites seeking nest entrances (Evans 1966b; O'Neill 2001).

Orientation and Homing

The ability of digger wasps to learn the location of their own nests, both in the general habitat and within aggregations, was a necessary prerequisite to the evolution of more complex forms of nesting behavior. Females could not go very far away to hunt after constructing a nest, or between trips to place multiple prey in cells, without having some ability to find their way back home (O'Neill 2001). The ability of wasps to locate their nests was remarked upon by earlier writers (e.g., Peckham and Peckham 1905). And the research of Tinbergen (1932, 1935) on *Philanthus* and Baerends (1941) on *Ammophila* demonstrated that digger wasps use visual cues in homing and orientation, cues learned during orientation flights. In the Bembicinae, orientation flights vary in form, being a series of high, oblique hovering flights interspersed with nest leveling in *Hoplisoides nebulosus,* "loops and figure eights" in *Gorytes canaliculatus,* "low and circling" flights in *Glenostictia scitula* (Evans 1966a), and slow, irregular spirals ascending to heights of three or more meters in *Bicyrtes quadrifasciatus* (Krombein 1955). However, *Microbembex monodonta* make "no conspicuous orientation flights during or following nest construction," yet easily find their concealed nest entrances amid many others within apparently homogeneous expanses of sand (Evans 1966a).

Evans (1966a) reviewed information on homing and orientation by Bembicinae available at the time. That information included the studies on *Bembix rostrata* by van Iersal (1952), van Iersal and van den Assem (1963), and Chmurzynski (1964), to which we can add more recent work on that species by Chmurzynski (1967), Tengö et al. (1990, 1996), and Schöne and Tengö (1991a,b). The earlier studies concentrated on the cues used by the wasps to find nests. Later studies by Tengö and his colleagues examined homing success in relation to distance from the nest and the stage in the nesting cycle. Homing success decreased linearly with distance of displacement from the nest, but increased as the nesting cycle progressed from digging through provisioning. Similar studies have also been done with *Argogorytes carbonarius* (Schöne et al. 1993, 1994). For a recent, brief review on homing and orientation in solitary wasps, see O'Neill (2001).

Oviposition and Provisioning

Oviposition and provisioning can be characterized in terms of the physical location of egg placement in the empty cell and in the timing and duration of provisioning relative to the laying and hatching of the egg and the spinning of the cocoon. In mass provisioners, females bring in provisions rapidly, either after laying an egg on the first prey in the cell or after provisioning is complete; the rate of provisioning proceeds independently of the developmental progress of the larva, which often does not hatch until after provisioning is complete. Evans (1966a) noted that apparently all Alyssontini and Gorytini mass provision cells and lay the egg on the last prey in the cell. Work since then confirms this, with the possible exceptions of *Gorytes tricinctus,* which lays the egg in the middle of the prey mass, and *Sphecius hogardii* and *Argogorytes hispanicus,* which lay the egg on the first prey in the cell; the latter species is also reported to provision cells over several days. Note that laying the egg early in the provisioning sequence was a necessary prerequisite for the evolution of progressive provisioning (Evans 1966a).

In the Stizini, the timing of oviposition and provisioning is divided along generic lines. All *Stizus* are mass provisioners that lay the egg on the first prey in the cell (Figure 1.5F), though some *Stizus* provision over several days—delayed or slow mass provisioning that may result in some prey being brought in after the egg has hatched. All *Bembecinus* are progressive provisioners that lay the egg in the empty cell. Evans (1966a) noted that six species of *Bembecinus* were known to lay their egg on a small pedestal of sand grains glued together, probably using secretions of the wasp (Figure 1.5G). Since then, this deft form of egg placement has been also seen in five further species (*B. cinguliger, B. egens, B. posterus, B. proximus,* and *B. strenuus*), and been confirmed for several subspecies of *B. hungaricus.*

In the Bembicini, *Microstictia* and *Stictiella* are mass provisioners, as are some *Bicyrtes (B. fodiens, B. variegata)* and *Bembix* (in North America, *B. hinei, B. stenebdoma;* in Australia, *B. cursitans, B. flavipes, B. kununurra, B. moma, B. mundurra, B. musca,* and *B. tuberculiventris)* (Evans and Matthews 1973). Many times, a female fails to fill the cell on the day of egg laying, perhaps because of prey scarcity or poor weather. But such delayed mass provisioning is especially common in *Bicyrtes* and *Stictiella,* so its occurrence may be independent of environmental conditions. The majority of Bembicini, however, are progressive provisioners, so that the first prey are not brought in until the egg is ready to hatch, and are then introduced daily until the larva is nearly fully grown. In some cases, when prey are very plentiful, the cell may be filled well before the larva is full-grown (truncated progressive provisioning). This is often the case in *Microbembex* and some *Bembix* (including, perhaps, some species listed above as mass provisioners). Clearly the distinction between mass and progressive provisioning is not as clear-cut as it was once believed to be. For further discussion, see Tsuneki (1956–58), Evans (1966a), Genise (1982f), and Asís et al. (1991).

The vast majority of Bembicinae lay the egg on the side of the thorax of the initial prey in the cell (Figure 1.5B, C, E, F), whether the prey is a fly, moth, bee, damselfly, or antlion; an exception is *Bicyrtes,* whose females deposit the egg semierect on the midventer of a bug (Hemiptera) (Figure 1.5D). Regardless of the exact position, there are indications that the pedestal prey of *Bembix* may be carefully chosen, rather than simply being the first prey that happens to have been brought into the cell. *Bembix moma,* for example, was seen to use a bee as an egg pedestal in all 17 cells examined by Evans and Matthews (1973), despite the fact that it also preyed on wasps and flies; the probability of this result occurring by chance in all 17 cells, given that bees made up ~74% of all prey, was only ~0.6%. The first prey may also be treated differently than others, as in the case of the *Bembix* species that kill the prey used as a pedestal, but paralyze the remaining prey in the cell. The explanation for this is, perhaps, that dead prey cannot squirm about and damage a delicate egg.

Like *Bembecinus,* some Bembicini lay the egg in the empty cell prior to provisioning. All species of *Microbembex* lay the egg in the empty cell in an erect position, toward the extremity of the cell, supported by a few sand grains. There are at least six *Bembix* that lay the egg erect in the empty cell: *B. citripes, B. melanaspis, B. multipicta, B. olivacea, B. texana,* and *B. troglodytes. Stictia carolina* and *Bembix occidentalis,* however, lay the egg flat in the bottom of the cell. In the two species of *Bembix* that make an elongate brood chamber, the egg is laid at the distal end of the empty chamber *(B. pruinosa)* or glued to a few sand grains near the extremity of the cell *(B. variabilis).*

Cell cleaning (or discarding of prey fragments) has been described in *Bembix niponica, B. texana, Rubrica nasuta,* and *Stictia heros,* as well as in several nonbembicine apoid wasps (Genise 1982b). While cleaning, the female backs from the nest in the morning before she hunts, and drags debris and uneaten parts of prey from the cell; she then deposits the materials some distance from the nest, usually after a short flight. In another species, *Bembix pallidipicta,* some females separate fresh prey from uneaten prey fragments by pushing the latter to the apex of the cell and partitioning them behind a barrier of sand. Perhaps cell cleaning plays a role in removing nest parasites, scavengers, and prey fragments that could become foci for mold growth. Or, given that all species with cell cleaning are progressive provisioners, maybe removal of prey fragments allows the female to better determine whether the larva will require additional fresh prey.

Number of prey per cell. The number of prey a mother provides to each offspring varies within and between species and genera, the maximum values recorded for each genus being spread over quite a range (Table 8.1). Prey numbers are easiest to determine for mass provisioners, but often difficult to determine for progressively provisioning species, especially those with cell cleaning. The number of prey provided can be as low as one per cell for some cells of *Sphecius* and *Exeirus,* which prey on large cicadas, two per cell in some cells of *Alysson cameroni,* or three per cell in *Ammatomus icarioides* and *Pseudoplisus ranosahae.* The highest value for prey per cell that we are aware of (74) came from our study of *Bembecinus quinquespinosus,* and that may have been from an incompletely provisioned cell.

Two factors that influence intraspecific variation in the number of prey per cell are offspring sex (daughters generally must receive a greater mass of prey on average) and the type of prey used. Sex-biased provisioning is undoubtedly a widespread phenomenon in sand wasps, and is a major proximate cause of sexual size dimorphism, with females typically being larger than males. It likely evolved because females require large bodies that can hold large eggs, and subdue and carry large prey (O'Neill 1985, 2001). Females of at least 12 species of solitary nest-provisioning wasps are known to provide more provisions for their daughters than for their sons (O'Neill 2001; O'Neill and O'Neill 2003). The list includes one sand wasp, *Sphecius speciosus,* whose females are nearly 2.5 times heavier than males (Coelho 1997). Dow (1942) found that daughters received 2.2–3.9 g of provisions, whereas sons were given 0.9–2.1 g; N. Lin (1979a) reported that all daughters got two cicadas, but most males got just one.

Evidence for the role of prey size relative to wasp size in determining the number of prey per cell is more extensive. Several examples, all cited earlier, include:

- *Argogorytes carbonarius,* which uses no more than three prey per cell when adult prey are provided, but as many as 27 when nymphs are used.
- *Austrogorytes bellicosus,* which may use as few as five *Eurymeloides pulchra* or as many as 27 prey of smaller species or earlier instars.
- *Gorytes tricinctus,* which provide as few as four prey when a large species of *Aphrophora* are used, but as many as ten when a small species are provided. *Gorytes canaliculatus* uses fewer prey per cell when cells are stocked with adult rather with nymphal leafhoppers.
- *Hoplisoides ater* and *H. nebulosus,* in which the number of prey per cell increases when small membracids are provided. *Hoplisoides ater* also tends to provision with a greater number of prey (10–34 per cell) than does another Cuban species, *Hoplisoides jaumei* (5–12), which tends to take fewer immatures.
- *Bembix variabilis,* for which perhaps twice as many prey are needed per cell when cells are provisioned with small Diptera, in comparison with when they are stocked with damselflies.

The number of prey used may vary among cells because of spatial and temporal changes in prey availability (as may be the case in any of the examples cited above) or because the size of prey taken varies among females of different size in the same species. For example, smaller females of *Bembecinus quinquespinosus* take a greater proportion of immature prey and smaller prey on average. Thus small females either have to provide fewer prey per offspring (thus having smaller offspring themselves) or provide a greater number of prey per cell (thus perhaps limiting the total number of cells they can provision during their lives). Size constraints on both female and male reproductive success are likely one of the major selection pressures favoring behavioral plasticity in digger wasps and other animals (O'Neill 2001; West-Eberhard 2003). Note, however, that Villalobos and Shelly (1996) found no relationship between female size and prey size for *Stictia heros.*

Estimates of Lifetime Reproductive Success

Directly estimating the lifetime reproductive success (LRS) of female sand wasps is difficult, because it entails prolonged studies of marked females whose nests then have to be excavated to achieve a total cell count. Ideally, offspring would also be followed to determine whether they survive until emergence, and then have their body size measured, because this may influence their own LRS (and, ultimately, that of their mothers). Not surprisingly, estimates of LRS are available for just a few species of solitary wasps (O'Neill 2001); and these either are based on counts of cell numbers or are indirect estimates based on yet other estimates of adult lifespan and the amount of time it takes to provision one cell. Even with the limited data available, one trend is obvious: the values for LRS for nest-provisioning species are low relative to most other insects, including many parasitic relatives of related families of Hymenoptera (O'Neill 2001). The difference, of course, is due to the time and effort nest provisioners put into digging and provisioning nests, as well as constructing closures and leveling mounds. How long does it take? Evans (1957b) observed that female *Bembix pallidipicta* took 4–6 days (mean = 4.3) for each unicellular nest. Martins et al. (1998) reported that *Bicyrtes angulatus* females take an average of 8.5 days from excavation to closure of their unicellular nests, and Matthews et al. (1981) found that female *Stictia maculata* took as long as 8 days to complete one unicellular nest.

Let's first consider some rough but reasonable estimates of LRS for four mass-provisioning species. Harris (1994) estimated that *Argogorytes carbonarius* females make 3–5 nests during their lives, with 6–9 cells per nest. Thus maximum lifetime reproductive success probably lies somewhere between 18 and 45 cells. We expect that females that achieved the upper part of that range would be rare (and lucky) and that the average value will be much lower because many females will fail. LRS is a bit lower for *Sphecius grandis* females, which have a mean adult lifespan of just 11 days and are capable of provisioning a total of no more than 20 cicadas during their lives (Hastings 1986). The number of prey used per nest cell has not been reported for this species, but its similar-sized relative *Sphecius speciosus* provides each offspring with 1–2 cicadas. If the range in prey number per cell is similar for two *Sphecius*, then *S. grandis* females probably produce no more than about 13 or so offspring during their lives if they average 1.5 cicadas per cell. This estimate is close to the *average* of ~16 cells per nest for one population of *S. speciosus* (Dambach and Good 1943), whose females are thought to dig just one nest in their lives. A still lower value for LRS was reported for another mass provisioner, *Bicyrtes angulatus*, whose females provisioned 1–7 unicellular nests during their lives (Martins et al. 1998).

LRS estimates for progressive provisioners tend to be lower than those for mass-provisioners. Larsson and Tengö (1989) estimated that *Bembix rostrata* females provision a maximum of no more than five cells during their lives. Martins (1993) found that female *Editha magnifica* took an average of over 12 days to dig and provision a nest, and that marked females lived from 12 to 78 days. Thus the potential maximum number of nests may be no more than about 6–7; in fact, five marked females in the study built just 2–4 unicellular nests in their lives. Similarly, Alexander et al. (1993) observed that 20 marked females of *Glenostictia pictifrons* took from 3 to 9 days (mean = 5.5) to dig and provision their unicellular nests. Females could thus provision an average of ~11 cells (range 7–20) if they managed to keep up the average pace for two months, but only 5.5 cells (range 3–10) if they lived for 30 days, as would seem more likely (our calculations). Similar estimates for *Bembix pallidipicta* (based on Evans's determination that females take 4–6 days to build a nest) would be ~10–15 cells if they lived for 60 days, but only ~5–8 cells if they lived for 30. (Note that even 30 days may be an overestimate for the reproductive life span of females.) All of the LRS estimates given above would also have to be adjusted downward to account for offspring mortality, the level of which is partly due to the mother's parental abilities (and so should be applied to her LRS ledger). Once significant evidence accumulates, one may also be able to test the hypothesis that the number of cells provisioned declines in the Bembicini if offspring survivorship increases with increasing levels of parental care. If so, the actual LRS values for mass and progressive provisioners may be more similar.

We should also expect that the structure and production capacity of the ovaries, and the size of the eggs of wasps, should be correlated with their reproductive lifestyle (Itô 1978; O'Neill 2001). As the above estimates of LRS suggest, most nest-provisioning sand wasps lay eggs at a relatively low rate because of the extended time invested in each nest cell. Consequently they rarely carry more than two mature oocytes in their ovaries at any time (Table 8.2); and those oocytes tend to be large relative to the mother's body size (Iwata 1964; Itô 1978; O'Neill 1985, 2001; Ohl and Linde 2003). Although more data are needed, the information in Table 8.2 suggests that progressively provisioning species carry fewer mature oocytes than do mass-provisioning species. All of this points to the general conclusion that female sand wasps, because of their short adult lives and extensive parental

Table 8.2 Number of ovarioles and number of mature oocytes in female Bembicinae. Table modified from Ohl and Linde (2003); all data by Ohl and Linde, unless otherwise specified. Because data on the number of oocytes in Ohl and Linde were given as mean ± standard deviation, we used those values to deduce the values reported here.

Species	Number of ovarioles per ovary	Maximum number of mature oocytes per female	Number of females examined
Mass provisioners			
Argogorytes mystaceus[a]	3	5	2
Hapalomellinus albitomentosus (Bradley)	3	5	1
Hoplisoides glabratus (Bohart)	3	2	1
Hoplisoides homonymus (Schultz)	3	2	1
Hoplisoides manjikuli Tsuneki	3	2	1
Bicyrtes variegatus	3	3	2
Progressive provisioners			
Bembecinus clypearis Bohart	3	1	1
Bembecinus hungaricus japonicus[a]	3	1	1
Bembecinus prismaticus (F. Smith)	3	1	1
Bembecinus quinquespinosus[b]	3	2	30
Microbembex argyropleura	3	1	3
Editha magnifica[c]	3		3
Microstictia femorata (Fox)	3	1	4
Bembix americana[b]	3	2	30
Bembix borrei thaiana Tsuneki[d]	3	2	2
Bembix dentilabris[d]	3	1	2
Bembix niponica[a]	3	1	2
Bembix troglodytes	3	1	1
Brood parasites			
Nipponysson rufopictus Yasumatsu and Maidl[d]	4	5	1
Nysson interruptus	4	5	1
Nysson tridens Gerstaecker	4	2	1
Nysson trimaculatus (Rossi)	4	5	4
Epinysson bellus (Cresson)	4	2	1
Zanysson texanus texanus (Cresson)	4	4	1
Foxia navajo Fox	4	4	1
Metanysson solani (Cockerell)	4	5	1
Stizoides renicinctus	4	5	1
Stizoides renicinctus[e]	4–5	6	30

[a]Iwata (1955); [b]O'Neill (1985); [c]Martins (1993); [d]Iwata (1960); [e]KMO and A. Pearce, unpublished.

care, are low-fecundity creatures. Later we will discuss the differences between the ovaries of nest-provisioning and brood parasitic Bembicinae.

Prey Type

Much could be said about hunting behavior and the kinds of prey taken by sand wasps. As a subfamily, Bembicinae prey on insects of least nine orders, two of which predominate (Table 8.3); in this discussion, we exclude the scavengers of the genus *Microbembex,* which do not experience the same set of constraints as wasps that hunt living prey. Homoptera are the sole prey of all Alyssontini and Gorytini, and are the only prey of all *Bembecinus* (Stizini) (except for a very few records of Diptera). The use of Homoptera by sand wasps is not unique among solitary aculeate wasps. Homoptera of the same families taken by sand wasps are also exploited by certain other crabronid wasps, including some Crabroninae (e.g., a few *Crossocerus*) and most Pemphredoninae (O'Neill 2001). The only wasps among the Alyssontini, Gorytini, and Stizini that do not prey on Homoptera are *Stizus,* which take grasshoppers, katydids, and mantids.

Diptera are the major prey of 9 of 14 genera of Bembicini for which records are available (Tables 8.3, 8.4), including its largest genus, *Bembix.* The second most common order taken by Bembicini is Lepidoptera, known to be prey of seven genera, including four that apparently prey on nothing else. Lepidoptera are also the prey of several other genera of solitary wasps, including *Ammophila* and *Podalonia* (Sphecidae) and the majority of Eumeninae (Vespidae) (O'Neill 2001). However, these wasps prey upon caterpillars, whereas Bembicini take adults. Hemiptera are known as prey of four genera of sand wasps, but only *Bicyrtes* specializes on them. Among other solitary wasps, *Astata* spp. are also predators of Hemiptera, especially Pentatomidae (Evans 1957b).

On the basis of existing phylogenetic hypotheses (Bohart and Menke 1976; Prentice 1998), it seems likely that use of Diptera, Hymenoptera, Lepidoptera, Neuroptera, and Odonata is an evolutionarily derived feature of the Bembicini. But the unresolved relationships among the genera discussed here (Prentice 1998) make it difficult at this time to further correlate patterns of prey use with phylogenetic patterns. One group of genera within the Bembicini that appears to form a monophyletic lineage is the "Stictiellina" (*Chilostictia, Glenostictia, Steniolia, Stictiella,* and *Xerostictia*) (Evans 1966a; Bohart and Horning 1971; Bohart and Menke 1976; Prentice

Table 8.3 Orders of insect prey reported for genera of sand wasps. X = taken by >50% of the species in the genus; x = taken by <50% of the species in the genus.

Tribe/genus	Odonata	Orthoptera	Mantodea	Homoptera	Hemiptera	Neuroptera	Lepidoptera	Diptera	Hymenoptera
Alyssontini									
Alysson				X					
Didineis				X					
Gorytini									
Clitemnestra				X					
Exeirus				X					
Sphecius				X					
Tanyoprimnus				X					
Ammatomus				X					
Argogorytes				X					
Harpactus				X					
Trichogorytes				X					
Austrogorytes				X					
Gorytes				X					
Pseudoplisus				X					
Sagenista				X					
Hoplisoides				X					
Hapalomellinus				X					
Stizini									
Stizus		X	x[a]						
Bembecinus				X				x[b]	
Bembicini									
Bicyrtes					X				
Hemidula								X	
Rubrica	x[c]						x	X	
Selman[d]								X	
Stictia				x			x	X	
Editha							X		
Trichostictia								X	
Zyzzyx							x	X	

Table 8.3 (continued)

Tribe/genus	Odonata	Orthoptera	Mantodea	Homoptera	Hemiptera	Neuroptera	Lepidoptera	Diptera	Hymenoptera
Bembicini *(cont.)*									
Stictiella							X		
Microstictia[d]							X		
Glenostictia		x			x			X	x
Xerostictia[d]				X		X			
Steniolia								X	
Bembix	x					x	x	X	x

[a]Observations to date suggest that species that prey on Orthoptera do not take Mantodea (and vice versa); [b]in just two species, a minority of prey were flies; [c]based on a single observation of a female carrying a dragonfly; [d]data available for one species only.

1998). Even here, no consistent pattern emerges, as these genera prey on a wide variety of insects. Thus with regard to prey use, the Stictiellina appears to be one of the most evolutionarily flexible lineages of the Bembicinae, along with certain groups of *Bembix.*

Of all sand wasp genera, *Bembix* exploits the widest variety of prey, something that has come to be fully appreciated only in the last 40 years. A similar form of diversification occurred in the crabronid genus *Crossocerus,* species of which variously prey on Diptera, Ephemeroptera, Homoptera, Lepidoptera, Mecoptera, Psocoptera, and Trichoptera (Bohart and Menke 1976). Of all sand wasps, Australian *Bembix* show the most radical evolutionary shifts in prey use. Perhaps one key to understanding why many *Bembix* have made evolutionary shifts to non-Diptera prey (Figure 8.3) is the recognition that even a species of *Bembix* that specializes on Diptera may use a wide variety of prey, including small and delicate Stratiomyidae and Dolichopodidae, robust fast-flying Calliphoridae, Sarcophagidae, and Tachinidae, and elongate, strong-flying Asilidae. The ability to exploit such a high diversity of dipteran prey may have made the transition to non-dipterous prey more likely when the opportunity arose. Certainly, it is more difficult to envision a switch to damselflies and butterflies happening as readily among the many bembicine specialists on small Homoptera. At a minimum, most of the Homoptera predators are too small (and perhaps

Table 8.4 Summary of prey records of New World Bembicini (orders in parentheses give non-Diptera orders among prey).

Genus (number of described species)	Diptera only	Diptera with occasional nondipterous prey	Diptera, but make regular use of other insect orders	Non-Diptera prey only
Bicyrtes (27)				11 species (Hemiptera)
Editha (7)				*E. adonis*, *E. fuscipennis*, *E. integra*, *E. magnifica* (Lepidoptera)
Glenostictia (21)	4 species		*G. scitula* (Homoptera, Hemiptera, Hymenoptera)	
Hemidula (2)	2 species			
Microstictia (11)				*M. minutula* (Lepidoptera)
Rubrica (4)	2 species	*R. nasuta* (Lepidoptera)		
Selman (1)	1 species			
Steniolia (15)	6 species			
Stictia (26)	9 species	*S. carolina* (Hemiptera, Lepidoptera)		
Stictiella (14)				*S. callista*, *S. emarginata*, *S. evansi*, *S. formosa*, *S. pulchella*, *S. serrata* (Lepidoptera)
Trichostictia (3)	2 species			
Xerostictia (1)				*X. longilabris* (Hemiptera, Neuroptera)
Zyzzyx (1)		*Z. chilensis* (Lepidoptera)		

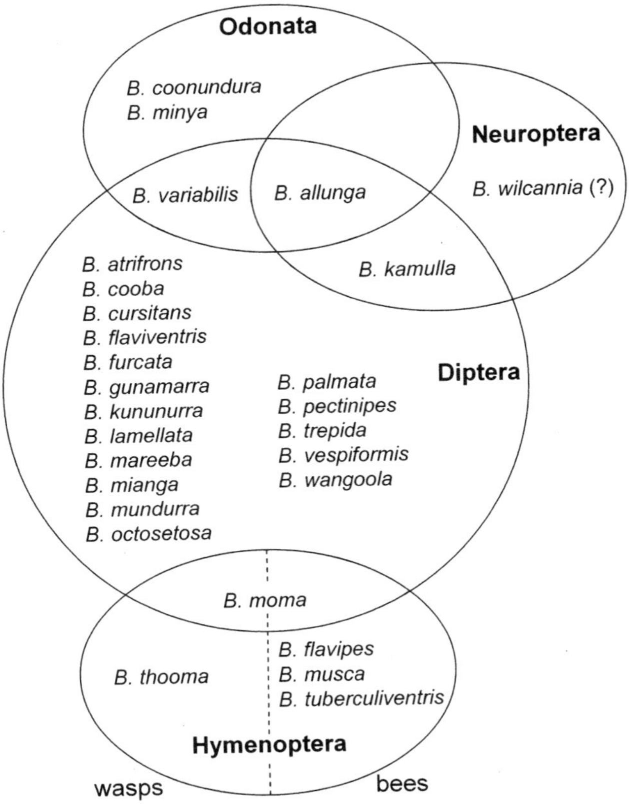

Figure 8.3. A representation of the known prey orders of 28 species of Australian *Bembix*. Modified from Evans and Matthews (1973). *B. wilcannia* record is based on single prey item.

too slow) to add such prey to their diets. Attacking stationary leafhoppers or spittlebugs on the host plants of prey is a very different matter from engaging in an aerial chase of a fly.

Evolutionary shifts in prey use may be favored by new opportunities that occur when novel prey appear in the habitat. Or they may be favored by environmental changes that make formerly favored prey less abundant. The degree of evolutionary flexibility seen within genera, most obviously *Bembix*, makes it clear that phylogenetic analyses of prey use need to include analyses below the genus level. Perhaps we can also gain some insight into evolutionary changes in prey use by considering species that include unusual prey types in their diets. Evans (2002) noted that if the ancestor of the Bembicini preyed on flies, then we should pay special attention to species from groups that normally use flies, but occasionally provision with other insects. He cited seven species of Bembicini that have been recorded rarely or occasionally taking prey outside of their "normal" range and another four that regularly use both flies and other insect orders. (We left several odd records off the list entirely that may have simpler explanations. Evans and Matthews (1973), for example, found a single cicadellid leafhopper among the dipteran prey within a nest of *Bembix variabilis*, suggesting that it may have been accidently deposited there by a female of a *Bembecinus* nesting at the same site. Records for *Bembix allunga* also include the wing of a cicada found in one nest cell, but Evans et al. (1982) thought that it was possible that it was extraneous material accidentally brought in within the grasp of an asilid fly used as prey.) To the records cited in Evans (2002), however, we have added several species of *Bembecinus* that show the opposite trend by including a few flies in their normal diet of homopterans.

1. *Zyzzyx chilensis* was reported by Janvier (1928) to take flies, but the wings of a skipper (Hesperiidae) were found in one nest.
2. *Bembix melanaspis*, along with the next three species listed, is a well-known predator of flies; the single odd record came from a report of the wings of a damselfly found in one nest.
3. *Bembix citripes* was reported including a small bee in one nest cell by Llano (1959).
4. *Bembix littoralis* was also found to include a single bee in one nest cell along with the requisite flies.
5. *Rubrica nasuta* has been reported taking flies in a few studies, but there are isolated records from several locations of their taking butterflies and moths, and a single record of a female carrying a dragonfly.
6. *Bembecinus cinguliger*, like other *Bembecinus*, preys mainly on Homoptera (94% of all records), but also included two species of Diptera in provisions (6%).
7. *Bembecinus oxydorcus* also included a few tephritid flies among its prey.
8. *Bembix variabilis* included only flies in its diet at 18 sites, only damselflies at 1 site, and both at 3 sites.
9. *Stictia carolina*, a well-known fly predator, was twice reported prey-

ing on skippers (in New Jersey and Oklahoma) and once on cicadas (in Oklahoma).

10. *Bembix kamulla* is reported taking flies (19% of all records) and antlions (81%).
11. *Bembix moma* regularly preys on flies (7% of all records), wasps (19%), and bees (74%).
12. *Bembix allunga* provisioned with flies (61% of all records), Neuroptera (30%), Odonata (7%), and Homoptera (2%).
13. *Glenostictia scitula* has been reported, unlike other members of its genus, taking Diptera, Hemiptera, Homoptera, Hymenoptera.

What can we learn from these records? Do they represent mistakes in prey capture, where the wasps failed to distinguish them from their typical prey (Evans et al. 1974)? Such might be the conservative explanation for numbers 1–7 on the list until further evidence appears. Numbers 6 and 7 on the list are of interest, however, because they represent two sympatric and apparently closely related species deviating in the same way from their ancestral diet. Perhaps something peculiar about these wasp species favored the inclusion of tephritids among their prey; alternatively, there may be something about the local tephritid species that makes them liable to be "confused" for homopterans (e.g., their size, patterns of movement, microhabitat). Either way, the situation could be similar to the very early stages in the evolution of other species that have made the transition to alternative prey types.

Because more information is available, numbers 8 and 9 may be better examples of wasps that are commonly expanding their diets when suitable alternatives are encountered. Still, it is difficult to envision what about damselflies, skippers, and cicadas makes them the likely alternatives. Finally, numbers 10–13 on the list clearly represent species that have fully made the transition to prey not used by their ancestors. But, again, why these particular non-Diptera prey rather than other insects? It must have something to do with the availability and abundance of these prey in the habitat in which the wasps hunt, as well as the nondipterans' similarity to normal Diptera prey and their susceptibility to prey capture techniques and venoms. Nevertheless, some of the transitions to non-Diptera are still quite puzzling, given the general lack of similarity in body form between, for example, damselflies and antlions, on the one hand, and the dipteran prey of *Bembix,* on the other. But that is our perception, and *Bembix* no doubt see their world differently than we do.

Several studies of sand wasps have documented intraspecific variation in prey use between and within populations. Perhaps the best explanation for geographic variation in prey use is concomitant variation in availability of potential prey. Geographic variation has been reported at a variety of taxonomic levels. The species of the cicada genus *Tibicen* used by *S. speciosus* vary latitudinally. The families of bees used by *Bembix tuberculiventris* also varied from north to south in Australia; in Northern Territory, females preyed solely on stingless bees of the genus *Trigona,* whereas further south, outside the range of stingless bees, they preyed on colletid and halictid bees. Even more striking was the sole use of Diptera by females in most populations of *Bembix variabilis,* when females were recorded preying only on Odonata at one site. This difference, however, could have been due to the time of year that the site was visited; at a few sites, both Diptera and Odonata were mixed within individual nest cells. Evans and Matthews (1973) suggested that damselflies were used when available because their larger size allowed fewer prey to be used per nest; they also noted that the females using damselflies constructed longer brood cells in order to accommodate the longer prey.

C. S. Lin (in Evans 1966a) reported that 77% of *Bembix nubilipennis* prey in a study near Austin, Texas, were house flies, whereas Evans reported that house flies made up just 7% of prey in his study in Indiana. This was likely due to a late-summer flush in the abundance of house flies in the vicinity of Austin. Opportunism is also evident in the seasonal variation in prey of *Rubrica nasuta,* which took flower flies of the genus *Eristalis* during January and February in Argentina, but almost entirely tabanids during late February and March. In Wyoming, *Bembix americana* apparently also concentrates on tabanids during years in which they are abundant. Finally, *Bembix occidentalis* provisioned solely with one species of otitid fly during the first two weeks of a study by G. E. Bohart and MacSwain (1939), before the wasps switched to flies from four other families, including a species of calliphorid fly that was apparently hunted at a newly established pig pen near the nest site. And the cicadellid prey taken by *Bembecinus hirtulus* varied among two sites in Australia.

Within a population, one often observes that some cells contain a certain narrow array of prey, likely the result of the short-term concentration by individual females on particularly profitable prey sources. Evans and Matthews (1973), for example, noted that one female of *Bembix variabilis* stocked her nest only with *Musca vetustissima* Walker (Muscidae), while another nearby female had provisioned solely with robber flies (Asilidae); at another site, 18 of 20 flies in one nest were all of the same genus of Muscidae *(Lispe)*. On the basis of variation in prey types present in different cells, Evans (1957b, 1966a) cites similar examples for all four of the nest-provisioning tribes of sand wasps, including *Alysson melleus, Bembecinus neglectus, Bembix amoena, B. belfragei, B. cinerea, B. nubilipennis, B. sayi, Bicyrtes quadrifasciata, Glenostictia scitula, Gorytes canaliculatus, Hoplisoides nebulosus, Clitemnestra bipunctata, Stictia carolina,* and *S. vivida.* Between-cell (or female) variation in prey type has also been mentioned in recent accounts of *Austrogorytes bellicosus, Bembecinus hungaricus, B. tridens, Bembix bubalus, Bicyrtes variegatus,* and *Rubrica nasuta.* Zolda and Holzinger (2002) reported that "three *B. tridens* females were observed provisioning with a single species of prey over a period of two weeks and even successive nests of individual wasps were provisioned with the same species of prey."

Prey records may also be biased toward one sex of insects, as in the case of the tendency to find excess females among the tabanid flies in some prey samples for *Bembix amoena, B. hinei,* and *Stictia carolina;* however, the opposite was true in one study of *Bembix americana* (Lane et al. 1986). And males made up 96% of *Trigona carbonaria* taken as prey by *Bembix musca,* which apparently hunted within mating swarms of the bee. A less extreme, but statistically significant, case of male prey bias was found in a large sample of the cicada prey of *Sphecius grandis* which included 55% males. This contrasts with a preponderance of females in several studies of *Sphecius speciosus* (i.e., 65% in Dow (1942) and 71% females in Dambach and Good (1943)). On the other hand, no sex bias was found for among prey of two species of *Bembecinus* by Zolda and Holzinger (2002). The fact that sand wasps sometimes exhibit sex-biased prey use is, perhaps, less interesting in itself than is the fact that it imposes sex-biased selection pressures on prey populations.

Thus if we take an accounting of the possible causes of variation of prey use within a species, present evidence suggests that much of the variation in published prey records is the result of large- and small-scale spatial variation, as well as seasonal and year-to-year variation, in the availability of different species, sexes, and developmental stages of potential prey. Some variation within populations is also attributable to the fact that larger females of a species can take larger prey than smaller females and therefore have a wider range of prey species and/or developmental stages available. Among sand wasps, this has been demonstrated for *Bembecinus quinquespinosus.* We expect that such a size constraint is common, especially in species that take prey that are large relative to their own size.

Finally, has competitive exclusion influenced the evolution of prey use in sympatric species of sand wasps? Evans (2002) hypothesized that competition resulting from the tendency among some *Bembix* to share nest sites could favor or at least reinforce the ability to use non-Diptera as prey. Testing this hypothesis can be tricky, because apparent resource partitioning among sympatric species could be due either to past competitive interactions favoring use of different prey, or else to other factors that have nothing to do with competition. We have discussed this problem with regard to the prey used by four sympatric species of beewolves (*Philanthus* spp.), among which there was little overlap in prey used when both size and taxa of prey were considered (O'Neill and Evans 1982). It is possible that the low degree of overlap was a consequence of the fact that the four beewolf species also differed in body size, and that body size evolved in response to factors other than competition. In fact, differences in prey use may have originally evolved in geographically distinct populations prior to sympatry. The same may apply to interpreting the low overlap in prey use by two sympatric species of *Bembecinus* in Europe, *B. hungaricus* and *B. tridens,* both of which are widely distributed (Zolda and Holzinger 2002).

Measuring Prey Diversity

Whether we are trying to understand the behavior, ecology, or evolution of a species, how would we best go about getting an accurate assessment of prey diversity in sand wasps? In particular, how would we estimate diversity in a way that would allow us make statistically valid comparisons between populations, species, genera, and tribes of sand wasps? The task is not straightforward, because of problems having to do with (1) the quality and size of the prey sample, (2) the appropriateness of using taxonomic categories in diversity estimates, and (3) assessment of the potential local prey base from which females take their prey. The statistical intricacies of characterizing diversity have been discussed by Magurran (2003).

One set of problems has to do with the quality of the prey sample, whether by sample we mean the number of wasp species studied or the number of prey recorded for each species. The best samples for a species would come from multiple females, in multiple locations, and would be obtained in different seasons and across different years. Some of the records we reported in earlier chapters are from just one or two females nesting in the same location. But how much is enough? Even after collecting over 3000 prey of *Philanthus sanbornii* Cresson during a five-year study, Stubblefield et al. (1993) were still adding new species to the list, which had grown to number 108. The large prey lists for *Bembix littoralis* (63 species) and *B. variabilis* (90 species), and the observed geographic variation in the lists attest to a similar situation in some sand wasps. For many groups of Bembicinae, variation in sample size and study duration among species makes quantitative comparisons of diet breadth of limited value. Exceptions include studies of extreme specialists (e.g, *Sphecius speciosus* or *Bembix flavipes*), or of cases where two species were studied simultaneously in one location.

Data exist to examine the value of large samples of prey for Gorytini and Australian *Bembix*. For the Gorytini, we could ask whether prey diversity differs among *genera*. But the numbers of families known to be prey of each genus of Gorytini could simply be a matter of the number of predator species examined, which ranged from 1 (many genera) to 15 *(Hoplisoides)*. If so, the more species studied in a genus, the more prey families should be found, due to increased sampling effort. In addition, geographic range expansion and adaptive radiation by a large genus may lead to an expansion of niche breadth as new prey types are encountered. It is undoubtedly true that more sampling will increase, at least up to a point, the prey family records of some genera. However, for the gorytine records summarized in Table 3.4, we found no significant correlation between the number of species studied in each genus and the number of prey families reported ($r = 0.39$, $P > 0.05$, $N = 19$ genera). So, in this case, sample size (number of species studied per genus) does not seem to have a major effect on estimates of prey diversity. In fact, two of the highest numbers of prey families recorded, nine for *Clitemnestra* and six for *Sagenista*, come from genera with few species.

The situation is different, however, for Australian *Bembix*, for which we could ask whether prey diversity differs among species rather than among genera. The number of dipteran prey recorded for Australian *Bembix* (among those that take at least some Diptera) ranges from 5 to 370 (Table

7.3). Among these 21 species, there is a significant correlation between the number of prey families documented and sample size for each species ($r = 0.80$, $P < 0.001$). Thus nearly two-thirds of the variation in the number of families recorded among species of Australian *Bembix* is perhaps explained by the size of the sample. The remaining variation is likely due to some combination of prey preferences of individual species of *Bembix*, local prey availability at the time each species was studied, the number of sites sampled, the duration of time over which sampling occurred, and the ability of each wasp species to capitalize on that prey availability. For example, small species of *Bembix* may be unable to pursue and subdue large flies, whereas large *Bembix* may be unable to efficiently handle small flies (which may in any case require too many foraging trips to be energetically viable as prey).

Because of problems with sample size, one must consider the extent of the known prey records before drawing the conclusion that a species of sand wasp specializes on a narrowly defined taxonomic group (i.e., a genus or species). We would perhaps be rash, for example, in concluding that *Sphecius hogardii* was a specialist on cicadas of the species *Uhleroides walkeri* based on the single study of this species by Genaro and de Varona (1998). On the other hand, multiple studies have documented that *Sphecius grandis* and *S. speciosus* prey (with one exception) only on cicadas of the genus *Tibicen*, so they do appear to be specialists at the generic level. Or is it just that no other cicadas within the appropriate size range consistently appear in their habitats? The natural prey of *Argogorytes carbonarius* in New Zealand was apparently just two species of the cercopid genus *Carystoterpa*, but the wasps have lately added *Philaenus spumarius*, a recent arrival from Europe.

A second set of problems encountered when trying to quantify prey diversity has to do with the use of taxonomically based measures of diversity to assess a wasp species' behavioral plasticity as a predator. Using the Linnaean classification system, we might conclude, for example, that a species that preys on ten families of flies has a wider diet than one that preys on five families. However, the level of morphological diversity used as a basis for classification by a taxonomist may be less meaningful to a wasp taking its prey on the wing. To the wasp, prey vary in their preferred habitats, visibility, body size and shape, flight speed, flight patterns and evasive maneuvers, hardness of exoskeleton, susceptibility to sting and venom, nutritive quality, and ease of transport. Any one of these factors could influence the evolutionary and behavioral responses that determine whether an insect species is included in a wasp's provisions. These qualities may correlate with our taxonomic categories, but only roughly sometimes. Should we conclude that a species of *Bembix* that preys on Calliphoridae, Sarcophagidae, and Muscidae has a broader diet than one that preys on Tabanidae and Dolichopodidae? Prey from the first three families listed may include a set of species that are relatively homogeneous with regard to body size and flight capability, whereas flies of the latter two families are very different from each other. And what about the equivalency of taxa? If we are using the number of prey genera in our measure of diversity, are genera of Tabanidae defined in the same way as genera of Sarcophagidae, or have entomologists classifying the former tended to be "lumpers" while those working with the former were "splitters"?

West-Eberhard (2003) used *Bembix variabilis* (predators of Diptera and Odonata) and *Bembix moma* (Diptera and Hymenoptera) as examples of associations between "intraspecific versatility" in behavior on the one hand (i.e., ability to use widely different prey) and adaptive radiation in behavior in the broader taxonomic group (i.e., adaptive radiation in prey use in the genus *Bembix*). She is probably correct, but to fully explore the hypothesis that high intraspecific behavioral plasticity is associated with the adaptive radiations in prey type in sand wasps, we need a measure of prey diversity that goes beyond taxonomic diversity and encompasses the biochemical, physiological, structural, and behavioral diversity of prey. All four forms of diversity, as well as interactions between the forms, will influence the benefits and costs associated with switching to novel prey types. As an illustration, consider how one would answer the following question. Does the vespid subfamily Eumeninae, all members of which prey on larval Lepidoptera and Coleoptera, exhibit lower prey diversity (and less of an adaptive radiation in prey use) than the single bembicine genus *Bembix*, which has been documented to take insects of seven orders? To answer this, we have to have some way of comparing the diversity of the two prey groups without just counting species, genera, or families. The adult prey taken by *Bembix* perhaps have greater behavioral and morphological diversity than the larval prey of eumenines, at least in terms of behaviors that affect vulnerability to capture. But the caterpillars used by eumenines may have a much greater biochemical diversity because of their defensive secretions. Is overcoming the behavioral and morphological diversity of prey a greater evolutionary hurdle than overcoming their biochemical diversity? The same type of question applies (though perhaps on a smaller scale) if we make prey diversity comparisons between, for example, *Bembix* and *Bembecinus,* or between the Bembicini and Gorytini.

Finally, is a wasp that takes a low diversity of prey incapable of preying on other insects, or are potential alternative prey just rare in its habitat? Recognizing that the best way to determine a wasp's prey preferences would be to compare the prey taken with the prey available, how do we assess the prey base available to a wasp species? One problem is that the sampling method used by wasps, which is based on their sensory and predatory capabilities, is different from the methods used by entomologists, such as sweep nets, vacuum samplers, pitfall traps, or sticky traps. The first question, then, is which sampling method to use? A sweep net or vacuum used to sample insects on plants is probably a more appropriate way to assess leafhopper availability than is a pitfall trap, which captures insects walking on the soil surface. And a sticky trap placed on a horse's hindquarters or next to a pile of what comes out of it might be the best available method to sample the flies taken by some *Bembix* and *Stictia.* For other prey, such as damselflies, the available resource base may be difficult to quantify, and for species that take a great diversity of prey and a wide variety of microhabitats an accurate assessment of prey availability is probably impossible. The second problem is how to employ the chosen method. For example, on which plants in the local environment do wasps hunt leafhoppers? On the basis of the known hosts of various homopteran prey, Zolda and Holzinger (2002) concluded that *Bembecinus hungaricus* hunted mainly on young trees, whereas the sympatric *B. tridens* hunts primarily on grasses, forbs, and shrubs. Because this was known only after prey had been identified, it would have been difficult to decide where to sample the potential prey base during their initial study. A third, sometimes intractable problem is that we may not know how accurately our sampling method estimates the relative abundances of different species of insects unless we have done an independent assessment (see Larson et al. [1999] for a study that attempted to do just such a thing).

Hunting and Prey Paralysis

Most studies of sand wasps center around nesting areas, where parenting and mating behaviors can be observed. *Microbembex monodonta* females forage for dead arthropods on the soil surface near their nests, at least until the area has been picked clean. Females of most species, however, generally forage outside of the nesting area, and it is often unknown where they hunt. But there are exceptions. In the Gorytini, *Sphecius speciosus* and *Exeirus lateritius* are known to take prey in trees, while there a number of observations of *Argogorytes mystaceus* and *A. carbonarius* attacking spittlebugs within spittle masses on plants. The study of the prey of *Bembecinus hungaricus* and *Bembecinus tridens* cited above illustrates the use of prey records, combined with knowledge of prey biology, to deduce hunting sites.

It is well known that many Bembicini hunt for flies that gather around livestock. Known examples include *Bembix americana, B. bidentata, B. moebii, B. olivacea, Rubrica nasuta; Stictia carolina* (the "horse guard"), *S. elegans,* and *S. signata;* Krombein and van der Vecht (1987) suggested that the source of prey for *Bembix glauca* in Sri Lanka was water buffalo. *Stictia signata* (Bates 1863), *R. nasuta* (Evans et al. 1974), *Bembix variabilis,* and *B. wangoola* (Evans and Matthews 1973) have been reported hunting biting flies on and around humans. And *Bembix* have been observed hunting around other centers of fly activity, such as livestock watering troughs, animal droppings, pigpens, and seaweed and dead fish on beaches. Similarly, *S. signata* hunts near garbage and vertebrate carrion, and *Stictia maculata* was observed hunting flies "in the vicinity of an open air privy" (Matthews et al. 1981).

Evans and Matthews (1973) hypothesized that *Bembix flavipes,* which specialized on stingless bees, hunts at bee hives, whereas *B. tuberculiventris,* which takes both bees and flies, hunts on flowers. *Bembix americana, Editha magnifica, Glenostictia scitula, G. pictifrons, Stictia carolina,* and *Stictiella formosa* have also been observed hunting and taking prey on flowers; *B. americana* appears to hunt in a wide variety of other locations, including within grasses and tabanid mating aggregations. Several studies of *Rubrica nasuta* revealed them engaged in restricted prey searches around feces and within patches of flowers, as well as in broader, continuous searches along dirt roads and within open vegetation.

There have been few detailed studies of the hunting behavior of sand wasps, though many studies have documented a few prey captures, usually observed at a distance where the stinging sequence could not be scrutinized. The most detailed behavioral studies of stinging by solitary wasps document the number, location, and sequence of placement of stings, the form and duration of paralysis, and the existence of lesions, especially those in the central nervous system of prey (reviewed in Piek 1985; Steiner

1986; O'Neill 2001). Other studies go further and examine the biochemistry and pharmacology of venoms. As far as we know, no comprehensive study has been done on any sand wasp, but among bembicines, we probably know most about stinging and paralysis by the cicada-killer wasps *Sphecius speciosus* (though there are some inconsistencies among different accounts). Reinhard (1929) described the stinging of cicadas by *Sphecius speciosus* as follows: "Recurving her needle-tipped abdomen, the wasp feels with it for a spot in the breastplate of the cicada. Deliberately, she plunges her sting up to its hilt, deep into the side of the struggling insect. The stricken cicada is quickly quiet." The point of insertion of the sting was later confirmed by microscopic examination of "over a score" of prey to be "a single [slit-like] puncture in the membranous articulation at the base of the front leg, either on the right or on the left side." Each cicada is apparently stung just once. Dambach and Good (1943), however, claim that the sting is inserted in the abdomen. We expect that Reinhard's description represents the more typical situation, because prey are more likely to be inactivated when stung near centers of locomotion in the thorax. Regardless of the point of entry, and perhaps it is variable in this species, the effect of the venom is rapid and long-lasting. According to Bringer (1996), paralyzed cicadas live longer (up to about 10 days) and lose body mass at a lower rate than cicadas that the researcher captured alive. Clearly, combined with the fact that the venom prevents the cicada from struggling during prey carriage and storage, the effects of the venom are highly adaptive for the wasp. And the venom of *Sphecius* is delivered by a robust sting (Prentice 1998), part of "the most highly developed sting apparatus within the whole family Sphecidae" (Radović 1985).

For any solitary wasp, an ideal venom would immediately paralyze the prey (providing for easier handling during prey transport) and keep it immobile (to prevent its thrashing about in the nest cell), while maintaining its nutritional value and water content prior to feeding by the larva. Reviews of the effects of bembicine venoms make it clear that the degree and duration of the effect is variable not only across the subfamily but within species. It is not unusual for prey to remain alive and twitching for several weeks following stinging (Steiner 1986). Evans (1966a) noted that the prey of Gorytini "does not normally recover from paralysis" and more recent work on *Gorytes canaliculatus* and *Sphecius speciosus* confirms this. *Bicyrtes* prey are said to be well paralyzed, but to remain alive and responsive to stimuli for several days. On the other hand, many or all prey of some

Bembix (e.g., *B. cinerea, B. dentilabris, B. texana*), as well as those of *Glenostictia scitula* and *Steniolia* spp., are apparently killed outright as a result of stinging; they appear to be stiff and dry soon after being placed in nests (Evans 1966a). Killing of prey has also been reported in species of *Alysson, Bicyrtes, Stictiella,* and *Zyzzyx.* But other prey in these same species are clearly just paralyzed, often so lightly that they may live for long periods and exhibit considerable movement when removed from cells. Evans (1966a), for example, notes that the lepidopteran prey of *Stictiella* are usually "deeply paralyzed or dead"; however, prey in one exhumed cell of *Stictiella formosa* were so incompletely paralyzed that, when they were removed from the cell, "they created pandemonium by flopping aimlessly but vigorously around on the ground" (Gillaspy et al. 1962).

Steiner (1986) points out that even among solitary wasp prey provisioned in the same cell, one often observes variation in the degree of paralysis: "some prey are found dead, others deeply paralyzed and a few even show various degrees of recovery and/or imperfect paralysis." Perhaps this results from variation in the pharmacological effect of venoms on different species or sizes of prey. Or maybe there is variation among prey types on how well the venom is delivered to its target tissue. Either way, future studies should make note not only of variation in degrees of paralysis but of how the variation correlates with prey species or prey size. Another source of variation may occur when females provision a cell so rapidly so that their supply of venom begins to run low. If so, prey brought in at the end of a rapid provisioning sequence should be more lightly paralyzed than those brought in earlier (or than those brought in over longer intervals). But some variation in the effect of stinging on prey in the same cell may be adaptive. *Bembix* females often kill the first prey in the cell (when it serves as a pedestal for the egg), but paralyze subsequent prey.

The sting apparatus of apoid wasps seems to have considerable systematic value (Prentice 1998) and its form varies among taxa using different types of prey (i.e., in size, curvature, and rigidity). As might be expected in a group that has dropped the use of its sting for capturing prey, the morphology of the sting of *Microbembex* differs from that of other Bembicinae. Prentice (1998) notes that the sting of *Microbembex* is "proportionally smaller than in other Bembecina" (i.e., the taxa covered here in Chapters 6 and 7), and that the apex of the sting shaft is "somewhat dull" (i.e., in contrast to the sharp penetrating point found on the sting shaft of predatory apoid wasps). Similarly, Radović (1985) describes the stylet of the sting apparatus of *Microbembex* as "rather tender and short." *Microbembex* also have a short furcula, the structure or apodeme to which the sting muscles attach (Prentice 1998), perhaps indicating that the sting apparatus has lost some of its power. However, Evans (1966a) noted that the sting and poison sacs of *Microbembex monodonta* are as large as would be expected for a wasp of its size. Radović (1985) hypothesized that the highly curved sting stylets of *Bembix, Rubrica, Stictiella,* and *Zyzzyx* may aid in subduing highly mobile prey. This has been disputed by Prentice (1998), who raises the examples of the genus *Miscophus* and the tribe Trypoxylini, which attack relatively slow-moving prey yet possess strongly curved stylets. Thus much work needs to done to link sting morphology with prey type and stinging behavior.

Prey Carriage

Once prey are subdued, bembicines carry them to nests using either their mandibles or their legs. Among sand wasps, mandibular prey carriage, which is considered to be an ancestral trait for apoid wasps, is used only by *Alysson* (and perhaps also *Didineis*) of the Alyssontini and *Clitemnestra* of the Gorytini. Alyssontini carry prey using a combination of flying and walking, the latter being used as females near their nests. Most Gorytini, and all Stizini and Bembicini, have pedal-type carriage, holding prey with their middle legs, sometimes with the aid of other legs, as they carry prey forward, usually in flight. *Pseudoplisus natalensis* use their hind legs to assist in the carrying of large spittlebug prey.

The use of hind legs is particularly evident in species of *Sphecius,* which apparently require them for stabilizing their exceedingly large and heavy prey. Several authors have remarked on the apparent difficulty experienced by female cicada-killers as they lug about their large prey. In order to get airborne with such prey, females have to first climb trees on foot before launching themselves homeward. The earlier suggestion that the hind tibial spurs of *Sphecius* evolved to assist in holding prey during prey carriage (Evans 1966a) may be incorrect (Coelho and Wiedman 1999). There has also been the suggestion that the long legs of *Exeirus,* the other cicada hunters among sand wasps, are adaptations for straddling prey while in transit to the nest (Evans 1966a), as is certainly the case for spider wasps (Pompilidae); in fact, *Exeirus lateritius* was once erroneously classified as a pompilid. Evans and Matthews (1973) note the intriguing possibility that certain abdominal structures of *Bembix thooma* have evolved to aid in carrying its prey (thynnine wasps). In females (but not males) of this species, sternite 2 has a "broad dish-like depression medially," sternite 6 has a "very high, arching median ridge on [its] apical two-thirds," and tergite 6 is "unusually slender." Unfortunately, prey carriage in this species has not been observed in detail.

Feeding by Adults

Not much new can be said about the feeding habits of adult sand wasps. Evans (1966a) noted then that adults obtain carbohydrates in the form of nectar from flowers, honeydew (the secretions of homopterans), and in *Exeirus and Sphecius,* plant sap. He noted that *Alysson* and *Clitemnestra* are often found on honeydew, but not flowers. The previous suggestion that *Bembecinus* do not visit flowers seems not generally true. We saw *Bembecinus quinquespinosus* feeding only at honeydew sources, F. W. Gess and S. K. Gess, however, observed *Bembecinus* feeding on flowers and extrafloral nectaries in South Africa, and Karl Krombein found them at flowers in Sri Lanka. *Bembicini* are common visitors to flowers, and genera with long tongues such as *Steniolia* can feed on flowers whose nectar is generally inaccessible to their short-tongued relatives. Recent additions to the (sparse) literature on adult Bembicini feeding on flowers include observations we cited earlier on *Bicyrtes angulatus* visiting the weed species *Waltheria americana, Editha magnifica* feeding on nectar at *Veronia* (the same plant on which many of the lepidopteran prey were also taken), and *Bembix promontorii* of both sexes visiting *Eucalyptus* flowers. With regard to the potential of bembicines as pollinators, two recent studies leave little cause for optimism (if you are a flower). Males of *Bembix dentilabris* commonly visit *Kallstroemia grandiflora* in Arizona, but by robbing nectar from a position beneath the blossom, they do not pick up or deposit pollen. The story is a little better for *Bembix amoena,* which, though inferior to bees and flower flies as a pollinator of onions in Utah, was the most effective among a large group of apoid wasps examined.

Evans (1966a) also brings together scattered observations of adult female sand wasps feeding on their prey. He cites observations of species of *Bembix, Harpactus, Hoplisoides, Microbembex,* and *Zyzzyx* feeding on the body fluids of prey, and records of *Stictia signata* actually capturing and eating mosquitoes whole. The rarity of such observations suggests that adult feeding on prey is not common, but it may increase in occurrence during times when nectar is scarce.

Brood Parasitic Bembicinae

When microbiologists attempt to establish host relationships, they follow a rigorous set of rules known as Koch's postulates. No pathologist would be taken seriously if he proposed that a particular organism caused a disease simply because its spores had been collected in the vicinity of sick patients. Strictly interpreted, Koch's postulates do not directly apply to studies of brood parasites. But a similar level of skepticism applied to wasp studies would demand that a researcher proposing a host relationship be able to rear the presumed brood parasite from the nest cells of the host. Although potential hosts are often cited, not all investigators search for the inconspicuous brood parasite eggs hidden within prey masses. Nor do many rear cell contents to expose the presence of brood parasites. Perhaps this is partly due to the emphasis researchers usually place on preserving and identifying prey found within cells. We might also add that because brood parasitic Nyssonini have likely evolved to be inconspicuous as a means of avoiding the attention of host females, wasp ethologists perhaps also find it difficult to detect their activities. So it is not particularly surprising that even though there are hundreds of brood parasitic species of Bembicinae, we have very little hard biological information. Nevertheless, as a general contention, it seems certain that all Nysonnini and all *Stizoides* (Stizini) are brood parasites. The six genera of Nyssonini for which we have some evidence of brood parasitism represent 86% of the species in the Nyssonini; and none of the remaining genera have more than 10 species each. However, some evidence regarding specific host relationships (Table 4.1) should be often considered no more than a working hypothesis and a first step to future studies.

Regardless of whether we are always confident in our suppositions about specific host relationships, it is clear that brood parasitic Bembicinae exhibit a number of traits related to their lifestyle. Female Nyssonini lack foretarsal rakes and scopae, because these are no longer needed for nest excavation (Ohl and Linde 2003). What digging female *Nysson* and *Stizoides* do is restricted to opening temporary nest closures and closing hosts' nests that they have attacked. Female Nyssonini also have reduced stings, because these are no longer required for hunting (Radović 1985). Like the brood parasitic and parasitoid Chrysididae, wasps of this tribe (with one exception) have a "heavily reinforced or 'armored' integument, presumably as defense against the stings or bites of their hosts" (Bohart and Menke 1976). The propodeal spines of Nyssonini, perhaps also used for defense, are also "reminiscent" of those in the Chrysididae.

The reproductive systems of brood parasitic bembicines reflect the fact that these wasps are unconstrained by the need to build and provision a nest. Ohl and Linde (2003) compared the reproductive systems of 131 species of apoid wasps, including nine brood parasitic species, eight Nyssonini and one *Stizoides;* in Chapter 5, we discussed further data on *Stizoides renicinctus.* Ovarioles in the ovaries of wasps represent parallel strands of developing oocytes (eggs); because only the oocytes at the distal ends of each ovariole are mature or nearly so, the number of ovarioles should correlate with the number of eggs ready to be laid. All nest-provisioning apoid wasps listed by Ohl and Linde had either six (most species) or four ovarioles (*Oxybelus* spp.). (Table 8.2 gives their data for Bembicinae, along with a few additions.) The only apoid wasps with eight ovarioles were the nine brood parasites in their sample. And we have since found that *S. renicinctus* females usually have up to 10 ovarioles. Among other solitary aculeate wasps, ovariole numbers exceeding six have been reported only for parasitoids and brood parasites of the families Mutillidae (up to 8) and Chrysididae (up to 108) (review in O'Neill 2001).

Furthermore, individuals in the eight species of Nyssonini on Ohl and Linde's list carried 2–5 mature oocytes (overall mean = 3.9); similarly, *Stizoides renicinctus* females carry up to 6 mature oocytes. This contrasts with seven nest-provisioning species of *Bembecinus* (Stizini) and *Bembix* (Bembicini), where the within-species means ranged from 1.0 to 1.5 (overall mean = 1.1). Finally, it would seem likely that a species that carried more mature oocytes would have to carry smaller mature oocytes because of space constraints (O'Neill 1985). Ohl and Linde assessed the relative size of the largest oocyte using the "egg index," the length of the longest mature oocytes divided by the thorax width (a measure first used by Iwata 1955). For seven species of Nyssonini, the egg index ranged from 0.36 to 0.90 (mean = 0.74). In a large sample of *S. renicinctus* females, the average was 0.73 (KMO and A. Pearce, submitted). For five nest-provisioning *Bembecinus* and *Bembix* it ranged from 0.67 to 1.01 (mean = 0.82). Thus, compared with brood parasitic Nyssonini, progressive- or delayed-provisioning *Bembecinus* and *Bembix* carry a greater number of relatively small oocytes. No one has directly measured the LRS of brood parasitic sand wasps. However, the structure of their reproductive systems suggests that the maximum value for offspring per lifetime among brood parasites must be higher than the maximum values for nest provisioners we cited earlier.

Natural Enemies

Insects of a variety of orders have been reported as brood parasites, parasitoids, and predators of the immature and adults stages of sand wasps, and have been implicated as important selective agents in the evolution of nest structure, nest closures, and accessory burrows as well as various avoidance behaviors (Evans 1966a,b; McCorquodale 1986; Spofford and Kurczewski 1990, 1992). Evans (1966a) also mentions the likelihood that birds prey on sand wasps, and we earlier noted our observations of horned lizards preying on male *Bembecinus.* Here we reiterate which groups are the major natural enemies of Bembicinae; more details on the manner in which various natural enemies attack sand wasps can be found in O'Neill (2001).

Table 8.5 provides an updated version of a similar table in Evans (1966a), although we have excluded (1) records cited then as having "an element of doubt" associated with them and (2) scavengers of the families Chloropidae, Otitidae, and Phoridae (which are technically not "natural enemies," because they often only feed on refuse in cells). The table lists several categories of natural enemies, including:

(1) *Brood parasites* (also sometimes simply called cleptoparasites, but see Field 1992), species whose larvae feed upon the host's nest provisions. The larvae of all brood parasites listed develop within the nest cell, but they differ among one another in their manner of gaining entry to nests. Along with bembicine brood parasites, adult females of most Chrysididae and certain genera of Miltogramminae (e.g., *Metopia* and *Phrosinella*) deposit their eggs directly within nest cells or somewhere in the host nest burrow. Some *Senotainia,* on the other hand, deposit a larvae (or an egg that immediately hatches) onto the prey as the prey-carrying wasp approaches her nest entrance. The egg or larva of the sand wasp may be killed directly (and perhaps eaten by the parasite larva), or the bembicine larva may simply starve due to loss of provisions. In either case, the majority of the parasites' food comes from the host's provisions.

Table 8.5 Major natural enemies of sand wasps. Updated and modified from Table 47 of Evans (1966a), with additions from research cited earlier in this book.

	Natural enemy			
Type of enemy	Order	Family	Genus	Host/prey genera
Brood parasites	Diptera	Sarcophagidae (Miltogramminae)	*Amobiopsis, Craticulina, Opsidia, Protomiltogramma, Taxigramma*	*Bembix*
			Hillarella	*Rubrica*
			Metopia	*Bembix, Bicyrtes, Gorytes, Sphecius*
			Pachygraphomyia	*Rubrica*
			Pachyophthalmus	*Hoplisoides*
			Phrosinella	*Bembix, Gorytes*
			Senotainia	*Bembix, Bicyrtes, Glenostictia, Hapalomellinus, Hoplisoides, Sphecius, Stictiella*
	Hymenoptera	Crabronidae	*Nysson, Epinysson*	*Argogorytes, Bembecinus, Gorytes, Harpactus, Hoplisoides, Lestiphorus, Oryttus*
		Chrysididae	*Elampus*	*Hoplisoides*
			Hedychridium	*Harpactus*
			Hedychrum, Holopyga	*Bembecinus*
Prey thieves		Crabronidae	*Bembix, Stictia*	*Bembix, Stictia*
		Formicidae[a]	Various	*Bembecinus, Bembix, Bicyrtes*[b]

Parasitoids of larvae	Diptera	Bombyliidae	*Exoprosopa*	*Bembix, Microbembex*
			Hyperalonia	*Rubrica*
			Ligyra	*Bicyrtes*
			Villa	*Bembix*
	Hymenoptera	Chrysididae	*Parnopes*	*Bembix, Glenostictia, Microbembex, Steniolia*
		Mutillidae	*Dasymutilla*	*Bembecinus, Bembix, Microbembex, Steniolia*
			Ephutomma	*Bembix*
			Mutilla	*Stizus*
			Smicromyrme	*Bembecinus*
Parasitoids of adults	Coleoptera	Rhipiphoridae	*Macrosiagon*	*Bembix, Stizus*
	Diptera	Conopidae	*Physocephala*	*Bembix*
	Strepsiptera	Stylopidae	*Paraxenos, Pseudoxenos*	*Bembecinus, Bembix, Sphecius, Stizus*
Predators of adults	Diptera	Asilidae	*Diogmites, Megaphorus, Proctacanthus, Promachus*	Various[b]
	Neuroptera	Myrmeleontidae		*Bembix*
	Hymenoptera	Crabronidae	*Bembix*	*Bembix*
			Philanthus[c]	*Alysson, Bicyrtes, Harpactus, Microbembex, Stictiella, Trichogorytes*
		Mutillidae	*Pseudophotopsis*	*Bembix*

[a]Either from within nests or at nest entrance; [b]probably common; [c]Evans and O'Neill (1988).

(2) *Prey thieves,* those species that steal prey and take them elsewhere. Perpetrators include conspecifics, other species of solitary wasps, and ants, all of which coopt prey to use them as their own provisions. The impact of ants as nest plunderers may be underestimated, especially because their activities may occur at times when researchers are not monitoring nest aggregations since wasps are not present above ground (e.g., in the evening). Ants also steal prey when females deposit them outside of nests prior to entering.

(3) *Parasitoids of larvae,* whose own larvae feed directly on larvae of sand wasps, sometimes (as in the case of the Mutillidae), after the wasp larva has spun its cocoon. *Parnopes* and *Mutilla* oviposit in the host cells, whereas *Dasymutilla* do so within the host cocoon. *Exoprosopa* of the Bombyliidae drop their eggs into open nest entrances.

(4) *Parasitoids of adults,* whose larvae feed on the bodies of adults, either killing them outright in some cases (e.g., Conopidae) or at least debilitating them (e.g., Strepsiptera). The effect of the latter may be so minimal that some stylopized females of *Bembix littoralis* and *Bembix variabilis* construct and provision nests (Evans and Matthews 1973). However, there is also evidence that strepsipterans have significant sublethal effects on the reproductive success of their hosts (Salt 1931; Bohart 1941).

(5) *Predators of adults,* which take their prey on the wing and on plants (asilids and *Philanthus*), on the ground (antlions), or within nests *(Pseudophotopsis).* The last would be the stuff of sand wasp nightmares if they had them: adults of *Pseudophotopsis continua* enter a *Bembix* nest at night, bite the adult in the neck, and feed on the wasp's body fluids (Mellor 1927).

Both *Gorytes canaliculatus* (Powell 1974) and *Alysson conicus* (O'Brien and Kurczewski 1982) prey upon leafhoppers parasitized with Dryinidae. This would generally be termed incidental cleptoparasitism, as in the case of ichneumon wasps parasitizing caterpillar prey of eumenines and other wasps (e.g., Krombein 1967). In the latter case, the ichneumons may complete development feeding on the prey, thereby depriving the nest-provisioner's larva. However, it is not known whether this is the case with the dryinids, which perhaps may be consumed along with the leafhopper; in fact, parasitized leafhoppers may also be easier to capture.

Reports of natural enemies often result from *ad lib* observations, and therefore little information is available on the level of mortality inflicted on sand wasp populations by specific natural enemies; and, as we said was the case for Nyssonini and *Stizoides*, many reported host records for the brood parasites and parasitoids in Table 8.5 are less than certain. But a few quantitative estimates have been made. Evans and Matthews (1973) found rates of stylopization to be 11.5% in *Bembix littoralis* and 7.6% in *B. variabilis*. In a more recent survey, Spofford and Kurczewski (1990, 1992) reported that 20% of 25 *Bicyrtes ventralis* nest cells were parasitized by the miltogrammine *Senotainia trilineata*, 17% of 24 *M. monodonta* nest cells were parasitized by *Senotainia* spp., and 9% of 23 *B. americana* nest cells were parasitized by the miltogrammine flies, including *Senotainia*. Even the value of 20% is lower than many values reported for bombyliid- and miltogrammine-caused mortality inflicted on some species of *Philanthus*, which range as high as 41% (Evans and O'Neill 1988). However, things are not always so rosy for sand wasps. Martins et al. (1998) reported observing an entire aggregation of 25 nests of *Bicyrtes angulatus* being decimated by termites and ants raiding nests (and an overall loss rate of 90% during their study—due to ants, termites, and flies). Cane and Miyamoto (1979) reported that 4 of 14 nests of *Bembix multipicta* were destroyed by plundering ants. Raiding by ants was deterred when the female constructed thick nest closures, but females also attempted to discourage the ants by picking them up in their mandibles and carrying them in flight up to a meter away. Larsson (1986) found that the ratio of ant (as well as miltogrammine fly) numbers to nests of *Bembix rostrata* declined with increasing nest densities, and suggested that this provides a major selective advantage to nesting in aggregations.

Male Behavior

Compared with the complex sequences of female behaviors involved in digging, provisioning, and disguising a nest while keeping track of location in the habitat, the behaviors typically displayed by male sand wasps are relatively simple. Most often, they occupy themselves by patrolling emergence and nesting areas in search of receptive virgins, otherwise visiting a few flowers during the day and sleeping off the remainder it (and the night) on plants, in holes in ground, or under rocks. Fortunately, there are exceptions, and as has long been observed, even the patrolling males can provide fascinating natural spectacles in the form of "sun dances." And because observers have generally focused their attention on nesting areas, we may have missed activities of males at other sites (see sections on *Bembix flaviventris, B. furcata,* and *B. moma* males in Chapter 7).

As of 1966, detailed information on male Bembicinae was restricted to Kullenberg's (1956) studies of sexual attractants used by orchids to attract male *Argogorytes,* N. Lin's (1963a) observations on territorial behavior in cicada-killers, and general descriptions of the "sun dances" of male sand wasps by Rau and Rau (1918), Evans (1957b), and others. However, a number of papers dedicated primarily to male sand wasps have since appeared on various Gorytini (Alcock 1975a; Hastings 1989a,b; Coelho and Holliday 2000; Coelho 2001), Stizini (O'Neill and Evans 1983; O'Neill et al. 1989; Asís et al. 2006), and Bembicini (Chmurzynski 1977; Schöne and Tengö 1981; Longair et al. 1987; Dodson and Yeates 1989; Larsson 1989a; Thomas and Nonacs 2002). In addition, other papers now often include at least brief descriptions of where and when males were active and what they were doing. There are also further reports on how orchids exploit the sexual eagerness of male *Argogorytes* (Ågren and Borg-Karlson 1984; Ågren et al. 1984; Borg-Karlson 1990).

In a review of male behavior in solitary wasps, O'Neill (2001) classified mating strategies according to when, where, and how males compete for mates. This classification scheme forms a convenient framework for discussing what we know about male sand wasps.

(1) *The time in the adult females' reproductive cycle when they are sexually receptive.* The best evidence that we have is that females of most solitary wasps are receptive at the time they emerge as adults and that they mate just once in their lives; see O'Neill (2001) for known exceptions, none of which include sand wasps. As a result, a male sand wasp that does not reach a female soon after she emerges will likely lose out to more timely competitors. For this reason, most solitary wasps, including Bembicinae, exhibit *protandry.* Protandry tends not to be absolute, but the average date of male sand wasps' emergence is almost always earlier than the average for females, with the result that most males are present during the peak period of female emergence. As noted by Evans (1966a), the number of active males typically declines within 2–3 weeks, and males may be absent during the latter half of a nesting season.

Hasting's (1989b) study of protandry in *Sphecius grandis* suggests why seasonal emergence periods of the two sexes overlap. In his three-year study, because the period over which females emerged was 2–5 times longer than the average adult male lifespan, males that emerged too early risked missing the peak of female emergence (as they did in one of the three years). Experiments with *Bembecinus quinquespinosus* demonstrated that, after mating once following emergence, females refused further sexual advances by males; they then left the emergence area to nest elsewhere, where they may never have come in contact with mate-seeking males again. We saw both sexes feeding on honeydew on sunflowers, but never saw mating attempts in those locations. So it seems that the best strategy for a male *B. quinquespinosus* is to emerge early (but not too early) and live a fast life during the period in which most virgin females are present. It would be interesting to determine whether the story is different in bivoltine species, where males of the first generation each year might have a chance to mate with females of both generations if they live long enough (Werren and Charnov 1978).

(2) *The location in the habitat where males find receptive females.* The fact that female sand wasps are sexually receptive at or soon after the time of emergence explains why males of most species search for females in the area where virgin females emerge or within the present year's nesting area; females begin to nest soon after emerging. In those species that nest in the same location each year, emergence and nesting areas are, of course, one and the same. Males of many species of *Bembix* gather in emergence/nesting areas, sometimes in very high densities (see Chapter 7). Other species of Bembicinae in which males rendezvous with females in nesting/emergence areas include:

Argogorytes mystaceus	*Sphecius grandis*
Bembecinus cinguliger	*Sphecius speciosus*
Bembecinus neglectus	*Steniolia elegans*
Bembecinus quinquespinosus	*Steniolia nigripes*
Bembecinus strenuus	*Stictia carolina*
Bembecinus tridens	*Stictia heros*
Bicyrtes angulatus	*Stictia signata*
Editha magnifica	*Stictia vivida*
Microbembex monodonta	*Stizus continuous*
Pseudoplisus ranosahae	*Stizus perrisi*
Rubrica nasuta	*Stizus pulcher*

Other locations at which the sexes meet include sleeping clusters (*Steniolia obliqua*), hunting sites (*Exeirus lateritius*), and landmark sites on hilltops *(Bembix furcata),* but such examples have so far rarely been seen and need further study.

(3) *How males compete for access to receptive females or locations where they are likely to appear.* The most basic distinction that is often made in the form of mating strategies is whether males compete by scramble competition (i.e., racing to sexually receptive females) or interference competition (e.g., fighting for direct access to females or to sites where females are likely to appear). The sun dances exhibited by *Bembix* and many other sand wasps are clear examples of scramble competition. Although contact may occur between males, these are most often best interpreted as males making close contact to determine the identity of a potential mate (rather than as attempts to drive competitors from the area). In fact, males of some species are often so intent on close investigation that they even attempt to copulate with conspecific males, and they often investigate inanimate objects. *Bembecinus quinquespinosus* have been seen attempting to copulate with rabbit droppings and lizard feces (O'Neill and Evans 1983), whereas male *Glenostictia satan* may mount fly puparia and various plant fragments (Longair et al. 1987). The ease with which males are attracted to objects that resemble females, either visually or chemically, has allowed certain orchids to evolve mechanisms for attracting males of *Argogorytes fargeii* and *Argogorytes mystaceus* (and other Hymenoptera) in order to deceive them into vectoring pollen. Nevertheless, even *Argogorytes* males are more attracted to conspecific females than they are to orchids (Kullenberg 1956).

The form of the search pattern is not the same for all species that exhibit sun dances; they vary in the height of flights above the ground, the pattern of flight, and the frequency with which males alight on the soil surface. As noted in our discussion of *Bembix* sun dances, much of the intra- and interspecific variation in patrolling flights may be due to such factors as the densities of searching males, local topography, and temperature. However, some of the variation may be due to interspecific variation in what is the best way to locate and intercept a virgin female. For example, given the density of competitors and the sensory capabilities of males, what path and flight speed during patrolling flights maximize the likelihood of intercepting and detecting a receptive female?

Searches for virgin females may extend to attempts to get at them at the very moment they break through the soil surface as they make their way upward after emerging from their cocoons. Patrolling males of *Stizus con-*

tinuous, Bembecinus neglectus, B. quinquespinosus, B. tridens, and *Bembix rostrata* often pause on the ground in emergence areas to dig at locations where females may be about to emerge. They often make errors, uncovering conspecific males or other species, but the behavior brings them one step closer to being the first to intercept a virgin. *Bembix rostrata* males can evidently detect emerging females that are several centimeters beneath the surface. In an experimental study, in which different insects (or parts of insects) were buried in emergence areas, searching males preferred virgin females to older females and decapitated females to males. Chemical and vibrational cues emanating from the body inform the males of the sex, status, and location of preemergent individuals (Schöne and Tengö 1981; Larsen et al. 1986). Digging males often attract competitors so that, when the female does emerge, a cluster of many males may form around her (so-called mating balls). Mating balls have been reported in all of the species listed above, as well as in *Stictia carolina* (C. S. Lin 1971), *Glenostictia satan,* and *Bembix americana antilleana.* Within the jostling mating balls, males are certainly engaging in interference competition as they attempt to mount a female. In *B. quinquespinosus,* interference continues as males pounce on a pair even after one male successfully mounts and attempts to carry off a female.

Are male sand wasps ever territorial? In the sand wasp literature, the term "territory" seems sometimes to be used simply to denote a fixed area occupied by a male, but more recent usage also incorporates defense of an area. We follow Brown's (1964) stricter definition of a territory as a "fixed area from which intruders are excluded by some combination of advertisement, threat, and attack." The definition implies both site specificity and mechanisms by which one male excludes others from a site. Even under a strict interpretation of the definition, male territoriality occurs in a large number of solitary wasp species (Evans and O'Neill 1988; O'Neill 2001). Among the nonbembicine apoid wasp species, males defend: (1) individual nests where females are actively provisioning (e.g., some *Trypoxylon;* Brockmann 1989), (2) resources other than nests, such as hunting sites (e.g., *Aphilanthops subfrigidus;* O'Neill 1990), (3) nesting areas, both compact and diffuse, which are also emergence sites when females nest in the same locations each year (e.g., many *Cerceris, Eucerceris, Philanthus;* Evans and O'Neill 1988), and (4) landmark sites distinguished as being either conspicuous physical features in the environment (e.g., *Hemipepsis ustulatus;* Alcock 1981) or scent-marked locations (e.g., *Philanthus bas-*

ilaris; O'Neill 1983a). Do any Bembicinae fit into these categories? We know of no examples among sand wasps of categories 1 and 2 above. However, the well-documented mating systems of *Sphecius grandis, S. speciosus,* and *Stictia signata* are clear examples of mating systems in category 3. In fact *S. speciosus* was the first hymenopteran for which territoriality was clearly demonstrated (N. Lin 1963a).

Other sand wasps may also have territorial males. *Rubrica nasuta* males, for example, each patrol a particular group of nests, flying about 20 cm above the ground and striking at other males that intrude; males return to perches between patrolling flights. Any male removed from his territory is promptly replaced by another male. *Editha magnifica* also defend clusters of nests at times; and males seem to engage in postcopulatory mate-guarding, forcing females into their nests and scraping soil into the entrance. There are also reports of male territoriality in emergence/nesting areas in *Stictia carolina, S. heros,* and *S. vivida,* though overt aggression involving physical contact among males has not been described. Dodson and Yeates (1989) report a similar form of behavior of *Bembix furcata* males as they defend hilltop landmark territories; this system may be similar to that described by Alcock (1979, 1980) for the pompilid wasp *Hemipepsis ustulata.*

Attempts to classify the mating systems of male sand wasps are often complicated, though they are made much more interesting by the existence of alternative mating tactics within populations. The following examples have been reported in the Bembicinae:

- *Bembecinus quinquespinosus:* Large males patrol for females within emergence areas, competing vigorously for females emerging from the ground. Small males, who are also darker in color, patrol in lower densities in the periphery of the emergence area, where they probably achieve lower average mating success.
- *Bembix rostrata:* Most males patrol the emergence area, but during periods of high temperature, some males hover over restricted areas that have been interpreted as territories (Larsson and Larsson 1989).
- *Editha magnifica:* Early in the season, when males far outnumber females, males patrol the nesting area. Later, when the sex ratio is more female-biased, some males defend clusters of nesting females.
- *Sphecius grandis:* Large males defend territories in the emergence area, where they intercept virgin females soon after they emerge from the ground. Small males wait in trees in the periphery of the emergence

area. Intermediate-sized males sometimes take over territories abandoned by the larger males (Hastings 1989a).

- *Stictia heros:* Males perform typical patrolling flights ("sun dances") during the morning. Later, a small number of males hover over fixed positions in the same area, where they pounce on or chase away intruding insects. Hovering males are larger than patrolling males, and are able to tolerate higher ambient temperatures.
- *Stictia signata:* Males patrol with a sinuous flight early in the day, then change to a more stationary hovering flight later in the day, positioning themselves about 4 m apart and interacting with neighboring males over a restricted area.

Certain of these alternative tactics appear to be expressions of conditional strategies. Those of *Bembecinus quinquespinosus, Sphecius grandis,* and *Stictia heros* are size-related conditional strategies. Small males of *B. quinquespinosus* would rarely succeed within the emergence areas, because they have a difficult time picking up and carrying virgin females in flight to avoid competitors. Small males of *S. grandis* cannot defend territories from larger competitors, so they await females outside of the emergence area. Small males of *S. heros* cannot adopt the more thermally stressful hovering tactic. Temperature may also be important in the *B. quinquespinosus* mating system: larger males are also lighter colored than small males, with more reflective exoskeletons, so they may be better able to tolerate conditions on the hot soil surface in the emergence area.

Sleeping

The last behavior to discuss is sleeping, which, though inherently less interesting than many of the complex behaviors covered to this point, does at least show some species-specificity. Different species sleep in different locations, gather together in clusters of different size and composition (in terms of both sexes and species), and adopt different postures (O'Neill 2001). The following is an updated version of a similar summary presented in Evans (1966a). Species for which new or confirming observations have been published since 1966 are marked with an asterisk.

A. Sleep on vegetation (both sexes)
   1. Solitarily or in small groups with no bodily contact: *Stizoides* (*S. renicinctus** often mixed with *Prionyx, Sphex,* or other species),

some *Bicyrtes* (*B. angulatus**), *Rubrica nasuta**, *Steniolia tibialis**, and probably many others.

2. In dense clusters with considerable bodily contact: *Bembecinus* (*B. cinguliger**, *B. neglectus**), *Steniolia, Stictiella,* some *Glenostictia, Zyzzyx, Rubrica.*

B. Males sleep in loose clusters on vegetation, females in their nests: *Glenostictia scitula, Sphecius speciosus* (probably).

C. Sleep in the ground

1. Both females and males together in nest burrows: *Clitemnestra.*
2. Females in their nests, males in short sleeping burrows: *Bembix* (*B. cursitans**, *B. lamellata**, *B. musca**, *B. palmata**, *B. trepida** males), *Bembecinus* (*B. comberi** and *B. luteolus** females). Of particular interest are the unusually long sleeping burrows of certain male Australian *Bembix* that may exceed 15 cm in length.
3. Both males and females in short sleeping burrows: *Microbembex* (*M. argentifrons**), some *Bicyrtes.*

D. Under rocks in tight clusters: *Bembecinus quinquespinosus**.

Sand Wasp Conservation

We conclude with a discussion of factors, other than endemic natural enemies, that may influence the size and stability of sand wasp populations, some of which may be threatened with extinction (Day 1991). As noted earlier, sand wasps tend to have relatively specific habitat requirements, in terms of soil and moisture content, soil surface integrity, surface vegetation density, subsurface root density, insolation and temperature at the soil surface, and prey availability.

In some cases, population declines may result from natural habitat changes that modify nesting habitat. Dune blowouts may become overgrown with vegetation and unsuitable for species of *Bembix* that prefer open habitats. Kula and Tyrner (2003) documented declines in digger wasp populations as vegetational changes in Bohemian forests reduced solar radiation reaching the soil surface. On the other hand, Lüps (1973) found that a *Bembecinus tridens* aggregation in Germany persisted in the face of encroaching vegetation; whether it did so over a longer period is unknown. In other cases, rapid population declines may follow specific weather events. During the early 1980s, we had been studying a large population of

Bembecinus quinquespinosus on the Pawnee National Grasslands in Colorado, when a prolonged period of heavy rains prior to adult emergence in July killed many adults in their cocoons. Through the late 1990s, we were unable to locate nesting areas of the species, though we occasionally collected a few adults on flowers. Such events may have a temporary effect on a population size or induce the wasps to relocate if nearby habitats are more suitable. Along these lines, one wonders about the fate of populations of *Bembix glauca* nesting on beaches along the southeastern coast of Sri Lanka inundated by the tsunami of late 2004. But weather-related events may sometimes change habitats in ways at least temporarily beneficial to wasps. *Bembecinus tridens* was among the most abundant sphecids recolonizing steppe habitat surveyed in 1994–1995 following flooding of the habitat of the Rhone River in 1993 (Zehnder and Zettel 1999). In assessing possible population declines, one must of course take into account that there is likely to be natural variation in the size of nesting aggregations (e.g., those reported for *Bembecinus neglectus* by Evans 1955).

Human-caused habitat changes that threaten wasp populations. Disturbances to critical features of sand wasp habitat may be permanent, as when a site is radically modified due to "development" (e.g., building of a mall), or temporary, as when soil in a habitat is frequently disrupted by grazing livestock, hiking humans, or off-road vehicles. In South Africa, S. K. Gess and F. W. Gess (1989) and F. W. Gess and S. K. Gess (1993) noted several effects of grazing on solitary wasp and bee populations. First, heavy grazing changed the composition of vegetation, in particular causing replacement of the shrub *Pentzia incana* (Thunb.) Kuntze by *Chrysocoma ciliata* L. *Pentzia* plays host to a greater diversity of herbivorous insects that provide prey for wasps, including species of *Bembecinus.* Second, grazing livestock also trample soil. The Gesses observed a rapid decline in the size of a *Bembix bubalus* aggregation due to trampling that modified the surface integrity of soil. We have observed at least temporary disruptions of this type in the western U.S. On the other hand, the keeping of large numbers of livestock and the consequent building up of species of biting flies and dung-feeding flies is probably a boon to the *Bembix* and *Stictia* that hunt around livestock. However, such a benefit would have to be balanced against the costs due to the disruption of soil and vegetation.

Sand wasps that may benefit from human activities. As with robins, house sparrows, rats, coyotes, and other species that thrive in human habitats, certain sand wasps appear to benefit from human activities (in ways beyond the setting aside of land reserves, which are undoubtedly good for many wasps). We say that they "appear to benefit," because it is possible that the species cited in the following examples would do even better in the total absence of humans. *Gorytes laticinctus, G. tricinctus,* and *Pseudoplisus natalensis* have been found nesting in flowerpots, and *Exeirus lateritius* will nest in gardens (Froggatt 1903). Cicada-killers *(Sphecius speciosus),* which are such common inhabitants of playgrounds and lawns that they are considered pests, seem less deterred by the presence of humans than humans are by them. *Bembix nubilipennis* has also long been known to inhabit sandlot baseball diamonds (Rau and Rau 1918), whereas *Bembix texana* and *Stictia carolina* will nest in parking lots. Dirt roadways and trails commonly attract female sand wasps, as well as other digger wasps. *Bembix trepida,* for example, commonly nested in the hard-packed sand of roads and paths in picnic areas of southeast Australia (Evans and Matthews 1973). In what are fairly typical cases, mixed-species aggregations of wasps (including Bembicinae) have been documented in such locations by Lüps (1973) in Germany and by Martins et al. (1998) in Brazil. Earlier in this chapter we cited the example of an aggregation of *Bembix americana comata* that had been present in a driveway in Alameda, California, for at least 15 years. One suspects, however, that most Americans would respond to such a phenomenon with a frantic call to their local extension agent, or with a good dose of whatever insecticide happened to be handy.

Evans (1975) monitored ground-nesting wasps colonizing a patch of habitat in Bedford, Massachusetts, bulldozed in the spring of 1972, with the resulting removal of vegetation and exposure of sandy soil. During the summers of 1972–1973, the 25 species of apoid wasps colonizing the bulldozed area from an adjacent habitat (which had been monitored in previous years) included four species of Bembicinae: *Bembix americana, Bicyrtes quadrifasciatus, Gorytes canaliculatus,* and *Hoplisoides nebulosus.* Similarly, *Bembecinus mexicanus* and *Bembix multipicta* were found nesting in heaps of ground limestone left from a road-building operation in Yucatan, Mexico (Evans 1966a). Krombein and van der Vecht (1987) also noted that *Bembix borrei* used sand piles at construction sites. Australian *Bembix* have on several occasions been found nesting in soil of sites recently disturbed by human activities. A *Bembix gunamarra* aggregation was found in a sandy field that had been bulldozed just two or three years earlier, whereas *Bembix moma* occupied a site along an irrigation ditch (Evans and Matthews 1973). Along the eastern subtropical coast of Australia, Evans et al. (1982) found *Bembix promontorii* females nesting in vertical sandbanks resulting from man-made excavations in dunes. As in the case of *B. borrei* just mentioned, *Bembix minya* nests were found in a pile of "builder's sand" that had been present for just seven days (Austin 1999). Finally, *Bembix lamellata* nests at one site were found in the soil eroded from a slag heap at an abandoned mine. Evans also found *Steniolia elegans* nests in the slag pile of an abandoned mine in Colorado. Clearly, not all anthropogenic changes that favor sand wasps are otherwise welcome. Air pollution damage that caused the opening up of spruce stands seemed to benefit some species of bees and wasps, including the bembicine brood parasite *Nysson spinosus* (Kula and Tyrner 2003). But even those of us smitten with sphecophilia would be awfully short-sighted to trumpet this as an important benefit of pollution.

Interactions with invasive species. Effects of invasive species (or invasive populations of *Homo sapiens*) on sand wasps can be positive or negative. Benefits may accrue when invasive species are added to the diet of a wasp, either as prey or as nectar sources. In New Zealand, for example, *Argogorytes carbonarius* now preys upon the spittlebug *Philaenus spumarius*, an invasive from Europe. Harris (1994) argues that the existence of this new and abundant prey has led to a shift to building nests with >2 cells, which could lead, we surmise, to greater numbers of cells provisioned if fewer new nests need to be initiated. Similarly, *Bembix* around the world have added various tramp species of flies to their diets, including flies of the genera *Lucilia* (Calliphoridae), *Musca, Muscina, Orthellia,* and *Stomoxys* (Muscidae). The exact geographic origins of the house fly *(Musca domestica)* are unknown, though it is likely not endemic to the New World or Australia (Marquez and Krafsur 2002; Krafsur et al. 2005). Nevertheless, it is preyed upon by *Rubrica nasuta* and many *Bembix* spp. in these regions. Similarly, *Orthellia caesarion* is a European species taken in North America by at least seven species of *Bembix*, whereas the European *Muscina assimilis* is taken in North America by *Bembix americana* and *B. pallidipicta.* It is unknown whether non-native flies have provided a boon to foraging female *Bembix*, or whether these opportunistic predators have just substituted one equivalent prey type for another. Similarly, wasps that take nectar from invasive plants may not be experiencing a net gain relative to native plants. Martins et al. (1998) stated that *Waltheria indica,* apparently of North American origin, acts as an important nectar source for *Bicyrtes angulatus* in Brazil. We can also add our own anecdotal evidence that we have often collected apoid wasps in high numbers on weeds such as Canada thistle (*Cirsium arvense* L.), white sweetclover (*Melilotus alba* Desr.), and leafy spurge (*Euphorbia esula* L.) in the Rocky Mountain region of the western United States.

On the other side of the ledger may be invasive species of insects that are natural enemies of sand wasps, or invasive species of weeds that compete with the host plants of native prey species. Both possibilities are speculative at this time, but one likely culprit is the imported fire ant *(Solenopsis invicta),* which is ubiquitous in the southern U.S. and whose congeners have been reported as nest marauders of *Bembecinus* and *Bembix. Solenopsis* also plunder nests in their native South America (see account of *Bicyrtes angulatus*).

Finally, it is possible that sand wasps themselves can be invasive, though as far as we know, no one has systematically accumulated records of major range shifts by Bembicinae. It may be that sand wasps, being soil-nesters, are less likely to be moved great distances by humans than are cavity-nesters and mud-nesters, which can be more easily transported while in diapause, within or attached to plant materials (Pagliano et al. 2001; Nishida and Beardsley 2002; Ćetković et al. 2004; Bogusch et al. 2005). At least one bembicine, *Hoplisoides semipunctatus,* has likely been introduced to the southern U.S. from its native South America, "since collections have been made near airfields" (Bohart 1997). Reports that *Gorytes tricinctus* and *Pseudoplisus natalensis* nest in the soil of potted plants suggests a possible route for the transport of bembicines to new areas. And the fact that sand wasps are often opportunistic predators may indicate that prey availability may not be a great barrier to establishment once a species reaches a new location.

Monitoring sand wasp populations. Whether habitat changes or species movements actually cause increases or decreases in sand wasp populations is a question that will be best answered with quantitative analyses of population size and distribution. However, accurate monitoring of the status of sand wasp populations is hindered by several factors. First is the growing scarcity of experts able to identify the many species involved, a problem caused by the retirement of expert taxonomists who are not always being replaced by younger entomologists (Day 1991). Second is the paucity of research funds, which are generally devoted to species of interest to the general public and governmental agencies. Outside of a group of dedicated cognoscenti and insect natural history buffs, sand wasps are perhaps short on the kind of charisma that causes people to (rightfully) fret about beached whales and injured eagles. Third is the practical difficulty of monitoring widely scattered aggregations that may change location from year to year (e.g., Evans et al. 1986) without experiencing consistent population declines. Finally, as Day (1991) and others have pointed out, we need to develop a consistent set of criteria for judging the vulnerability of a species to extinction. To what density must a population decline to indicate a critical situation for a population? How long must it be at a low level before we should worry? How far apart must suitable habitat fragments be before natural recolonization of a locally extinct population becomes unlikely in the short term? How do species differ in their ability to recover from near brushes with extinction, whether they be local, regional, or global? Do population declines create genetic bottlenecks that significantly reduce the probability of population recovery?

The list of aculeate Hymenoptera on European Red Data Lists compiled by Day (1991) includes over 40 species in 11 genera of Bembicinae: *Alysson* (5 species), *Argogorytes* (1), *Bembecinus* (1), *Bembix* (3), *Didineis* (1), *Gorytes* (12), *Harpactus* (8), *Hoplisoides* (1), *Lestiphorus* (2), and *Nysson* (8). The list also contains Chrysididae that are known or suspected natural enemies of Bembicinae: *Hedychridium integrum* Dahlbom, *Hedychridium roseum* (Rossi), *Hedychrum chalybaeum* Dahlbom, and *Holopyga chrysonota* (Foerster). Some of these wasps are classified as rare and vulnerable in specific parts of Europe, others may already be locally extinct. As of 2006, no sand wasp had made it onto the U.S. Endangered Species List, which would be good news if none deserved to be on it. With their specific habitat requirements, many sand wasps are just the kinds of species that could be susceptible to habitat disruption. But there have not been enough long-term quantitative habitat surveys and population censuses that include sand wasps. Among other surveys, the lists of wasp faunas in Jackson Hole, Wyoming (Evans 1970), the Yukon and Northwest Territories of Canada (Steiner 1973), several areas of South Africa (F. W. Gess 1981; F. W. Gess and S. K. Gess 1991), Sri Lanka (Krombein 1984, 1985; Krombein and van der Vecht 1987), the California Channel Islands (Rust et al. 1985), and the pine barrens in upstate New York (Kurczewski 1998) are invaluable (along with museum collections), because they can at least tell us what species occupied a particular site at a particular time. But nonquantitative, short-term surveys of sites affected by humans will not be enough in themselves to provide arguments that measures should be taken to preserve wasp populations. Researchers familiar with the fauna of their local sites may at least provide the first warning. The question remains, however, whether it is likely that concern for a rare sand wasp will ever halt the use of sand dunes for off-road vehicles in the United States, or that a grassroots movement will form to reintroduce *Bembix tarsata* into Austria, where it is thought to be extinct (Dollfuss 1983).

Sand wasps as biodiversity indicators. It is perhaps an occupational hazard common to many ecologists with deep devotion to a particular set of organisms that they want to see their pet taxon as a critical functional component of ecosystems. Many candidates have therefore been put forward for key indicator taxa, whose abundance could be monitored and used either as an alarm bell for ecological disturbance or as a sign of good conservation practices. So too for the champions of apoid wasps, with whom we share taxon-loyalty. Tscharnke et al. (1998), for example, put forward cavity-nesting solitary wasps as bioindicators of "community structure and function."

Gayubo et al. (2005) recently made a cogent argument that apoid wasps offer "guarantees of being good indicators of biodiversity, both for predicting the diversity of other groups of animals and for all the species of a given area." They base this suggestion on seven criteria independently proposed by Pearson (1994) as qualities of useful biodiversity indicators, defined as "a group of taxa . . . whose diversity reflects some measure of diversity . . . of other higher taxa in a habitat or landscape." Gayubo and his colleagues argue that useful indicator groups should be easy to sample and have a well-established classification system (as is the case for apoid wasps even given several minor disputes about phylogenetic relationships). They should have a well-known biology, so that we know their functional role in an ecosystem (for solitary wasps, many comprehensive reviews exist documenting their prey and habitat requirements). The abundance of reliable bioindicators should reflect habitat characteristics that are important to a wide diversity of other organisms (which in this case includes prey, nectar sources, natural enemies, and species with shared niche requirements). Because of their requirements for certain types of nest sites, wasps may also be indicators of certain general habitat characteristics such as soil disturbance and vegetation cover. Organisms proposed as indicators should contain a mix of generalist species with wide geographic distribution (e.g., *Bembecinus tridens, Bembix americana*) and specialist species that are sensitive to habitat changes (e.g., the wasps on the European Red Data List).

Gayubo et al. do admit that apoid wasps are probably weak candidates relative to the criterion that they have direct economic importance (e.g., as biocontrol agents), and that monitoring bees would solve the problem. It is certainly true that apoid wasps include no really important biocontrol agents, but they may well have some economic benefit as natural enemies, considering the wide diversity of insects on which they prey (O'Neill 2001). For example, *Bicyrtes spinosus* preys on hemipteran pests of beans, squashes, and tomatoes, whereas *Sphecius, Stizus, Stictia, Bembix,* and the multitude of leafhopper-hunting bembicines prey on groups of insects that include pests. Thus if Gayubo et al. are correct, sand wasps and their relatives not only enhance our understanding of insect behavior and evolution, but may give hints as to the state of the broader environment. But even if they fail in the latter sense, they have performed admirably as subjects of ethology.

Appendix

Research Wish List

Despite over a century of research on sand wasps, we should emphasize that the summary presented in this book reveals major gaps in our understanding of the biology of the Bembicinae. In this appendix we therefore present a "wish list" detailing some of the questions researchers might concentrate on prior to the next review of the biology of the Bembicinae written, perhaps, by professors Anna Lysson and Glenn O'Stictia. Our list is idiosyncratic in the sense that it reflects our own interests, but we think that it at least broadly covers the kinds of questions that generally intrigue insect ethologists. We firmly believe that any well-done research on sand wasps is welcome, including basic descriptive studies of populations that have already received attention. At a minimum, this type of work provides data on temporal variation, or lack thereof, in behavior. But despite our conviction that descriptive studies are invaluable, we also strongly promote quantitative and experimental approaches and encourage the integration of phylogenetic analyses with comparative behavioral studies. In addition, researchers should be alert to the possibilities of using particular models and theories, where appropriate, to guide their studies (e.g., central place foraging models, producer-scrounger models, sex allocation models, kin selection models, Hamilton's selfish-herd hypothesis). And so to some of our favorite questions.

Nest-site selection and nest construction. How do females choose nest sites at different spatial scales? What soil and microclimatic characteristics are analyzed during nest-site selection? How does nest-site selection and nest structure affect nesting success and offspring development? Is cell depth a facultative response to the depth of moisture in the soil?

Provisioning and prey choice. Where do females hunt? How do they locate and capture prey? What are the pharmacological effects of venoms? What are the relative roles of prey preference, prey defenses, and prey availability in determining diet? Can central place foraging models be applied to understand prey choice? Are mass provisioning and progressive provisioning in its different forms obligate or facultative features of a species? What is the adaptive value of cell cleaning? Are prey carriage flights modified in the presence of parasites? Are male and female cells differentially provisioned? Do small and large females (or younger and older females) exhibit different patterns of sex allocation? Do the quality and quantity of available prey influence sex allocation?

Social interactions. How do wasps interact in aggregations, perhaps influencing nest density and the growth of aggregations during a season? Does the density of active nests remain relatively constant during a season, even as the number of nests in a aggregation changes? Do females gain direct benefits from aggregating at nest sites, as opposed to aggregating simply because potential nest sites are rare? Do bembicines treat kin and nonkin differentially? How common is intraspecific and interspecific prey theft? Are such behaviors widespread in a population or do certain females specialize in theft? Do thieves and nonthieves have equal reproductive success? Is the balance between hunting for prey ("producing") and prey theft ("scrounging") subject to frequency-dependent natural selection?

Brood parasitism. What are the real hosts of Nyssonini and *Stizoides?* What cues do brood parasites use to locate host nests? How do they attack host nests? What is the effect of brood parasitism on individual hosts and populations as a whole? How does the lifetime reproductive success of brood parasites compare with that of their nest-provisioning relatives?

Natural enemies. What features of nests, such as their location, structure, closures, and accessory burrows, deter parasitoids and how do they accomplish this? What cues, either chemical, visual, or textural, alert natural enemies to the presence of a nest entrance? As suggested by Matthews (1991), do mound-leveling and mound-building disrupt chemical cues used by brood parasites and ants to locate nest entrances? Does the per capita rate of nest parasitism or predation decrease with increasing nest density?

Mating strategies. How often do females mate and at what times in their lives do they become receptive? How do males locate potential mates and recognize receptive females? How do males compete for mates? Are males truly aggressive toward one another, or does apparent aggression represent investigatory contacts? How does the operational sex ratio (a good indicator of the level of competition experienced by males) change over time and in space within populations? Is success in mate competition related to body size? Do alternative mating tactics exist within populations, expressed either by different males in different locations, or by the same males at different times? Are forms of precopulatory, copulatory, and post-copulatory courtship evident?

Phylogenetic studies. Can phylogenetic analyses, particularly those using molecular methods, be used to enhance our understanding of evolutionary transitions between different behavioral strategies in sand wasps (e.g., transitions to new prey types or nesting strategies)? Can we relate adaptive radiation in behavior among sand wasps to plasticity in behavior at the individual level (West-Eberhard 2003)?

Faunal studies and conservation. Which populations and species of sand wasps are in danger of extinction? What long-term trends in population size can we detect in populations that may be stressed by human activities? Can we integrate plans to conserve sand wasp populations into plans to conserve entire habitats and their associated fauna and flora? How can we convince the general public that sand wasps matter?

References

Ågren, L., and A.-K. Borg-Karlson. 1984. Responses of *Argogorytes* (Hymenoptera: Sphecidae) males to odor signals from *Ophrys insectifera* (Orchidaceae). Preliminary EAG and Chemical Investigation. *Nova Acta Regiae Societatis Scientiarum Upsaliensis,* ser. 5, C, 3: 15–25.

Ågren, L., B. Kullenberg, and T. Sensenbaugh. 1984. Congruences in pilosity between three species of *Ophrys* (Orchidaceae) and their hymenopteran pollinators. *Nova Acta Regiae Societatis Scientiarum Upsaliensis,* ser. 5, C, 3: 15–25.

Alayo, D. P. 1969. Estudios sobre los Himenópteros de Cuba. III. Subfamilia Nyssoninae (Familia Sphecidae). *Poeyana,* ser. A, 59: 1–34.

Alayo Soto, R. 1989. Observaciones en himenópteros cubanos. III. Hallazgo de una población *Bembix americana antilleana* Evans et Matthews (Sphecoides: Nyssonidae). *Poeyana* 363: 1–4.

Alcock, J. 1972. Variation in cell number in a population of *Bembix americana* (Hymenoptera: Sphecidae). *Psyche* 79: 158–164.

——— 1973. Notes on an aggregation of digger wasps in Seattle, Washington (Hymenoptera). *Wasmann Journal of Biology* 31: 323–336.

——— 1975a. The behaviour of western cicada killer males, *Sphecius grandis* (Sphecidae, Hymenoptera). *Journal of Natural History* 9: 561–566.

——— 1975b. Notes on the behavior of three Argentine sphecids (*Microbembex uruguayensis, Tachytes fraternus,* and *T. amazonus*). *Pan-Pacific Entomologist* 51: 195–200.

——— 1975c. The behavior of some bembicine wasps of southern Arizona (Hymenoptera: Sphecidae, *Microbembex, Glenostictia, Xerostictia*). *Southwestern Naturalist* 20: 337–342.

——— 1979. Behavioural consequences of size variation among males of the territorial wasp *Hemipepsis ustulata* (Hymenoptera: Pompilidae). *Behaviour* 71: 322–355.

——— 1980. Notes on the reproductive behaviour of some Australian solitary wasps (Hymenoptera: Sphecidae, *Tachysphex* and *Exeirus*). *Journal of the Australian Entomological Society* 19: 259–262.

——— 1981. Lek territoriality in the tarantula hawk wasp *Hemipepsis ustulata* (Hymenoptera: Pompilidae). *Behavioral Ecology and Sociobiology* 8: 309–317.

——— 2005. *An Enthusiasm for Orchids: Sex and Deception in Plant Evolution.* Oxford: Oxford University Press.

Alcock, J., and G. Gamboa. 1975. The nesting behavior of some sphecid wasps of Arizona, including *Bembix, Microbembex,* and *Philanthus. Journal of the Arizona Academy of Sciences* 10: 154–165.

Alcock, J., and A. F. Ryan. 1973. The behavior of *Microbembex nigrifrons* (Hymenoptera: Sphecidae). *Pan-Pacific Entomologist* 49: 144–148.

Alcock, J., E. M. Barrows, G. Gordh, L. J. Hubbard, L. Kirkendall, D. W. Pyle, T. L. Ponder, and F. G. Zalom. 1978. The ecology and evolution of male reproductive behaviour in bees and wasps. *Zoological Journal of the Linnaean Society* 64: 293–326.

Alexander, B. A. 1992. An exploratory analysis of cladistic relationships within the superfamily Apoidea, with special reference to sphecid wasps. *Journal of Hymenoptera Research* 1: 25–61.

Alexander, B. A., R. L. Minckley, and D. Yanega. 1993. Nesting biology of *Glenostictia pictifrons* (F. Smith) (Hymenoptera: Sphecidae: Bembicini). *Journal of the Kansas Entomological Society* 66: 108–120.

Amarante, S. T. P. 1996. An interesting hunting tactic of *Stictia punctuata* [*sic*] (Fabr.) and other observations on the hunting behavior of Bembicini wasps (Sphecidae: Nyssoninae). *Sphecos* 30: 9–10.

——— 2002. A synonymic catalog of the neotropical Crabronidae and Sphecidae (Hymenoptera: Apoidea). *Arquivos de Zoologia Sao Paulo* 37: 1–139.

Arens, L. E., and E. L. Arens. 1953. The behavior of the wasp *Stizoides tridentatus* F. (Hymenoptera, Sphecidae). *Entomologicheskoe Obozrenie* (Moscow) 33: 190–193. (In Russian.)

Arnold, G. 1945. The Sphecidae of Madagascar. Cambridge: Cambridge University Press.

Asís, J. D., S. F. Gayubo, and J. Tormos. 1991. Notes on the natural history of *Stizus continuus ibericus* Beaumont (Hymenoptera: Sphecidae). *Journal of Natural History* 25: 1331–37.

——— 1992. Data on the nesting behaviour of five European *Bembix* and descriptions of the mature larvae of *B. merceti* and *B. rostrata* (Hymenoptera, Sphecidae). *Deutsche Entomologische Zeitscrift* 39: 221–231.

Asís, J. D., J. Tormos, and S. F. Gayubo. 2004. Nesting behaviour and provisioning in *Bembix merceti* and *Bembix zonata* (Hymenoptera: Crabronidae). *Journal of Natural History* 38: 1799–1809.

——— 2006. Territorial dynamics and contest behaviour in the solitary wasp *Stizus continuus* (Hymenoptera: Apoidea: Crabronidae). *Behaviour* 143: 83–104.

Asís, J. D., J. Tormos, and R. Jimenez. 1988. Contribution to the study of the biology of *Stizus continuus* (Hymenoptera: Sphecidae). *Entomological News* 99: 199–206.

Augener, H. 1927. Ein weiterer mitteleuropäischer Hymenopterenwitter aus der Fam. Sphegidae. *Archiv für Naturgeschichte,* Abteilung A 91: 164–167.

Austin, A. D. 1999. Use of Odonata as prey by sand wasps, *Bembix* spp. (Hymenoptera: Sphecidae). *Australian Entomologist* 26: 77–82.

Baerends, G. P. 1941. Fortpflanzungverhalten und Orientierung der Grabwespe *Ammophila campestris* Jur. *Tijdschrift Entomologie* 84: 68–275.

Balduf, W. V. 1941. Take off by prey laden wasps (Hymen: Pompilidae?, Sphecidae). *Entomological News* 52: 91–92.

Bates, H. W. 1863. The Naturalist on the River Amazons. 2 vols. London: John Murray.

Begum, A., S. K. Bose, and M. A. Howlander. 1989. Ethological observations on some solitary wasps (Hymenoptera: Sphecidae) in Dhaka, Bangladesh. *Bangladesh Journal of Zoology* 17: 123–129.

Beirne, B. 1956. Leafhoppers (Homoptera: Cicadellidae) of Canada and Alaska. *Canadian Entomologist* 138, supplement no. 2.

Benno, P. 1966. Enige aantekeningen bij de fenologie van *Lestiphorus bicinctus* (Rossi) en zijn koekoekswesp, *Nysson trimaculatus* (Rossi). *Entomologische Berichten* 26: 7–11.

——— 1977. De verspreiding van *Argogorytes* en hun respektievelijke koekoekwespen (Nysson) in Nederland (Hymenoptera: Sphecidae; Nyssoninae). *Entomologische Berichten* 37: 153–156.

Benson, R. B. 1934. *Bembix regnata* Parker, preying on butterflies. *Proceedings of the Royal Entomological Society of London* 9: 75–76.

Bernard, F. 1934. Observations sur les proies de quelques Hyménoptères. *Bulletin de la Société Entomologique de France* 39: 247–250.

——— 1935. Hyménoptères prédateurs des environs de Fréjus. *Annales de la Société Entomologique de France* 104: 31–72.

Bodkin, G. E. 1917. "Cowfly tigers," an account of the hymenopterous family Bembecidae in British Guiana. *Journal of the Board of Agriculture of British Guiana* 10: 119–125.

Bogusch, P., P. Liška, J. Lukaš, and A. Dudich. 2005. Spreading and summary of the knowledge of the invasive sphecid wasp *Sceliphron curvatum* (Smith 1870) in the Czech Republic and Slovakia (Hymenoptera: Apocrita, Sphecidae). *Linzer Biologische* Beiträge 37: 215–221.

Bohart, G. E., and J. W. MacSwain. 1939. The life history of the sand wasp, *Bembix occidentalis beutenmuelleri* Fox and its parasites. *Bulletin of the Southern California Academy of Sciences* 38: 84–98.

Bohart, G. E., W. P. Nye, and L. R. Hawthorn. 1970. Onion pollination as affected by different levels of pollinator activity. *Bulletin of the Utah Agricultural Experimental Station,* no. 482: 1–57.

Bohart, R. M. 1941. A revision of the Strepsiptera with special reference to the species of North America. *University of California Publications in Entomology* 7: 91–159.

——— 1967. New genera of Gorytini (Hymenoptera: Sphecidae: Nyssoninae). *Pan-Pacific Entomologist* 43: 155–161.

——— 1968a. A synopsis of North American species of the genus *Oryttus. Proceedings of the Biological Society of Washington* 81: 431–438.

——— 1968b. New species of *Pseudoplisus* from North and Central America. I. The *P. phaleratus* group. *Journal of the Kansas Entomological Society* 41: 494–501.

——— 1969. New species of *Pseudoplisus* from North and Central America. II. *Journal of the Kansas Entomological Society* 42: 392–405.

——— 1976. A review of the nearctic species of *Crabro* (Hymenoptera: Sphecidae). *Transactions of the American Entomological Society* 102: 229–287.

——— 1980. A review of the North American species of *Dienoplus* (Hymenoptera, Sphecidae). *Pan-Pacific Entomologist* 56: 63–70.

——— 1983. New species of *Glenostictia* (Hymenoptera: Sphecidae). *Proceedings of the Entomological Society of Washington* 85: 485–492.

——— 1984. A revision of the genus *Austrogorytes* Bohart (Hymenoptera: Sphecidae). *Australian Journal of Zoology* 32: 391–412.

——— 1993. Notes on *Microbembex* with new species from Texas, Mexico, and El Salvador (Hymenoptera, Sphecidae, Nyssoninae). *Journal of the Kansas Entomological Society* 66: 274–279.

——— 1996a. A review of the *Bembecinus* (Hymenoptera: Sphecidae: Stizini) in North and Central America. *Proceedings of the Entomological Society of Washington* 98: 517–526.

——— 1996b. A review of the genus *Bicyrtes* (Hymenoptera: Sphecidae, Nyssoninae, Bembicini). *Insecta Mundi* 10: 139–152.

——— 1997. A review of the genus *Hoplisoides* Gribodo (Hymenoptera: Sphecidae: orytini) in North America. *Proceedings of the Entomological Society of Washington* 99: 645–660.

——— 2000. A review of Gorytini in the Neotropical Region (Hymenoptera: Sphecidae: Bembicinae). *Contributions on Entomology, International* 4: 111–259.

Bohart, R. M., and J. E. Gillaspy. 1985. California sand wasps of the subtribe Stictiellina. *Bulletin of the California Insect Survey* 27: 1–89.
Bohart, R. M., and D. S. Horning, Jr. 1971. California Bembicine sand wasps. *Bulletin of the California Insect Survey* 13: 1–49.
Bohart, R. M., and A. S. Menke. 1976. *Sphecid Wasps of the World: A Generic Revision.* Berkeley: University of California Press.
Bohart, R. M., and L. Stange. 1977. *Liogorytes joergenseni* (Brèthes), a cicada killer in Argentina (Sphecidae, Nyssoninae). *Pan-Pacific Entomologist* 52: 313.
Bohart, R. M., and A. Willink. 1989. The genus *Microbembex* in Argentina (Hymenoptera: Sphecidae). *Transactions of the American Entomological Society* 114: 295–310.
Bondar, G. 1930. Vespas que protegem os animaes domesticos contra as muscas. *Correio Agricola* 8: 179–181.
Borg-Karlson, A.-K. 1990. Chemical and ethological studies of pollination in the genus *Ophrys* (Orchidaceae). *Phytochemistry* 29: 1359–87.
Borg-Karlson, A.-K., and J. Tengö. 1980. Pyrazines as marking volatiles in Philanthine and Nyssonine wasps (Hymenoptera: Sphecidae). *Journal of Chemical Ecology* 6: 827–835.
Borror, D. J., C. A. Triplehorn, and N. F. Johnson. 1989. *Introduction to the Study of Insects.* 6th ed. Philadelphia: Saunders College Publishing.
Bradbury, K., and W. Creswell. 1979. Observations of pseudocopulation between *Argogorytes mystaceus* (L.) (Hym., Sphecidae) and *Ophrys insectifera* (L.) (Orchidaceae) in Derbyshire. *Entomologist's Monthly Magazine* 114: 127–129.
Brauns, H. 1911. Biologisches über südafricanische Hymenopteran. *Zeitschrift für Wissenschaftliche Insektenbiologie* 7: 90–92.
Bringer, S. E. 1996. Effects of envenomization by cicada killers *(Sphecius speciosus)* on the longevity and physiology of annual cicadas (*Tibicen* spp.). M.S. thesis, Western Illinois University.
Brockmann, H. J. 1979. Nest-site selection in the great golden digger wasp, *Sphex ichneumoneus* L. (Sphecidae). *Ecological Entomology* 4: 211–224.
——— 1989. Mate conflict and male behaviour in a solitary wasp, *Trypoxylon (Trypargilum) politum* (Hymenoptera: Sphecidae). *Animal Behaviour* 37: 232–255.
——— 1997. Cooperative breeding in wasps and vertebrates: The role of ecological constraints. In J. C. Choes and B. J. Crespi, eds., *The Evolution of Social Behavior in Insects and Arachnids,* pp. 347–371. Cambridge: Cambridge University Press.
Brothers, D. J. 1999. Phylogeny and evolution of wasps, ants, and bees (Hymenoptera, Chrysidoidea, Vespoidea, and Apoidea). *Zoologica Scripta* 28: 233–249.

Brothers, D. J., and J. M. Carpenter. 1993. Phylogeny of Aculeata: Chrysidoidea and Vespoidea (Hymenoptera). *Journal of Hymenoptera Research* 2: 227–302.

Brown, J. 1975. *The Evolution of Behavior.* New York: W. W. Norton.

Callan, E. McC. 1976. Observations on the nesting behavior and prey of Gorytinae wasps in Trinidad (Hymenoptera, Sphecidae). *Psyche* 83: 324–335.

——— 1981. Nesting behavior and prey of *Argogorytes* Ashmead (Hymenoptera: Sphecidae). *Journal of the Washington Academy of Sciences* 70: 160–165.

——— 1991a. Nesting behaviour and prey of *Pseudoplisus* Ashmead (Hymenoptera: Sphecidae). *Journal of Natural History* 25: 1611–15.

——— 1991b. Nesting behaviour and prey of sand wasps in Trinidad (Hymenoptera: Sphecidae: Nyssoninae). *The Entomologist* 110: 134–138.

Cane, J. H., and M. M. Miyamoto. 1979. Nest defense and foraging ethology of a neotropical sand wasp, *Bembix multipicta* (Hymenoptera: Sphecidae). *Journal of the Kansas Entomological Society* 52: 667–672.

Cazier, M. A., and E. G. Linsley. 1974. Foraging behavior of some bees and wasps at *Kallstroemia grandiflora* flowering in southern Arizona and New Mexico. *American Museum Novitates,* no. 2546, 20 pp.

Ćetković, A., I. Radović, and L. Dorović. 2004. Further evidence of the Asian mud-daubing wasps in Europe (Hymenoptera: Sphecidae). *Entomological Science* 7: 225–229.

Chapman, R. N., C. E. Mickel, J. R. Parker, G. E. Miller, and E. G. Kelly. 1926. Studies in the ecology of sand dune insects. *Ecology* 7: 416–426.

Chmurzyński, J. A. 1964. Studies on the stages of spatial orientation in female *Bembex rostrata* (Linné, 1758) returning to their nests (Hymenoptera, Sphegidae). *Acta Biologia Experimentale* 24: 103–132.

——— 1967. On the role of relations between landmarks and the nest-hole in the proximate orientation of female *Bembex rostrata* (Linné) (Hymenoptera, Sphegidae). *Acta Biologia Experimentale* 27: 221–254.

——— 1977. Stimuli eliciting sexual pursuit in the digger wasp *Bembex rostrata* (L.) males (Hymenoptera, Sphegidae). I. The choice of an adequate method for the ethometry survey. *Acta Neurobiologica Experimentale* 37: 27–56.

Coelho, J. R. 1997. Sexual size dimorphism and flight behavior in cicada killers *(Sphecius speciosus). Oikos* 79: 371–375.

——— 1998. An acoustical and physiological analysis of buzzing in cicada killer wasps *(Sphecius speciosus). Journal of Comparative Physiology* A 183: 745–751.

——— 2001. Behavioral and physiological thermoregulation in male cicada killers *(Sphecius speciosus)* during territorial behavior. *Journal of Thermal Biology* 26(2): 109–116.

——— 2002. Spurred on to greater depths. *Natural History* 111: 20–22.

Coelho, J. R., and C. W. Holliday. 2000. Effects of size and flight performance on intermale mate competition in the cicada killer, *Sphecius speciosus* Drury (Hymenoptera: Sphecidae). *Journal of Insect Behavior* 14: 345–351.

Coelho, J. R., and K. Wiedman. 1999. Functional morphology of the hind tibial spurs of the cicada killer (*Sphecius speciosus* Drury). *Journal of Hymenoptera Research* 8: 6–12.

Cowan, D. P. 1991. The solitary and presocial Vespidae. In K. C. Ross and R. W. Matthews, eds., *The Social Biology of Wasps,* pp. 33–73. Ithaca, N.Y.: Cornell University Press.

Cullen, E. 1957. Adaptations of the kittiwake to cliff nesting. *Ibis* 99: 275–302.

Dambach, C. A., and E. Good. 1943. Life history and habits of the cicada killer *(Sphecius speciosus)* in Ohio. *Ohio Journal of Science* 43: 32–41.

Day, M. C. 1991. Towards the conservation of aculeate Hymenoptera in Europe. *Convention on the Conservation of European Wildlife and Natural Habitats, Nature and Environment Series* 51: 1–44.

Deleurance, E. P. 1944. Contributions à l'étude biologique de la Camargue. *Bulletin Muséum d'Histoire Naturelle de Marseille* 4: 56–61.

Dennis, D. S., and R. J. Lavigne. 1975. Comparative behavior of Wyoming robber flies (Diptera: Asilidae). *University of Wyoming Science Monographs* 30: 1–68.

Dodson, G. N., and D. K. Yeates. 1989. Male *Bembix furcata* Erichson (Hymenoptera: Sphecidae) behaviour on a hilltop in Queensland. *Pan-Pacific Entomologist* 65: 172–175.

Dollfuss, H. 1983. Rote Liste gefährdeter Grabwespenarten (Sphecidae, Hymenoptera). In J. Gepp, ed., *Rote Liste gefährdeter Tiere Österreichs,* pp. 76–82. Vienna: Bundesministerium für Gesundheit und Umweltschutz.

——— 1994. Rote Liste gefährdeter Grabwespen (Hymenoptera, Sphecidae). In J. Gepp, ed., *Rote Liste gefährdeter Tiere Österreichs, Grüne Reihe des Bundesministeriums für Umwelt, Jugend und Familie,* vol. 2, pp. 95–104. Graz: Styria Medien Service.

Dow, R. 1942. The relation of the prey of *Sphecius speciosus* to the size and sex of the adult wasp (Hym.: Sphecidae). *Annals of the Entomological Society of America* 34: 310–317.

——— 1976. Nesting of *Stizus iridis* at the type locality (Hymenoptera: Nyssonidae). *Proceedings of the Entomological Society of Washington* 78: 65–66.

Eason, P. K., G. A. Cobbs, and K. G. Trinca. 1999. The use of landmarks to define territorial boundaries. *Animal Behaviour* 58: 85–91.

El-Banna, S. Z., A. Abo-Ghalia, and F. Gilbert. 1999. Territoriality and foraging behavior of some solitary wasps in Egypt. *Egyptian Journal of Biology* 1: 96–103.

El-Sayed, A. M. 2006. The Pherobase: Database of Insect Pheromones and Semiochemicals. http://www.pherobase.com (accessed 10 April 2006).

Elliott, N., F. Kurczewski, S. Claflin, and P. Salbert. 1979. Preliminary annotated list of the wasps of San Salvador Island, the Bahamas, with a new species of *Cerceris* (Hymenoptera: Tiphiidae, Scoliidae, Vespidae, Pompilidae, Sphecidae). *Proceedings of the Entomological Society of Washington* 81: 352–365.

Evans, F. C. 1985. In praise of natural history. *Bulletin of the Ecological Society of America* 66: 455–460.

Evans, H. E. 1953. Comparative ethology and the systematics of spider wasps. *Systematic Zoology* 2: 155–172.

——— 1955. An ethological study of the digger wasp *Bembecinus neglectus,* with a review of the ethology of the genus. *Behaviour* 7: 287–303.

——— 1957a. Ethological studies on digger wasps of the genus *Astata* (Hymenoptera: Sphecidae). *Journal of the New York Entomological Society* 65: 159–185.

——— 1957b. *Studies on the Comparative Ethology of Digger Wasps of the Genus Bembix.* Ithaca, N.Y.: Cornell University Press.

——— 1959. Studies on the larvae of digger wasps (Hymenoptera, Sphecidae). Part V. Conclusion. *Transactions of the American Entomological Society* 85: 137–191.

——— 1963. *Wasp Farm.* Garden City, N.Y.: Anchor Press.

——— 1964. Further studies on the larvae of digger wasps (Hymenoptera, Sphecidae). *Transactions of the American Entomological Society* 90: 235–299, pl. 8–19.

——— 1966a. *The Comparative Ethology and Evolution of the Sand Wasps.* Cambridge, Mass.: Harvard University Press.

——— 1966b. The accessory burrows of digger wasps. *Science* 152: 465–471.

——— 1968. Notes on some digger wasps that prey upon leafhoppers. *Annals of the Entomological Society of America* 61: 1343–44.

——— 1969. *Life on a Little-Known Planet.* Guilford, Conn.: Lyons Press.

——— 1970. Ecological-behavioral studies of the wasps of Jackson Hole, Wyoming. *Bulletin of the Museum of Comparative Zoology, Harvard* 140: 451–509.

——— 1973. Notes on the nesting behavior of *Steniolia elegans* (Hymenoptera: Sphecidae). *Great Basin Naturalist* 33: 29–30.

——— 1975. Digger wasps as colonizers of new habitats (Hymenoptera: Aculeata). *Journal of the New York Entomological Society* 82: 259–267.

——— 1976a. Nesting behavior of *Trichogorytes cockerelli* (Ashmead) (Hymenoptera, Sphecidae, Nyssoninae). *Entomological News* 87: 33–37.

——— 1976b. Nesting behavior of *Microbembex hirsuta* Parker, with notes on related species (Hymenoptera: Sphecidae). *Proceedings of the Entomological Society of Washington* 78: 185–189.

——— 1976c. Bembicini of Baja California Sur: Notes on nests, prey, and distribution. *Pan-Pacific Entomologist* 52: 314–320.
——— 1977. Extrinsic vs. intrinsic factors in the evolution on insect sociality. *Bioscience* 27: 613–617.
——— 1978. A solitary wasp that preys upon lacewings (Hymenoptera: Sphecidae; Neuroptera: Chrysopidae). *Psyche* 85: 81–84.
——— 1982a. Two new species of Australian *Bembix* sand wasps, with notes on other species of the genus (Hymenoptera, Sphecidae). *Australian Entomological Magazine* 9: 7–12.
——— 1982b. Nesting and territorial behavior of *Philanthus barbatus* Smith (Hymenoptera: Sphecidae). *Journal of the Kansas Entomological Society* 55: 571–576.
——— 1984. *Insect Biology: A Textbook of Entomology.* Reading, MA: Addison-Wesley.
——— 1985. *The Pleasures of Entomology: Portraits of Insects and People Who Study Them.* Washington, D.C.: Smithsonian Institution Press.
——— 1989. The mating and predatory behavior of *Mellinus rufinodus* (Hymenoptera: Sphecidae). *Pan-Pacific Entomologist* 65: 414–417.
——— 1990. New Australian species and records of the *promontorii* group of the genus *Bembix* F. (Hymenoptera: Sphecidae: Nyssoninae). *Journal of the Australian Entomological Society* 29: 27–30.
——— 1993. *Pioneer Naturalists: The Discovery and Naming of North American Plants and Animals.* New York: Henry Holt and Company.
——— 1997. *The Natural History of the Long Expedition to the Rocky Mountains, 1819–1820.* Oxford: Oxford University Press.
——— 2001. *A Naturalist's Years in the Rocky Mountains.* Boulder, Colo.: Johnson Books.
——— 2002. A review of prey choice in bembicine sand wasps (Hymenoptera: Sphecidae). *Neotropical Entomology* 31: 1–11.
——— 2005. *The Man Who Loved Wasps: A Howard Ensign Evans Reader,* ed. M. A. Evans. Boulder, Colo.: Johnson Books.
Evans, H. E., and M. A. Evans. 1983. *Australia: A Natural History.* Washington, D.C.: Smithsonian Institution Press.
——— 1991. *Cache La Poudre: The Natural History of a Rocky Mountain River.* Boulder: University Press of Colorado.
Evans, H. E., and J. E. Gillaspy. 1964. Observations on the ethology of digger wasps of the genus *Steniolia* (Hymenoptera: Sphecidae: Bembicini). *American Midland Naturalist* 72: 257–280.
Evans, H. E., and C. S. Lin. 1956. Studies on the larvae of digger wasps (Hymenoptera, Sphecidae). Part II. Nyssoninae. *Transactions of the American Entomological Society* 82: 35–66.

Evans, H. E., and R. W. Matthews. 1968. North American *Bembix,* a revised key and suggested grouping. *Annals of the Entomological Society of America* 61: 1284–1299.

——— 1971. Nesting behavior and larval stages of some Australian Nyssonine sand wasps (Hymenoptera, Sphecidae). *Australian Journal of Zoology* 19: 273–310.

——— 1973. Systematics and nesting behavior of Australian *Bembix* sand wasps (Hymenoptera, Sphecidae). *Memoirs of the American Entomological Institute,* no. 20.

——— 1974. Observations on the nesting behavior of South American sand wasps (Hymenoptera). *Biotropica* 6: 130–134.

——— 1975. The sand wasps of Australia. *Scientific American* 233: 108–115.

Evans, H. E., and K. M. O'Neill. 1986. The reproductive and nesting biology of *Bembecinus nanus strenuus* (Mickel) (Hymenoptera, Sphecidae). *Proceedings of the Entomological Society of Washington* 88: 628–633.

——— 1988. *The Natural History and Behavior of North American Beewolves.* Ithaca, N.Y.: Cornell University Press.

Evans, H. E., and W. L. Rubink. 1978. Observations on the prey and nests of seven species of *Cerceris* (Hymenoptera: Sphecidae). *Great Basin Naturalist* 38: 59–63.

Evans, H. E., and M. J. West-Eberhard. 1970. *The Wasps.* Ann Arbor: University of Michigan Press.

Evans, H. E., M. A. Evans, and A. Hook. 1982. Observations on the nests and prey of Australian *Bembix* sand wasps (Hymenoptera: Sphecidae). *Australian Journal of Zoology* 30: 71–80.

Evans, H. E., R. W. Matthews, and E. McC. Callan. 1974. Observations on the nesting behavior of *Rubrica surinamensis* (DeGeer) (Hymenoptera, Sphecidae). *Psyche* 81: 334–352.

Evans, H. E., K. M. O'Neill, and R. P. O'Neill. 1986. Nesting site changes and nocturnal clustering in the sand wasp *Bembecinus quinquespinosus* (Hymenoptera: Sphecidae). *Journal of the Kansas Entomological Society* 59: 280–286.

Evans, M. A., and H. E. Evans. 1970. *William Morton Wheeler, Biologist.* Cambridge, Mass.: Harvard University Press. 380 pp.

Fabre, J. H. 1986. *Souvenirs Entomologiques.* 3 ser. Paris: Ch. Delegrave.

Felton, J. C. 1987. A new prey record for the solitary wasp *Gorytes laticinctus* (Lepeletier) (Hymenoptera: Sphecidae). *Entomologische Berichten* 47: 89–90.

Ferton, C. 1908. Notes détachées sur l'instinct des Hyménoptères mellifères et ravisseurs (4e Série). *Annales de la Société Entomologique de France* 77: 535–586.

——— 1911. Notes détachées sur l'instinct des Hyménoptères mellifères et ravisseurs (7e Série). *Annales de la Société Entomologique de France* 80: 351–412.

Field, J. 1992. Intraspecific parasitism as an alternative reproductive tactic in nest-building wasps and bees. *Biological Review* 67: 79–126.

Finnamore, A. T. and C. D. Michener. 1993. Superfamily Apoidea. *In* H. Goulet and J. T. Huber, eds., *Hymenoptera of the World: An Identification Guide to Families*, pp. 279–357. Ottawa: Agriculture Canada Publications.

Fontenelle, J. C. R., and R. P. Martins. 2002. Hunting behavior by the sand wasp *Rubrica nasuta* (Christ 1791) (Hymenoptera Sphecidae). *Tropical Zoology* 15: 187–196.

Fox, W. J. 1894a. Studies among the fossorial Hymenoptera. I. Synopsis of the North American species of *Alysson*. *Entomological News* 5: 86–89.

——— 1894b. Studies among the fossorial Hymenoptera. II. Synopsis of the N. Am. species of the genus *Didineis* Westmael. *Entomological News* 5: 126–128.

Fraizer, T. 1997. A dynamic model of mating behavior in digger wasps: The energetics of male-male competition mimic size- dependent thermal constraints. *Behavioral Ecology and Sociobiology* 41: 423–434.

Froggatt, W. W. 1903. Cicadas ("locusts") and their habits. *Agricultural Gazette New South Wales* 14: 334–341, 418–425.

Gayubo, S. F., and F. Sanza. 1986. Esfecidofauna de la Margen Derecha dela Cuenca Alta del Duero (Hymenoptera: Sphecidae). *Acta Salamanticensia* 74: 1–115.

Gayubo, S. F., J. A. González, J. D. Asís, and J. Tormos. 2005. Conservation of European environments: The Spheciformes wasps as biodiversity indicators (Hymenoptera: Apoidea: Ampulicidae, Sphecidae, and Crabronidae). *Journal of Natural History* 39: 2705–2714.

Genaro, J. A. 1994. Conducta de nidification de *Ochleroptera jamaica* (Hymenoptera: Sphecidae). *Caribbean Journal of Science* 30: 270–272.

——— 1995. Comportamiento de *Bembix americana antilleana* durante la nidificación (Hymenoptera: Sphecidae). *Revista de Biologia Tropical* 43: 271–275.

——— 1999. Interactions among species (Diptera, Hymenoptera) in *Stictia signata* nests (Hymenoptera: Sphecidae). *Revista de Biologia Tropical* 47: 535–538.

Genaro, J. A., and C. Juarrero de Varona. 1998. Comportamiento de *Sphecius hogardi* durante la nidificación (Hymenoptera: Sphecidae). *Caribbean Journal of Science* 34: 323–324.

Genaro, J. A., and C. Sánchez Alonso. 1990. *Microbembex argentifrons* (Cresson), una especie necrófaga (Hymenoptera: Sphecidae). *Poeyana* 405: 1–6.

Genise, J. F. 1979a. Comportamiento de nidificación de *Bicyrtes variegata* (Ol.) y *B. discisa* (Tasch.) (Hymenoptera, Sphecidae). *Revista de la Sociedad Entomológica Argentina* 38: 123–126.

——— 1979b. Interacción social en una agrupación de *Rubrica nasuta* (Christ.) (Hymenoptera, Sphecidae, Nyssoninae). *Physis: Revista Argentina de Ciencias Naturales* (C)38: 29–30.

——— 1980. Selección de presas en *Rubrica nasuta* (Christ) (Hymenoptera, Sphecidae). *Ecosur* 7: 15–18.

——— 1981a. Estudios sobre el comportamiento de Bembicini neotropicales II. *Selman notatus* (Tasch.), *Trichostictia guttata* (Tasch.) y *Editha integra* (Burmeister) (Hymenoptera, Sphecidae). *Revista de la Sociedad Entomológica Argentina* 40: 249–262.

——— 1981b. Observaciones sobre el compartamiento de nidificación de *Stictia arcuata* (Taschenberg). *Physis: Revista Argentina de Ciencias Naturales,* (C)40: 15–17.

——— 1982a. Estudios sobre el comportamiento de nidificación de Bembicini neotropicales V. *Hemidula singularis* (Taschenberg) (Hymenoptera, Sphecidae). *Revista de la Sociedad Entomológica Argentina* 41: 77–84.

——— 1982b. Comportamiento de desecho de presas en *Rubrica nasuta* (Christ) y revision de esta actividad en Sphecidae (Insecta, Hymenoptera). *Physis: Revista Argentina de Ciencias Naturales* (C)40: 101–109.

——— 1982c. Estudios sobre el comportamiento de nidificación de Bembicini neotropicales III. *Stictia flexuosa* (Taschenberg) y algunas reflexiones sobre le diferenciacion de nichos ecológicos en Bembicini (Hymenoptera, Sphecidae). *Physis: Revista Argentina de Ciencias Naturales* (C)41: 79–85.

——— 1982d. Estudios sobre el comportamiento de nidificación de Bembicini neotropicales IV. *Bembix citripes* (Taschenberg) (Hymenoptera, Sphecidae). *Physis: Revista Argentina de Ciencias Naturales* (C)41: 93–96.

——— 1982e. Estudios sobre el comportamiento de nidificación de Bembicini neotropicales VI. *Zyzzyx chilensis* (Eschscholz) (Hymenoptera, Sphecidae). *Revista de la Sociedad Entomológica Argentina* 41: 289–298.

——— 1982f. Estudios sobre el comportamiento de Bembicini neotropicales I. *Bicyrtes simillima* (Smith) y *Bicyrtes discisa* (Tasch.) (Hymenoptera, Sphecidae), con una revisión de lo tipos de aprovisionamiento en Bembicini. *Physis: Revista Argentina de Ciencias Naturales* (C)40: 93–99.

——— 1985. Estudios sobre el comportamiento de nidificación de Bembicini neotropicales VII. *Trichostictia vulpina* (Handlirsch) (Hymenoptera, Sphecidae). *Physis: Revista Argentina de Ciencias Naturales* (C)43: 1–4.

——— 1989. Estudios sobre el comportamiento de nidificación de Bembicini neotropicales. IX. *Hemidula burmeisteri* Willink (Hymenoptera, Sphecidae). *Revista de la Sociedad Entomológica Argentina* 45: 275–277.

Gess, F. W. 1981. Some aspects of an ethological study of the aculeate wasps and bees of a karroid area in the vicinity of Grahamstown, South Africa. *Annals of the Cape Provincial Museums, Natural History* 14: 1–80.

——— 1986. Three new species of southern African *Bembix,* a new synonymy, and biological notes on other species of the genus (Hymenoptera: Sphecidae: Nyssoninae). *Annals of the Cape Provincial Museums, Natural History* 16: 137–160.

Gess, F. W., and S. K. Gess. 1975. Ethological studies of *Bembecinus cinguliger* (Smith) and *B. oxydorcus* (Handl.) (Hymenoptera: Sphecidae), two southern African turret-building wasps. *Annals of the Cape Provincial Museums, Natural History* 18: 151–160.

——— 1991. A preliminary survey of the aculeate wasps and the bees of the lower reaches of the Nossob River Valley, Kalahari Gemsbok National Park, South Africa. *Koedoe* 34: 77–88.

——— 1993. Effects of increasing land utilization on species representation and diversity of aculeate wasps and bees in the semi-arid areas of southern Africa. In J. LaSalle and I. D. Gould, eds., *Hymenoptera and Biodiversity,* pp. 83–113. London: Natural History Museum.

Gess, S. K., and F. W. Gess. 1989. Notes on the nesting behaviour in *Bembix bubalus* Handlirsch in southern Africa with emphasis on nest sharing and reaction to nest parasites (Hymenoptera: Sphecidae). *Annals of the Cape Provincial Museums, Natural History* 18: 151–160.

Ghazoul, J., and P. G. Willmer. 1994. Endothermic warm-up in two species of sphecid wasp and its relation to behaviour. *Physiological Entomology* 19: 103–108.

Gillaspy, J. E. 1963. The genus *Stizoides* (Hymenoptera: Sphecidae: Stizini) in North America, with notes on the Old World Fauna. *Bulletin of the Museum of Comparative Zoology* 128: 369–391.

——— 1983. New taxa of bembicine Sphecidae (Hymenoptera). *Southwestern Naturalist* 8: 279–291.

Gillaspy, J. E., H. E. Evans, and C. S. Lin. 1962. Observations on the behavior of digger wasps of the genus *Stictiella* (Hymenoptera: Sphecidae), with a partition of the genus. *Annals of the Entomological Society of America* 55: 559–566.

Goodman, W. G. 1970. Observations on the behavior and biology of *Microbembex californica* Bohart. *Pan-Pacific Entomologist* 46: 207–209.

Gordon, D. M. 2000. Plants as indicators of leafcutter bee (Hymenoptera: Megachilidae) nest habitat in coastal dunes. *Pan-Pacific Entomologist* 76: 219–233.

Gould, S. J. 2001. Introduction. In C. Zimmer, ed., *Evolution: Triumph of an Idea,* pp. ix–xiv. London: William Heinemann.

Gourlay, E. S. 1964. Notes on New Zealand insects and records of introduced species. *New Zealand Entomologist* 3: 45–51.

Grandi, G. 1961. Studi di un Entomologo sugli Imenotteri superiori. *Bollettino dell'Istituto di Entomologia della Università di Bologna* 25: i–xv, 1–659.

Greene, H. W. 2005. Organisms in nature as a central focus for biology. *Trends in Ecology and Evolution* 20: 23–27.

Griswold, T. L. 1983. A new species of *Bembix* from Lower California (Hymenoptera: Sphecidae). *Pan-Pacific Entomologist* 59: 103–107.

Gwynne, D. T. 1978. Male territoriality in the bumblebeewolf, *Philanthus bicinctus* (Mickel) (Hymenoptera, Sphecidae): Observations on the behaviour of individual males. *Zeitschrift für Tierpsychologie* 47: 89–103.

——— 1980. Female defense polygyny in the bumblebeewolf, *Philanthus bicinctus* (Hymenoptera: Sphecidae). *Behavioral Ecology and Sociobiology* 7: 213–225.

Hamilton, W. D. 1964. The evolution of social behavior. *Journal of Theoretical Biology* 7: 1–52.

——— 1971. Geometry for the selfish herd. *Journal of Theoretical Biology* 31: 195–311.

Hamm, A. H., and O. W. Richards. 1930. The biology of the British fossorial wasps of the families Mellinidae, Gorytidae, Philanthidae, Oxybelidae, and Trypoxylidae. *Transactions of the Entomological Society of London* 78: 95–131.

Handlirsch, A. 1895. Nachträge und Schlusswort zur Monographie der mit *Nysson* und *Bembex* verwandten Grabwespen. *Sitzungsberichte der Kaiserlichen Akademie der Wissenschaften. Mathematisch-Naturwissenschaftliche Classe. Abteilung I* 104: 801–1079.

Harris, A. C. 1994. *Sphecidae (Insecta, Hymenoptera): Fauna of New Zealand, Number 22.* Lincoln,New Zealand: Manaaki Whenua Press.

Hartman, C. 1905. Observations on the habits of some solitary wasps of Texas. *Bulletin University of Texas* 65.

Hastings, J. 1986. Provisioning by female western cicada killer wasps, *Sphecius grandis* (Hymenoptera: Sphecidae): Influence of body size and emergence time on individual provisioning success. *Journal of the Kansas Entomological Society* 59: 262–268.

——— 1989a. The influence of size, age, and residency status on territory defense in male western cicada killer wasp (*Sphecius grandis*, Hymenoptera: Sphecidae). *Journal of the Kansas Entomological Society* 62: 363–373.

——— 1989b. Protandry in western cicada killer wasps (*Sphecius grandis*, Hymenoptera: Sphecidae): An empirical study of emergence time and mating opportunity). *Behavioral Ecology and Sociobiology* 25: 255–260.

Heinrich, B. 1993. *The Hot-blooded Insects.* Cambridge, Mass.: Harvard University Press.

Hennig, W. 1966. *Phylogenetic Systematics.* Urbana: University of Illinois Press.

Hilchie, G. J. 1982. Evolutionary aspects of geographical variation in color and prey in the beewolf species *Philanthus albopilosus* Cresson. *Quaestiones Entomologicae* 18: 91–126.

Hölldobler, B., and E. O. Wilson. 1990. *The Ants.* Cambridge, Mass.: Harvard University Press.

Holliday, C. W., and J. R. Coelho. 2006. Improved key to the New World species of *Sphecius* (Hymenoptera: Crabronidae). *Annals of the Entomological Society of America* 99: 793–798.

Hook, A. 1981a. Nesting biology of *Tanyoprymnus moneduloides* and *Ammatomus icarioides. Annals of the Entomological Society of America* 74: 409–411.

——— 1981b. Hesperiidae as prey of *Stictia carolina. Journal of the New York Entomological Society* 89: 202–203.

Itami, H. 1967. Biology of *Alysson pertheesi* Gorski in Japan. *The Life Study* (Fukui) 11: 4–5.

Itô, Y. 1978. *Comparative Ecology.* Cambridge: Cambridge University Press.

Iwata, K. 1936. On the habits of *Stizus* and *Bembix* which occur in Japan. *Kontyū* 10: 233–250.

——— 1937. On the habits of *Harpactus laevis* Latreille. *Kontyû* 11: 404–409.

——— 1955. The comparative anatomy of the ovary in Hymenoptera. Part I. Aculeata. *Mushi* 29: 17–34.

——— 1960. Comparative anatomy and overy of the Hymenoptera. Supplement on Aculeata with descriptions of ovarian eggs of certain species. *Acta Hymenopterologica* 1: 205–211.

——— 1964. Egg gigantism in subsocial Hymenoptera, with ethological discussion on tropical bamboo carpenter bees. *Nature and Life in Southeast Asia* 3: 399–433.

——— 1965. The comparative anatomy of the ovary in Hymenoptera (records on 64 species of Aculeata in Thailand with descriptions of ovarian eggs). *Mushi* 38: 101–109.

——— 1976. *Evolution of Instinct: Comparative Ethology of the Hymenoptera.* New Delhi: Amerind Publishing Company (translation of 1971 Japanese edition).

Jacobs, H. J., and J. Oehlke. 1990. Beitrage zur Insektenfauna der DDR: Hymenoptera: Sphecidae. 1. *Nachtrag Beiträge zur Entomologie* 40: 121–229.

Janvier, H. 1928. Recherches biologiques sur les prédateurs du Chile. *Annales des Sciences Naturelles. Zoologie (Série 10)* 11: 67–207.

——— 1974. Una colonia de *Argogorytes hispanicus* (Merc. 1906) en Málaga. *Graellsia* 27: 67–77.

Joos, B., and T. M. Casey. 1992. Flight energetics of the cicada killer wasp. *American Zoologist* 32: 54A (abstract).

Kaiser, W. 1995. Rest at night in some solitary bees—a comparison with the sleep-like state of honey bees. *Apidologie* 26: 213–230.

Karsai, I. 1989. Factors affecting diurnal activities of solitary wasps (Hymenoptera: Sphecidae and Pompilidae). *Entomologia Generalis* 14: 223–232.

Kazenas, V. L. 2001. *Fauna i biologiya royushchikh os (Hymenoptera, Sphecidae) Kazakhstana i Sredney Azii (Fauna and biology of sphecid wasps (Hymenoptera, Sphecidae) of Kazakhstan and Central Asia).* Almaty: Kazgos INTI.

Kazenas, V. L., and P. A. Esenbekova. 1995. Tsikadkovyie (Homoptera, Auchenorrhyncha)—dobycha royushchikh os (Hymenoptera, Sphecidae) v Kazakhstane—leafhoppers (Homoptera, Auchenorrhyncha)—prey od digger wasps (Hymenoptera, Sphecidae) in Kazakhstan. *Selevinia* 3: 79–81.

Kazenas, V. L., and V. I. Tobias. 1993. Night sleeping aggregations of digger wasps (Hymenoptera, Sphecidae) in southeastern Kazakhstan. *Entomological Review* 71: 13–16.

Kifune, T. 1988. Additional records of stylopized Sphecoidea (Hymenoptera) from the Yaeyama Islands, with description of the male of *Paraxenos orientalis*. *Kontyū* 56: 303–306.

Kifune, T., and Y. Hirashima. 1987. Three new species of the genus *Paraxenos* (Strepsiptera: Stylopidae) parasitic on *Bembix* (Hymenoptera: Sphecidae) of Sri Lanka and Australia in the collection of the Smithsonian Institution (Notulae Strepsipterologicae–XVII). *Esakia* 25: 155–160.

Kifune, T., and S. Yamane. 1985. Two new species of the genus *Paraxenos* (Strepsiptera, Stylopidae) and records of stylopized Sphecidae and Eumenidae (Hymenoptera) from the Ryukyus, Japan (Studies on the Japanese Strepsiptera IX). *Kontyū* 53: 49–58.

Kimsey, L. S., and R. B. Kimsey. 1981. A new species of *Bembix* from the California desert (Hymenoptera: Sphecidae). *Journal of the Kansas Entomological Society* 54: 486–488.

Kimsey, L. S., R. B. Kimsey, and C. A. Toft. 1981. Life history of *Bembix inyoensis* in Death Valley (Hymenoptera: Sphecidae). *Journal of the Kansas Entomological Society* 54: 665–672.

Krafsur, E. S., M. A. Cummings, M. A. Endsley, J. G. Marquez, and J. D. Nason. 2005. Geographic differentiation in the house fly estimated by microsatellite and mitochondrial variation. *Journal of Heredity* 96: 502–512.

Krombein, K. V. 1955. Some notes on the wasps of Kill Devil Hills, North Carolina, 1954 (Hymenoptera, Aculeata). *Proceedings of the Entomological Society of Washington* 57: 145–160.

——— 1959. Biological notes on some ground-nesting wasps at Kill Devil Hills, North Carolina, 1958, and some additions to the faunal list (Hymenoptera, Aculeata). *Proceedings of the Entomological Society of Washington* 61: 193–199.

——— 1967. *Trap-nesting Wasps and Bees: Life Histories, Nests, and Associates.* Washington, D.C.: Smithsonian Press.

——— 1972. Miscellaneous prey records of solitary wasps. VI. Notes on some species from Greece (Hymenoptera: Aculeata). *Proceedings of the Entomological Society of Washington* 74: 383–385.

——— 1984. Biosystematic studies of Ceylonese wasps. XIII: A monograph of the Stizinae (Hymenoptera: Sphecoidea, Nyssonidae). *Smithsonian Contributions to Zoology,* no. 388, 37 pp.

——— 1985. Biosystematic studies of Ceylonese wasps. XV: A monograph of the Alyssoninae, Nyssoninae, and Gorytinae (Hymenoptera: Sphecoidea, Nyssonidae). *Smithsonian Contributions to Zoology,* no. 414, 43 pp.

Krombein, K. V., and J. van der Vecht. 1987. Biosystematic studies of Ceylonese wasps. XVII: A revision of Sri Lankan and South Indian *Bembix* Fabricius (Hymenoptera: Sphecoidea: Nyssonidae). *Smithsonian Contributions to Zoology,* no. 451, 30 pp.

Krombein, K. V., and A. Willink. 1951. The North American species of *Bembecinus* (Hymenoptera, Sphecidae, Stizini). *American Midland Naturalist* 44: 699–713.

Krombein, K. V., P. D. Hurd, D. R. Smith, and B. D. Burks. 1979. *Catalog of Hymenoptera in America North of Mexico.* Vol. 2: *Apocrita (Aculeata).* Washington, D.C.: Smithsonian Institution Press.

Kruuk, H. 2004. *Niko's Nature: The Life of Niko Tinbergen and His Science of Animal Behaviour.* Oxford: Oxford University Press.

Kula, E., and P. Tyrner. 2003. Hymenoptera (Aculeata) of spruce stands in the air-pollution region of Northern Bohemia—Hymenoptera (Aculeata). *Journal of Forest Science* 49: 200–207.

Kullenberg, B. 1956. Field experiments with chemical sexual attractants on aculeate Hymenoptera. *Zoologiska Bidrag från Uppsala* 31: 253–354.

Kurczewski, F. E. 1998. Comparison of sand nesting wasps (Hymenoptera) from two pine barrens in upstate New York. *Entomological News* 109: 247–251.

Kurczewski, F. E., and H. F. Boyle. 2006. Nesting behavior, ecology, seasonal and geographic variation of the sand wasp, *Stictiella emarginata* (Hymenoptera, Sphecidae). *Canadian Field-Naturalist* 119: 6–15.

Kurczewski, F. E., and E. J. Kurczewski. 1971. Host records for some species of Nyssoninae. *Journal of the Kansas Entomological Society* 44: 334–337.

Lane, R. S., J. R. Anderson, and E. Rogers. 1986. Nest provisioning and related activities of the sand wasp, *Bembix americana comata* (Hymenoptera: Sphecidae). *Pan-Pacific Entomologist* 62: 258–268.

Larsen, O., G. Gleffe, and J. Tengö. 1986. Vibration and sound communication in solitary bees and wasps. *Physiological Entomology* 11: 287–296.

Larson, D. P., K. M. O'Neill, and W. P. Kemp. 1999. Evaluation of the accuracy of sweep sampling in determining grasshopper (Orthoptera: Acrididae) community composition. *Journal of Agricultural and Urban Entomology* 16: 207–214.

Larsson, F. K. 1986. Increased nest density of the digger wasp *Bembix rostrata* as a response to parasites and predators (Hymenoptera: Sphecidae). *Entomologia Generalis* 12: 71–75.

——— 1989a. Temperature-induced alternative male mating tactics in a tropical digger wasp. *Journal of Insect Behavior* 2: 849–852.

——— 1989b. Mating patterns in six insect species: Effects of weather and population density. Ph.D. dissertation, Uppsala University, Sweden.

——— 1990. Thermoregulation and activity patterns of the sand wasp *Steniolia longirostra* (Say) (Hymenoptera: Sphecidae) in Costa Rica. *Biotropica* 22: 65–68.

——— 1991. Some take it cool, some like it hot—a comparative study of male mate searching tactics in two species of Hymenoptera (Colletidae and Sphecidae). *Journal of Thermal Biology* 16: 45–51.

Larsson, F. K., and K. Larsson. 1989. Body size and nesting behaviour of the sand wasp *Stictia heros* (Hymenoptera: Sphecidae) in Costa Rica. *Journal of Tropical Ecology* 5: 117–120.

Larsson, F. K., and J. Tengö. 1989. It is not always good to be large: some female fitness components in a temperate digger wasp, *Bembix rostrata* (Hymenoptera: Sphecidae). *Journal of the Kansas Entomological Society* 62: 490–495.

Lavigne, R. J., and F. R. Holland. 1969. Comparative behavior of eleven species of Wyoming robber flies (Diptera: Asilidae). *University of Wyoming Science Monograph* 18: 1–61.

Leclercq, J., and M. Leclercq. 1970. Deux *Bembix* (Hymenoptera Sphecidae, Nyssoninae) attirés un piège et par les mammifères victimes de leurs proies. *Bulletin des Recherches Agronomiques de Gembloux*, n.s., 5: 267–270.

Lin, C. S. 1971. Bionomics of *Stictia carolina* at Lake Texoma, with notes on some neotropical species (Hymenoptera: Sphecidae). *Texas Journal of Science* 23: 275–286.

Lin, N. 1963a. Territorial behavior in the cicada killer wasp *Sphecius speciosus* (Hymenoptera: Sphecidae). *Behaviour* 20: 115–133.

——— 1963b. Observations of suspected density-dependent fighting between females of the cicada killer wasp *Sphecius speciosus*. *Bulletin of the Brooklyn Entomological Society* 58: 121–123.

——— 1966. Copulatory behavior of the cicada killer wasp, *Sphecius speciosus*. *Animal Behaviour* 14: 130–131.

——— 1967. Role differentiation in copulating cicada killer wasps. *Science* 157: 1334–35.

——— 1978. Contributions to the ecology of the cicada killer, *Sphecius speciosus* (Hymenoptera: Sphecidae). *Journal of the Washington Academy of Sciences* 68: 75–82.

——— 1979a. Differential prey selection for the sex of offspring in the cicada killer *Sphecius speciosus* (Hymenoptera: Sphecidae). *Proceedings of the Entomological Society of Washington* 81: 269–275.

——— 1979b. The weight of cicada killer wasps, *Sphecius speciosus*, and the weight of their prey. *Journal of the Washington Academy of Sciences* 69: 159–163.

Llano, R. J. 1959. Observaciones biológicas de insectos bonaerenses. *Suplemento de la Revista de Educacion de la Provincia de Buenos Aires*, no. 12. 136 pp.

Lomholdt, O. 1975. The Sphecidae (Hymenoptera) of Fennoscandia and Denmark. *Fauna Entomologica Scandinavica* 4: 1–224.

Longair, R. W., J. H. Cane, and L. Elliott. 1987. Male competition and mating behavior within mating aggregations of *Glenostictia satan* Gillaspy. *Journal of the Kansas Entomological Society* 60: 264–272.

Lüps, P. 1969. Die Grabwespe *Bembecinus tridens* (Fabricius) (Hymenoptera, Sphecidae). *Jahrbuch des Naturhistorischen Museums der Stadt Bern* 3: 201–215.

——— 1973. Beobachtungen an *Bembecinus tridens* (Fabricius) (Hymenoptera, Sphecidae). *Bulletin de la Société Entomologique Suisse* 46: 131–139.

Macedo, J. F., and R. P. Martins. 1998. Potencial da erva daninha *Waltheria americana* (Sterculiaceae) no manejo integrado de pragas e polinizadores: visitas de abelhas e vespas. *Anais de Sociedade Entomologica do Brasil* 27: 29–40.

Magurran, A. E. 2003. *Measuring Biological Diversity*. New York: Blackwell.

Malloch, J. R., and S. A. Rohwer. 1930. New forms of sphecoid wasps of the genus *Didineis* Wesmael. *Proceedings of the United States National Museum* 77: 1–7.

Marquez, J. G., and E. S. Krafsur. 2002. Gene flow among geographically diverse house fly populations (*Musca domestica* L.): A worldwide survey of mitochondrial diversity. *Journal of Heredity* 93: 254–259.

Martin, W. F., and R. F. Martin. 1990. Reproduction of the sand wasps *Stictia signata* (L.) and *Bicyrtes variegata* (Olivier) (Hymenoptera: Sphecidae) on the Caribbean coast of Quintana Roo, Mexico. *Pan-Pacific Entomologist* 66: 71–78.

Martins, E. P. 1996. *Phylogenies and the Comparative Method in Animal Behaviour.* New York: Oxford University Press.

Martins, R. P. 1993. The biology of *Editha magnifica* (Perty 1834) (Hymenoptera Sphecidae). *Tropical Zoology* 6: 109–123.

Martins, R. P., L. A. Soares, and D. Yanega. 1998. The nesting behavior and dynamics of *Bicyrtes angulata* (F. Smith) with a comparison to other species in the genus (Hymenoptera: Sphecidae). *Journal of Hymenoptera Research* 7: 165–177.

Matthews, R. W. 1991. Evolution of social behavior in the sphecid wasps. In K. G. Ross and R. W. Matthews, eds., *The Social Biology of Wasps*, pp. 570–602. Ithaca, N.Y.: Cornell University Press.

Matthews, R. W., and H. E. Evans. 1971. Biological notes on two species of *Sericophorus* from Australia. *Psyche* 77: 413–429.

——— 1974. Notes on the behavior of three species of *Microbembex* (Hymenoptera: Sphecidae) in South America. *Journal of the Georgia Entomological Society* 9: 79–85.

Matthews, R. W., and J. R. Matthews. 2005. Biological notes on ground-nesting digger wasps from western Nebraska. *Prairie Naturalist* 37: 61–71.

Matthews, R. W., R. A. Saunders, and J. R. Matthews. 1981. Nesting behavior of the sand wasp *Stictia maculata* (Hymenoptera: Sphecidae) in Costa Rica. *Journal of the Kansas Entomological Society* 54: 249–254.

McCorquodale, D. B. 1986. Digger wasp (Hymenoptera: Sphecidae) provisioning flights as a defence against a nest parasite, *Senotainia trilineata* (Diptera: Sarcophagidae). *Canadian Journal of Zoology* 64: 1620–1627.

McKamey, S. H. 2006. Checklist of leafhopper species, 1758–1955 (Hemiptera: Membracoidea: Cicadellidae and Myserslopiidae) with synonymy and distribution [Catalogue of the Homoptera, Fascicle 6, Abridged]. http://www.sel.barc.usda.gov/selhome/leafhoppers/HTML/html_main.html (accessed 13 April 2006).

Mellor, J. E. M. 1927. A note on the mutillid *Ephutoma continua* Fabr. and of *Bembix mediterranea* Hdl. in Egypt. *Bulletin Societe Royale d'Egypte* 20: 69–79.

Melo, G. A. R. 1999. Phylogenetic relationships and classification of the major lineages of Apoidea (Hymenoptera), with emphasis on the crabronid wasps. *Natural History Museum, University of Kansas, Scientific Papers* 12: 1–55.

Menke, A. S. 1997. Family-group names in Sphecidae (Hymenoptera: Apoidea). *Journal of Hymenoptera Research* 6: 243–255.

Menke, A. S., and F. Fernández C. 1996. Claves illustradas para las subfamilias, tribus y géneros de esfécidos neotropicales (Apoidea: Sphecidae). *Revista Biologia Tropical,* supplement 2, 44: 1–68.

Michener, C. D. 2000. *The Bees of the World.* Baltimore: Johns Hopkins University Press.

Myartseva, S. N. 1965. Royushchiye osy (Hymenoptera, Sphecidae) nizoviy Murgaba. In G. S. Medvedev and A. N. Luppova, eds., *Nasekomyiye nizoviy Murgaba (vostochnoya Turkmeniya): Fauna, ekologiya, khozaystvennoye znacheniye*, pp. 74–99. Ashhabad: Turkmenskoye Izdatelstvo.

——— 1976. Zametki po biologii maloizuchennykh sfetsid Turkmenii

(Hymenoptera, Sphecidae). In A. O. Tashliev, ed., *Ekologiya i khozaystvennoye znachenye nasekomych Turkmenii,* pp. 72–82. Askhabad: Ylym.

Nazarova, S., and S. B. Baratov. 1981. The predatory wasps (Hymenoptera: Sphecidae, Vespidae) of Tadzhikistan and their role in regulating the numbers of Tabanidae. *Entomological Review* 60: 95–101.

Nielsen, E. T. 1945. Moeurs des *Bembex. Spolia Zoologica Museum Hauniensis* 7: 1–174.

Nishida, G. M., and J. W. Beardsley. 2002. A review of the insects and related arthropods of Midway Atoll. Records of the Hawaii Biological Survey for 2000. Bishop Museum Occasional Papers 68: 25–69.

Noss, R. F. 1996. Are the naturalists dying off? *Conservation Biology* 10: 1–3.

O'Brien, M. F., and F. E. Kurczewski. 1979. Observations on the nesting behavior of *Alysson conicus* Provancher (Hymenoptera: Sphecidae). *Proceedings of the Entomological Society of Washington* 81: 435–437.

——— 1982. Further observations on the ethology of *Alysson conicus* Provancher (Hymenoptera: Sphecidae). *Proceedings of the Entomological Society of Washington* 84: 225–231.

Ohl, M. 1999. A revision of *Stizoides* Guérin-Méneville, 1844: Taxonomy, phylogenetic relationships, biogeography, and evolution (Hymenoptera: Apoidea: "Sphecidae"). *Mitteilungen aus dem Museum für Naturkunde in Berlin, Zoologische Reihe* 75: 63–169.

——— 2005. A review of the Nearctic digger-wasp genus *Microstictia* Gillaspy, 1963, with description of two new species from Arizona and Florida (Hymenoptera: Apoidea: Crabronidae). *Systematics and Biodiversity* 2: 341–353.

Ohl, M., and C. Bleidorn. 2006. The phylogenetic position of the enigmatic family Heterogynaidae based on molecular data, with description of a new, nocturnal species (Hymenoptera: Apoidea). *Systematic Zoology* 31: 321–337.

Ohl, M., and D. Linde. 2003. Ovaries, ovarioles, and oocytes in apoid wasps, with special reference to cleptoparasitic species (Hymenoptera: Apoidea, "Sphecidae"). *Journal of the Kansas Entomological Society* 76: 147–159.

O'Neill, K. M. 1979. Territorial behavior in males of *Philanthus psyche* (Hymenoptera, Sphecidae). *Psyche* 86: 19–43.

——— 1983a. Territoriality, body size, and spacing in males of the beewolf *Philanthus basilaris* (Hymenoptera: Sphecidae). *Behaviour* 86: 395–421.

——— 1983b. The significance of body size in territorial interactions of male beewolves (Hymenoptera: Sphecidae, *Philanthus*). *Animal Behaviour* 31: 404–411.

——— 1985. Egg size, prey size, and sexual size dimorphism in digger wasps (Hymenoptera: Sphecidae). *Canadian Journal of Zoology* 63: 2187–93.

——— 1990. Female nesting behavior and male territoriality in *Aphilanthops subfrigidus* Dunning (Hymenoptera: Sphecidae). *Pan-Pacific Entomologist* 66: 19–23.

——— 2001. *Solitary Wasps: Natural History and Behavior.* Ithaca, N.Y.: Cornell University Press.

O'Neill, K. M., and H. E. Evans. 1982. Patterns of prey use in four sympatric species of *Philanthus* (Hymenoptera: Sphecidae) with a review of prey selection in the genus. *Journal of Natural History* 16: 791–801.

——— 1983. Alternative male mating tactics in *Bembecinus quinquespinosus* (Hymenoptera: Sphecidae): Correlations with size and color variation. *Behavioral Ecology and Sociobiology* 14: 39–46.

——— 1999. Observations on nests and prey of *Podalonia valida* (Hymenoptera: Sphecidae). *Proceedings of the Entomological Society of Washington* 101: 312–315.

O'Neill, K. M., and R. P. O'Neill. 1988. Thermal stress and microhabitat selection in territorial males of *Philanthus psyche* (Hymenoptera: Sphecidae). *Journal of Thermal Biology* 13: 15–20.

——— 2003. Sex allocation, nest structure, and prey of *Isodontia mexicana* (Saussure) (Hymenoptera: Sphecidae). *Journal of the Kansas Entomological Society* 76: 447–454.

O'Neill, K. M., H. E. Evans, and L. Bjostad. 1991. Territorial behavior in males of three North American species of bumble bees (Hymenoptera: Apidae, *Bombus*). *Canadian Journal of Zoology* 69: 604–613.

O'Neill, K. M., H. E. Evans, and R. P. O'Neill. 1989. Phenotypic correlates of mating success in the sand wasp *Bembecinus quinquespinosus* (Hymenoptera: Sphecidae). *Canadian Journal of Zoology* 67: 2557–68.

Packer, L. 1987. The history of *Alysson lunicornis* (Fabricius) (Hym., Sphecidae) in Great Britain. *Entomologist's Monthly Magazine* 123: 37–40.

Pagliano, G., and A. Alma. 1997. Ricerche etologiche su Gorytini e Alyssonini (Hymenoptera Sphecidae) parassitoidi di Auchenorryncha (Rhynchota Homoptera). *Rivista Piemontese di Storia Naturale* 18: 173–181.

Pagliano G., P. L. Scaramozzino, and F. Strumia. 2001. Introduction and spread of four Aculeate Hymenoptera in Italy, Sardinia, and Corsica. In A. D. Austin and M. Dowton, eds., *Hymenoptera. Evolution, Biodiversity, and Biological Control,* pp. 290–296. Collingwood, Australia: CSIRO Publishing.

Parker, J. B. 1925. Notes on the nesting habits of *Bembix comata* Parker. *Proceedings of the Entomological Society of Washington* 27: 189–195.

Pearson, D. L. 1994. Selecting indicator taxa for the quantitative assessment of biodiversity. *Philosophical Transactions of the Royal Society of London,* ser. B 345: 75–79.

Peckham, G. W., and E. G. Peckham. 1905. *Wasps: Social and Solitary.* Boston: Houghton, Mifflin and Company.

Pfennig, D. W., and H. K. Reeve. 1989. Neighbor recognition and context-dependent aggression in a solitary wasp, *Sphecius speciosus* (Hymenoptera: Sphecidae). *Ethology* 80: 1–18.

——— 1993. Nepotism in a solitary wasp as revealed by DNA fingerprinting. *Evolution* 47: 700–704.

Philippi, T., and W. G. Eberhard. 1986. Foraging behavior of *Stictia signata* (Hymenoptera: Sphecidae). *Journal of the Kansas Entomological Society* 59: 604–608.

Piek, T. 1985. Insect venoms and toxins. In G. A. Kerkut and L. I. Gilbert, eds., *Comprehensive Insect Physiology, Biochemistry, and Pharmacology*, pp. 595–636. New York: Pergamon.

Pimenta, H. R., and R. P. Martins. 1999. The natural history of the neotropical sand wasp *Rubrica nasuta* (Christ 1791) (Hymenoptera Sphecidae) in Brazil. *Tropical Zoology* 12: 273–288.

Post, D. C. 1981. Observations on female nesting and male behavior of *Stictia signata* (Hymenoptera: Sphecidae) in Brazil. *Revista de Biologia Tropical*, 29: 105–113.

Powell, J. A. 1974. On the nesting behavior and larval development of the sand-wasp *Gorytes canaliculatus* (Packard) in California (Hymenoptera: Sphecidae). *Journal of the Kansas Entomological Society* 47: 1–7.

Prentice, M. A. 1998. The comparative morphology and phylogeny of apoid wasps (Hymenoptera: Apoidea). Ph.D. dissertation, University of California, Berkeley.

Pulawski, W. J. 1985. *Harpactes* Shuckard, 1837, the valid name for the genus currently called *Dienoplus* Fox, 1894 (Hymenoptera: Sphecidae). *Systematic Entomology* 10: 59–63.

——— 1997. A review of New Guinean *Ochleroptera* Holmberg 1903 (Hymenoptera: Sphecidae). *Journal of Hymenoptera Research* 6: 256–262.

——— 2006a. Catalog of the Sphecidae sensu lato (= Apoidea excluding Apidae). California Academy of Sciences. http://www.calacademy.org/research/entomology/Entomology_Resources/Hymenoptera/sphecidae/Genera_and_species_PDF/introduction.htm.

——— 2006b. Bibliography of the Sphecidae sensu lato (= Apoidea excluding Apidae). http://www.calacademy.org/research/entomology/Entomology_Resources/Hymenoptera/sphecidae/bibliography.pdf.

Radović, I. T. 1985. Morphology and adaptive value of the sting apparatus of digger wasps (Hym. Sphecidae). *Acta Entomologica Jugoslavica* 21: 61–74.

Rau, P. 1938. Additional observations on the sleep of insects. *Annals of the Entomological Society of America* 31: 540–556.

Rau, P., and N. Rau. 1918. *Wasp Studies Afield.* Princeton, N.J.: Princeton University Press.

Reinhard, E. G. 1929. *The Witchery of Wasps.* New York: Century Co.

Richards, O. W. 1937. Results of the Oxford University Expedition to British Guiana, 1929. Hymenoptera, Sphecidae, and Bembecidae. *Transactions of the Royal Entomological Society of London* 86: 101–118.

——— 1971. The habits of *Editha magnifica* (Perty) (Hym., Sphecidae, Bembicini). *Entomologist's Monthly Magazine* 107: 112–113.

Riley, C. V. 1892. The larger digger wasp. *Insect Life* 4: 248–252.

Roberts, L. W., and B. H. Wilson. 1967. Predation on horse flies by two bembicine wasp species in certain areas of southern Louisiana. *Journal of Economic Entomology* 60: 412–415.

Rodeck, H. G. 1931. Nesting habits of *Bembicenus godmani* (Cameron). *Journal of the Colorado and Wyoming Academy of Sciences* 1: 61.

Romel, K. E., and M. D. Dykstra. 1991. Insects and related pest of lawns, forage grasses, and legumes: Lawns. *Canadian Agricultural Insect Pest Review* 69: 11–12.

Rubink, W. L. 1977. Nesting behavior and prey records of *Hoplisoides splendidulus* (Bradley). *Southwestern Entomologist* 2: 1–3.

——— 1978. The use of edaphic factors as cues for nest-site selection by sand wasps (Hymenoptera: Sphecidae). Ph.D. dissertation, Colorado State University.

——— 1982. Spatial patterns in a nesting aggregation of solitary wasps: Evidence for the role of conspecifics in nest-site selection. *Journal of the Kansas Entomological Society* 55: 52–56.

Rubink, W. L., and K. M. O'Neill. 1980. Observations on the nesting behavior of three species of *Plenoculus* Fox (Hymenoptera: Sphecidae). *Pan-Pacific Entomologist* 56: 187–196.

Rust, R., A. Menke, and D. Miller. 1985. A biogeographic comparison of the bees, sphecid wasps, and mealybugs of the California Channel Islands (Hymenoptera, Homoptera). In A. Menke and D. Miller, eds., *Entomology of the California Channel Islands,* pp. 29–59. Santa Barbara: Santa Barbara Museum of Natural History.

Salt, G. 1931. A further study of the effects of stylopization on wasps. *Journal of the Experimental Zoology* 59: 133–166.

Sánchez, C. S., and J. A. Genaro. 1992a. Observaciones sobre la conducta de nidificación de *Hoplisoides ater* y *H. jaumei* (Hymenoptera: Sphecidae). *Ciencias Biológicas* 25: 150–154.

——— 1992b. Observaciones sobre la conducta de nidificación en esfécidos de Cuba (Hymenoptera). *Bicyrtes spinosa* (Fabr.). *Poeyana, Academia de Ciencias de Cuba,* no. 411. 8 pp.

Schmidt, K. 1979. Materialen zur Aufstellung einer Roten Listen der Sphecidae (Grabwespen) Baden-Württembergs. I. Philanthinae and Nyssoninae—Verhoff. *Naturschutz Land-schaftspflege Baden-Württembergs* 49/50: 271–369.

Schöne, H., and J. Tengö. 1981. Competition of males, courtship behaviour and chemical communication in the digger wasp *Bembix rostrata* (Hymenoptera, Sphecidae). *Behaviour* 77: 44–66.

——— 1991a. Insolation, air temperature and behavioural activity in the digger wasp *Bembix rostrata* (Hymenoptera: Sphecidae). *Entomologia Generalis* 17: 259–264.

——— 1991b. Homing in the digger wasp *Bembix rostrata* (Hymenoptera, Sphecidae)—release direction and weather condition. *Ethology* 87: 160–164.

Schöne, H., A. C. Harris, H. Schöne, and P. A. Mahalski. 1993. Homing after displacement in open or closed containers by the digger wasp *Argogorytes carbonarius* (Hymenoptera, Sphecidae). *Ethology* 95: 152–156.

Schöne, H., H. Schöne, W.-D. Kühme, and L. Kühme. 1994. Arena headings, vanishing bearings, and homing times by *Argogorytes carbonarius* (Hymenoptera, Sphecidae) after displacement in open or closed containers. *Ethology* 98: 291–297.

Schöne, H., J. Tengö, D. Kühme, H. Schöne, and L. Kühme. 1993. Homing with or without sight of surroundings and sky during displacement in the digger wasp *Bembix rostrata* (Hymenoptera Sphecidae). *Ethology, Ecology and Evolution* 5: 549–552.

Sheehan, W. 1984. Nesting biology of the sand wasp *Stictia heros* (Hymenoptera: Sphecidae: Nyssoninae) in Costa Rica. *Journal of the Kansas Entomological Society* 57: 377–386.

Soares, L. A., and R. P. Martins. 1995. The spinning and structure of the cocoon in *Rubrica nasuta* (Christ.) (Hymenoptera: Sphecidae). *Journal of the Kansas Entomological Society* 68: 234–238.

Spofford, M. G., and F. E. Kurczewski. 1990. Comparative larvipositional behaviours and cleptoparasitic frequencies of Nearctic species of Miltogrammini (Diptera: Sarcophagidae). *Journal of Natural History* 24: 731–755.

——— 1992. Counter-cleptoparasitic behaviours of species of Sphecidae (Hymenoptera) in response to Miltogrammini larviposition (Diptera: Sarcophagidae). *Journal of Natural History* 26: 993–1012.

Stange, L. A. 2000. The cicada-killers of Florida (Hymenoptera: Sphecidae). Florida Department of Agriculture and Consumer Services, *Entomology Circular*, no. 402.

Steiner, A. L. 1973. Solitary wasps from subarctic North America. II. Sphecidae form the Yukon and Northwest Territories, Canada: Distribution and ecology. *Quaestiones Entomologicae* 9: 13–34.

——— 1986. Stinging behaviour of solitary wasps. In T. Piek, ed., *Venoms of the Hymenoptera: Biochemical, Pharmacological, and Behavioural Aspects*, pp. 63–160. New York: Academic Press.

Stubblefield, J. W. 1984. Genus *Stizus* Latreille in North America (Hymenoptera: Nyssoninae). *Bulletin of the Museum of Comparative Zoology* 150: 397–428.

Stubblefield, J. W., J. Seger, J. W. Wenzel, and M. M. Heisler. 1993. Temporal, spatial, sex-ratio, and body size heterogeneity of prey species taken by the beewolf *Philanthus sanbornii* (Hymenoptera: Sphecidae). *Philosophical Transactions of the Royal Society of London B* 339: 397–423.

Tashiro, H. 1987. *Turfgrass Insects of the United States and Canada.* Ithaca, N.Y.: Cornell University Press.

Tengö, J., H. Schöne, and J. Chmurzyński. 1990. Homing in the digger wasp *Bembix rostrata* (Hymenoptera, Sphecidae) in relation to sex and stage. *Ethology* 86: 47–56.

Tengö, J., H. Schöne, W. D. Kühme, H. Schöne, and L. Kühme. 1996. Nesting cycle and homing in the digger wasp *Bembix rostrata* (Hymenoptera Sphecidae). *Ethology Ecology and Evolution* 8: 207–211.

Thomas, J. F., and P. Nonacs. 2002. Size dimorphism and male aggregation behavior in the sand wasp, *Steniolia nigripes* (Sphecidae: Bembicinae). *Sociobiology* 40: 317–323.

Thornhill, R., and J. Alcock. 1983. *The Evolution of Insect Mating Systems.* Cambridge, Mass.: Harvard University Press.

Tinbergen, N. 1932. Über die Orientierung des Bienenwolfes (*Philanthus triangulum* Fabr.). *Zeitschrift für Vergleichende Physiologie* 16: 305–335.

——— 1935. Über die Orientierung des Bienenwolfes. II. Die Bienenjagd. *Zeitschrift für Vergleichende Physiologie* 21: 699–716.

Toft, C. A. 1987a. Population structure and survival in a solitary wasp (*Microbembex cubana:* Hymenoptera, Sphecidae, Nyssoninae). *Oecologia* 73: 338–350.

——— 1987b. Activity budgets, habitat use, and body size in two coexisting species of sand wasps (*Microbembex:* Sphecidae, Hymenoptera). *Ecological Entomology* 12: 427–438.

Trivers, R. L. 1972. Parental investment and sexual selection. In B. Campbell, ed., *Sexual Selection and the Descent of Man,* pp. 136–179. Chicago: Aldine.

Tscharnke, T., A. Gathmann, and I. Stefan-Dewenter. 1998. Bioindication using trap-nesting bees and wasps and their natural enemies: Community structure and interactions. *Journal of Applied Ecology* 35: 708–719.

Tsuneki, K. 1956–58. Ethological studies on *Bembix niponica* Smith, with emphasis on the psychobiological analysis of behavior inside the nest (Hymenoptera, Sphecidae). Parts I–III. *Memoirs of the Faculty of Liberal Arts, Fukui University,* ser. 2, 6: 77–172; 7: 1–116; 8: 1–78.

——— 1965a. Nesting biology of *Argogorytes mystaceus grandis* Gussakovskij (Hymenoptera, Sphecidae). *The Life Study* (Fukui) 9: 41–42.

——— 1965b. The nesting biology of *Stizus pulcherrimus* F. Smith (Hym., Sphecidae) with special reference to the geographical variation. *Etizenia, Occasional Publications of the Biological Laboratories of Fukui University* 10: 1–21.

——— 1969. Gleanings on the bionomics of the East-Asiatic non-social wasps (Hymenoptera). IV. Some species of Bembicini, Stizini, Gorytini, Mellinini and Alyssonini. *Etizenia, Occasional Publications of the Biological Laboratories of Fukui University,* no. 10: 19 pp.

——— 1976. The biology of *Stizus pulcherrimus* (F. Smith) in Japan (Hym., Sphec., Nyssoninae). *Special Publications of the Japan Hymenopterist Association* 1: 1–24 (+ 3 plates).

——— 1982. Successive observations on the life of *Gorytes tricinctus* Perez, with special reference to the cocoon building behavior. *Special Publications of the Japan Hymenopterist Association* 22: 1–22.

Tsuneki, K., and T. Iida. 1969. The biology of some species of the Formosan Sphecidae: With descriptions of their larvae (Hymenoptera). *Etizenia, Occasional Publications of the Biological Laboratories of Fukui University,* no. 37: 1–21.

Tyler, T. L. 1986. Sleeping habits of *Steniolia tibialis* Handlirsch (Hymenoptera: Sphecidae) in Sierra County, California. *Pan-Pacific Entomologist* 62: 226.

van Iersel, J. J. A. 1952. On the orientation of *Bembex rostrata* L. *Transactions of the Ninth International Congress of Entomology* 1: 384–393.

van Iersel, J. J. A., and J. van den Assem. 1963. Aspects of orientation in the digger wasp *Bembex rostrata. Animal Behaviour,* supplement no. 1: 145–162.

Villalobos, E. M., and T. E. Shelly. 1994. Observations on the mating behavior of male *Stictia heros* (Hymenoptera: Sphecidae). *Florida Entomologist* 77: 99–104.

——— 1996. Intraspecific nest parasitism in the sand wasp *Stictia heros* (Fabr.) (Hymenoptera: Sphecidae). *Journal of Insect Behavior* 9: 105–119.

Waldbauer, G. P., J. G. Sternberg, and C. T. Maier. 1977. Phenological relationships of wasps, bumblebees, their mimics and insectivorous birds in an Illinois sand area. *Ecology* 58: 583–591.

Wcislo, W. T. 1981. The roles of seasonality, host synchrony, and behaviour in the evolutions and distributions of nest parasites in Hymenoptera (Insecta), with special reference to bees (Apoidea). *Biological Review* 62: 515–543.

——— 1984. Gregarious nesting of a digger wasp as a "selfish herd" response to a parasitic fly (Hymenoptera: Sphecidae; Diptera: Sarcophagidae). *Behavioral Ecology and Sociobiology* 15: 157–160.

——— 1998. Sexual dimorphism of wasp antennal structure in relation to parasitic and non-parasitic behavior (Hymenoptera: Sphecidae). *Journal of Hymenoptera Research* 7: 178–181.

Weaving, A. J. S. 1989a. Habitat selection and nest construction behaviour in some Afrotropical species of *Ammophila* (Hymenoptera: Sphecidae). *Journal of Natural History* 23: 847–871.

——— 1989b. Notes on nesting behaviour and prey of two southern African species of *Stizus* (Hym., Sphecidae, Nyssoninae). *Entomologist's Monthly Magazine* 125: 157–167.

Werren, J. H., and E. L. Charnov. 1978. Facultative sex ratios and population dynamics. *Nature* 272: 349–350.

West-Eberhard, M. J. 2001. The importance of taxon-centered research in biology. In M. J. Ryan, ed., *Anuran Communication,* pp. 3–7. Washington, D.C.: Smithsonian Institution.

——— 2003. *Developmental Plasticity and Evolution.* Oxford: Oxford University Press.

Wheeler, W. M., and R. Dow. 1933. Unusual prey of *Bembix. Psyche* 40: 57–59.

Will, H. C. 1935. An epidemic of the giant sand wasp *Sphecius speciosus* (Drury) at Narrows, Virginia (Hymen.: Sphecidae). *Entomological News* 46: 135–136.

Williams, F. X. 1928. Studies in tropical wasps—their hosts and associates (with descriptions of new species). *Bulletin. Reports of Work of the Experiment Station of the Hawaiian Sugar Planters' Association. Entomological Series* 19: 1–179.

Williams, G. C. 1966. *Adaptation and Natural Selection.* Princeton, N.J.: Princeton University Press.

Willink, A. 1947. Las especies argentinas de "Bembicini" (Hym.: Sphecidae: Nyssoninae). *Acta Zoologica Lilloana* 4: 509–651.

Willmer, P. G. 1985. Thermal ecology, size effects, and the origins of communal behavior in *Cerceris* wasps. *Behavioral Ecology and Sociobiology* 17: 151–160.

Wilson, E. O. 1992. *The Diversity of Life.* New York: W. W. Norton and Company.

Yasumatsu, K., and H. Masuda. 1932. On a new hunting wasp from Japan. *Annals of the Society of Hist. Nat. Fukuokensis* 1: 53–65. (In Japanese, English summary).

Zehnder, G., and J. Zettel. 1999. Succession and zonation in the flood plain area Rottensand (Pfynwald VS; Switzerland). I. Colonization of a flooded area by digger wasps (Hymenoptera, Sphecidae). *Mitteilungen-der-Schweizerischen-Entomologischen- Gesellschaft* 72: 123–137.

Zolda, P., and W. E. Holzinger. 2002. Prey use in two sympatric species of the digger wasp genus *Bembecinus* (Hymenoptera: Apoidea: Sphecidae). *Entomologia Generalis* 26: 101–106.

Zolda, P., J. Ortel, and W. Waitzbauer. 2001. Observations on the nesting behaviour of the digger wasp *Bembecinus hungaricus* in Austria (Hymenoptera: Sphecidae). *Entomologia Generalis* 25: 107–114.

Index

Numbers in **bold type** indicate page numbers for individual species accounts.